Prof Silliman
with the compliments
of the author

MINERALOGY AND CHEMISTRY:

ORIGINAL RESEARCHES

BY PROF. J. LAWRENCE SMITH,

OF LOUISVILLE.

LOUISVILLE, KY.:

PRINTED BY JOHN P. MORTON AND COMPANY.

1873.

CONTENTS.

MEMOIR ON EMERY.

Communicated to the Academy of Sciences of the French Institute, July, 1850.

PART FIRST.

ON THE GEOLOGY AND MINERALOGY OF EMERY FROM OBSERVATIONS MADE IN ASIA MINOR.

Of all the mineral substances employed in the arts few have offered so little opportunity for geological examination as emery, and conseqnently our knowledge of it in this particular is very limited.

Aware of the importance of the study of this substance *in situ*, both in a scientific and practical point of view, I did not lose the opportunity afforded by my late position under the Turkish Government to develop certain facts that came under my notice the latter part of the year 1846. Prior to that period emery (which term is here used, as in the arts, to express that mixed granular corundum employed for abrasion), although known to exist in many places in greater or less abundance, was supplied to the arts almost entirely from the island of Naxos in the Grecian Archipelago. So true is this that the proprietors of the mines in that island controlled completely the price of this mineral. The emery from Naxos frequently went under the name of Smyrna emery, from the fact of its coming to us from that port, where it is originally carried from the island for future exportation.

Prior to 1846 the existence of emery was not remarked in Asia Minor or any of the contiguous islands except that of Samos, which fact is alluded to in Tournefort's travels in the seventeenth century. In the latter part of 1846 I arrived in Smyrna, and was shown specimens, which I recognized as emery, that came from a place about twenty miles north of Smyrna;

they had been first discovered through the agency of a knife-grinder of the country, who had been in the habit of using it to charge his wheels with. The importance of this circumstance to the Turkish Government, as well as to the arts (emery being at that time sold at a most exorbitant price), induced me to return to Smyrna in the early part of 1847 for the purpose of examining the supposed locality of this mineral. On this second visit other localities were made known to me that an English merchant by the name of Healy had succeeded in bringing to light.

The first locality toward which I directed my examination was that of Gumuch-dagh, a mountain about twelve miles east of the ruins of Ephesus. Before, however, arriving there I discovered this mineral imbedded in a calcareous rock in a valley twenty miles south of Smyrna, called Allahmân-Bourgs. The position not being very favorable for the study of the geology of this substance, my route was continued to the place originally fixed upon. Obtaining guides at the village of Gumuch, I commenced the examination of the mountain, which is composed of bluish marble resting on mica-slate and gneiss. On the very summit of the mountain the emery was found scattered about and projecting above the surface of the soil. After examining the extent of the formation and satisfying myself that it was there *in situ*, I returned to Constantinople and made a report to the Ottoman Government. Although I gave no notice to the scientific world of the result of my examination, the editor of the *Journal de Constantinople* inserted a small note in his journal, in May, 1847, to the following effect: "It is some time since M. Lawrence Smith, American mineralogist, discovered at Magnesia, near to Gumuch-Kuey, an emery-mine, of which he brought specimens to Constantinople. The government have sent to the place a commission composed of Mr. Smith and some of the officers of the imperial powder-works to examine thoroughly into the importance of this mine, and according to the report that will be made the government will decide on the steps to be taken with reference to it," etc.

This circumstance, unimportant in itself, has subsequently become of great value to secure to me the priority of the discovery and examination of emery *in situ* in Asia Minor, and also to show that I have been instrumental in the development which has been subsequently given to this emery in a commercial point of view. Since the first discovery other localities have been ascertained by me, all of which will be alluded to in this memoir.

LOCALITIES OF EMERY IN ASIA MINOR AND THE NEIGHBORING ISLANDS.

Gumuch-dagh.—In going from Ephesus east to Gouzel-Hissar (the ancient Tralles) we pass by the ruins of the ancient city of Magnesia on the Miandre, and near to this latter is a beautiful valley, celebrated for its figs, in which is situated the village of Gumuch at the foot of a mountain bearing the same name. It was here that the emery formation was first examined. All the rocks of the surrounding country appear to belong to the old series; the limestone is entirely devoid of fossils and metamorphic in its character; it rests on the older schists, of which mica-schist appears the most abundant, and this again farther to the north was traced in contact with gneiss. The limestone is of a light-blue, passing into a coarse-grained marble, and on the south side the rock by its decay leaves in many places precipices of considerable elevation that add much to the picturesque appearance of the region.

The emery is found in different places in the Gumuch Mountain; the place, however, to which it is traced in greatest abundance is on a part of the summit about three miles from the village of Gumuch, and some fifteen hundred or two thousand feet above the level of the valley; it overlooks the magnificent plain of the Miandre, whose curiously tortuous course is seen as if traced on a map. The emery lies scattered on the surface in the greatest profusion, in angular fragments of a dark color, and large masses of several tons' weight are seen projecting above the surface; in penetrating the soil the emery is found imbedded in it, and a little farther down it is come to in the rock. In fact, by breaking the marble that projects above the surface at this spot, we are sure to find nodules of the mineral.

Sometimes the emery forms almost a solid mass several yards in length and breadth. One of these places, opened for the purpose of exploring, is about ten or twelve yards square, and all the rock taken out is emery; the spaces between the blocks are filled with an earth highly charged with oxide of iron. In some places the masses are consolidated by carbonate of lime of infiltration, which must not be confounded with the emery in its original gangue (the marble), in which it is found in nodules sometimes round and at other times fissured so as to represent angular fragments. In no place does it present any thing like a vein, nor has it signs of stratification. The largest mass at this locality that I saw unbroken must weigh from thirty to forty tons.

Attached to this mineral, more especially in the fissures and on the surface, are several minerals that will be alluded to hereafter.

Kulah. — This locality of emery is the second in importance in Asia Minor. It is a town situated about a hundred and fifty miles from Gumuch and twenty miles from the ancient city of Philadelphia (one of the seven churches). It is near the river Hermes, and on that interesting volcanic district of Asia called Catacecaumene, or the burnt country, resembling in many respects the volcanic region of Auvergne. The rocks forming the base of this region are of the older metamorphic series, covered to a greater or less depth by lava of different volcanic periods, which has flowed from the numerous craters that form the prominent feature of this region. The most common rocks in the mountain ranges about Kulah are white granular limestone, mica-slate, hornblende-schist, gneiss, and granite; the last four are seen more conspicuously in the mountain two or three miles to the south, which have not been subjected to volcanic action; the limestone overlies these rocks.

Before arriving at the place where I examined the emery (about two miles to the northeast of Kulah), an outcropping of gneiss was seen and subjected to the closest scrutiny without discovering the slightest trace of corundum; and I will here remark that although I have found several thin layers of mica-schist engaged in the marble, in no instance was there any trace of corundum in it.

The marble in this region is very compact, of great hardness, and I may also add of great purity. I can not say whether this hardness is traceable to a greater depth than that to which it has felt the influence of the superimposed lava.

Here again the emery was found on the surface, but not in such abundance as at Gumuch-dagh, and moreover the soil is not as deep as in the latter place. The emery as seen in the marble at Kulah is capable of being studied with the greatest satisfaction, particularly as two or three places in the rock have been quarried.

Adula.—Not far from this town, which is about twelve or fifteen miles east of Kulah, I have also discovered emery; only, however, in very small quantity.

Manser.—About twenty-four miles north of Smyrna emery is found in small quantity in the soil. In this, as well as in the former place, white granular limestone is found.

Island of Nicaria, Grecian Archipelago.—I have also been able to examine thoroughly the emery of this island, which promises to be of importance to the arts. It is only within about twelve months that it has been brought to light. The mineral of this locality presents some peculiar features, which will be alluded to hereafter. The geology is the same as that of the other localities already alluded to; namely, when found in contact with the rock it is always with limestone.

Island of Samos.—This locality has furnished me with only a few nodules imbedded in the soil, with a little calcareous rock attached to the surface.

Island of Naxos.—This old and well-known locality is here alluded to simply because it has furnished me with specimens, the examination of which forms a part of this memoir. It is found in large blocks mixed with a red soil and also imbedded in white marble. It is taken principally from the north and east sides of the island; the best comes from Vothrie, nine miles from the shore, and is embarked at Sulionos. Another good locality is at Apperanthos, seven miles from the shore, and it is embarked at a small port called Moutzona. In the south of the island it is found near Yasso. It is in such abundance on this island that, notwithstanding the immense quantity carried off, it is not yet found necessary to quarry it from the rock.

CONCLUSIONS WITH REFERENCE TO THE GEOLOGY OF EMERY.

The localities at Gumuch-dagh and Kulah are those which afforded me the best means of studying the geology of emery, although in every instance I have found it associated with the old limestone overlying mica-slate, gneiss, etc.

It is imbedded either in the earth that covers the limestone or in the rock itself, and exists in masses from the size of a pea to that of several tons' weight, generally angular, sometimes rounded, and when in the latter form they do not appear to have become so by attrition.

The masses in the soil possess but little interest for the geologist, as they may have been left there by the decomposition of the rock or been transported from other positions; still the latter is difficult of supposition in reference to what is found at Gumuch-dagh, for here it is only on the summit and not on the sides of the mountain that the emery has been traced. But having had the means of studying the emery and rock in contact, I have come to the firm conclusion that the *emery has been formed and consolidated in the limestone in which it is found*, and that it has not been detached from older rocks, as granite, gneiss, etc., and lodged in the limestone at the period of its formation. My reasons for so thinking are the following:

1. In no instance could the closest investigation of the older rocks of these localities that are below the limestone furnish the slightest indication of the existence of emery there; and moreover the masses of emery in the limestone never had fragments of another rock attached to them. A few thin layers of mica-slate were found in the limestone, but they were not in contact with the emery nor contained any traces of corundum. I dwell thus much on this point because in my specimens the calcareous rock in connection with the emery is under two forms—that of the original rock, and that formed by the infiltration of calcareous water in the fissures which exist near the surface.

2. The limestone immediately in contact with the emery differs almost invariably in color and composition from the mass of the rock, and at Kulah, where the marble forming the rock is remarkably pure (as evinced by analysis), the part in contact with the emery is of a dark-yellow color, resembling spathic iron, and contains a large portion of alumina and oxide of iron. The thickness of this interposing coat between the emery and the marble is variable; but what is certain, it passes gradually into white marble, so that their crystalline structures

run into each other, showing that they are one and the same rock. Had the masses of emery been broken from an older rock and imbedded in the marble at its formation, there is no reason why the contact should not always be direct and immediate without this transition from ferro-aluminous limestone to pure marble. What we see is just what should be expected in ferruginous and aluminous minerals forming and separating themselves from a limestone not yet consolidated.

This kind of separation between the emery and the marble has been highly useful in the facility that it has indirectly afforded for exploring this mineral. It has been stated that at all the localities under consideration, but principally at Gumuch and Naxos, the emery exists in great abundance, detached from the rock, in a red earth; now this earth is simply the result of the decomposition of this heterogeneous calcareous envelope, which from its nature is easy of disaggregation by the influence of atmospheric agents. Had the emery been in immediate contact with the marble, we could hardly have expected this spontaneous separation in so great a quantity. I have in some instances seen small nodules of emery in small cavities in the rock, but perfectly detached.

3. The immense mass alluded to as covering several square yards of surface is another evidence of the emery having been formed in the limestone, for this mass does not consist of a single piece, but of a number of different sizes, not lying together irregularly, but with their contiguous surfaces more or less parallel, although removed a little distance from each other; in fact, it is just what we would expect in a large mass that for some cause or other had been fissured in various directions.

4. Yet another circumstance to be remarked in connection with this part of the subject is that in the examination of the surface of contact between the emery and the rock we do not always see it marked by a distinct outline; but the minerals constituting the emery, as well as those associated with it, are more or less disseminated in the limestone at the point of contact. The value of this argument is better understood on examining the specimens in my possession.

Enough having been said to prove that the emery under consideration was formed within the limestone in which it is found, I will allude to the process of segregation which has given rise to this formation.

It would appear that the substances eliminated from the calcareous rock were silica, alumina, and oxide of iron, and that these three, in the exercise of homogeneous and chemical attractions, have given rise to the minerals which constitute and are associated with emery. In my collection there is a specimen exhibiting this fact in a remarkable manner. It is a nodule, showing emery in the center, with two concentric layers, the inner of *chloritoid* and the outer of *emerylite;* the latter was in contact with the limestone.

Emery—Mixture of corundum (alumina a little hydrated) and oxide of iron.

Chloritoid—Silica 24, alumina 40, oxide of iron 28, water 7.

Emerylite—Silica 30, alumina 50, lime 13, water 5.

It is seen that in commencing from the external surface, in which direction we must regard the consolidation of the nodule, the larger portion of silica eliminated has combined with a large portion of alumina and some lime to form a peculiar mineral; next, the remainder of the silica combines with an additional quantity of alumina and considerable oxide of iron to form another mineral; and finally, the remaining alumina and oxide of iron crystallize separately. Facts of this kind in geology are not infrequent, but they are always highly interesting and worthy of remark.

In concluding the geological considerations of emery with reference to the localities in Asia Minor and the neighboring islands, I would remark that at some future time, when the observations become extended, it will doubtless be found that the emery forms the geognostic mark of extensive calcareous formations in that part of the world, just as the flints do in the chalk of Europe.

MINERALOGICAL POSITION OF EMERY.

Emery is considered by some as corundum; others suppose it represented by some rock or other, not always the same, in which corundum is disseminated in greater or less quantity; others again consider it a mixture of corundum and oxide of iron. I am of opinion that the latter is the most correct manner of regarding this substance.

Emery, properly speaking, is not a simple mineral, but a mechanical mixture of granular corundum and oxide of iron, in which the former usually predominates. It has not the aspect of corundum disseminated in a rock, for it is found in distinct masses of different dimensions and of great hardness, and when broken giving way in the directions of fissures, which exist commonly in the mass. Most frequently there is no other evidence of the presence of corundum in emery but its hardness. The oxide of iron present is always under the form of magnetic oxide more or less mixed with oligiste; sometimes it is titaniferous. There are other minerals associated with the emery, all of which will be described hereafter.

The aspect of this substance differs more than is supposed, for until lately the emery brought from Naxos has been the criterion by which to judge others. The localities that I have discovered furnish me with specimens showing considerable difference not only as regards color, but also in the structure.

The *Naxos emery* is of a dark-gray with a mottled surface, and with small points of a micaceous mineral disseminated in the mass. It frequently contains bluish specks or streaks, which are easily recognized as being pure corundum.

The *Gumuch-dagh emery* is commonly of a fine grain, and dark-blue bordering on black, not unlike certain varieties of magnetic iron-ores. With this variety we frequently find pieces of corundum of some size. The interior of the mass is tolerably free from the micaceous specks found in that of Naxos.

The *Kulah emery* is usually coarse-grained, and much darker than that of Gumuch-dagh, its external surface resembling sometimes that of chromate of iron.

The *Nicaria emery* in many instances presents a schistose or lamellated structure to a very remarkable degree, so much so that certain specimens might pass for gneiss. The color is dark-blue and somewhat mottled, like that of Naxos. There is also much that is quite compact found in the same locality. The lamellated variety contains an abundance of a micaceous mineral, which in this instance appears to have determined its structure.

The *Samos emery*, as yet found only in small quantities, and in the form of nodules, is uniformly of a dark-blue color, sometimes of a coarse-grained and at other times of a fine-grained structure, not unlike certain varieties of very compact blue limestone.

Fracture.—The fracture of emery is tolerably regular, and the surface exposed is granular, of an adamantine aspect; it is exceedingly difficult to break when not traversed by fissures or not of a lamellated structure, as much of that from Nicaria. When reduced to powder it varies in color from that of a dark-gray to black. The color of its powder affords no indication of its commercial value.

The powder examined under the microscope shows the distinct existence of the two minerals, corundum and oxide of iron, which appear inseparable, as the smallest fragment contains the two together.

Magnetism.—As it is natural to suppose, all specimens of emery affect more or less the magnetic needle; in some the magnetism is barely perceptible, in others it amounts to strong polarity.

Odor.—Emery when moistened always affords a very strong argillaceous odor—even the most compact varieties.

Specific gravity.—The different varieties do not vary much in their specific gravity, it being always in the neighborhood of 4. The specific gravity of various specimens will be given on a following page.

Hardness.—The hardness of emery is its most important property, as to it is due the value of this substance in the arts. For this reason I have devoted much time and attention to the determination of it. In a mineralogical sense, its hardness is not difficult to determine; for if we try different varieties of emery by scratching agate or other hard substance, the effect will naturally be very nearly the same, for in every case it will be some point of corundum that has produced the scratch. If, however, we happen not to rub a point of corundum against the agate, no effect will be produced on the latter, but the emery will yield. As this method leads to no practical result, I have sought out another, which may properly be called one for determining the *effective hardness of emery and corundum*, and is as follows.

Fragments are broken from the piece to be examined, and crushed in a diamond mortar with two or three blows of a hammer, then thrown into a sieve (the one employed had four hundred holes to the square centimetre); the portion passing through is collected, and that remaining on the sieve is again placed in the mortar and two or three blows given, then thrown into the sieve; the operation is repeated until all the emery has passed through the sieve. The object of giving but two or three blows at a time is to avoid crushing any of the emery to too fine a powder.

Thus pulverized, it is intimately mixed and a certain portion of it is weighed (as I operated with a balance sensible to a milligramme, the quantity used never exceeded a gramme). To test the effective hardness of this, a circular piece of glass about four inches in diameter and a small agate mortar are used. The glass is first weighed and placed on a piece of glazed paper; the pulverized emery is then thrown on it little by little, at each time rubbing it against the glass with the bottom of the agate mortar.

The emery is brushed off the glass from time to time with a feather, and when all the emery has been made to pass once over the glass it is collected from the paper and made to pass through the same operation, which is repeated three or four times. The glass is then weighed, after which it is subjected to the same operation as before, the emery being by this time reduced to an impalpable powder. This series of operations is continued until by repeated weighing the loss sustained by the glass is reduced to a few milligrammes. The total loss in the glass is then noted, and when all the specimens of emery are submitted to this operation under the same circumstances we get an exact idea of their relative hardness.

The blue sapphire of Ceylon was pulverized and experimented with in this way; it furnished me with a unit of comparison by which to compare the results obtained. This operation is long but certain, and for the harder varieties of emery it is necessary to repeat the rubbing six or seven times, and it requires nearly two hours for completion.

The results that I have obtained are interesting, and have furnished me with the means of forming conclusions that I could not otherwise have come at.

Glass and agate have not been chosen for this experiment without a certain object, as experiments were first made with two pieces of agate, with two pieces of glass, and with metal and glass. The agates were found too hard, as they crushed the emery without producing hardly any abrasive effect; the others were found not to crush the emery sufficiently, making the experiment tedious and long. With the glass and agate we have a hard substance which crushes the emery, and in a certain space of time reduces it to such an impalpable state that it has no longer any sensible effect on the glass, and on the other hand the glass is soft enough to lose during this time sufficient of its substance to allow of accurate comparative results. In the employment of this method in the arts it would not be necessary to go to the sapphire for a standard of comparison; any good emery would answer the purpose quite as well.

It must be understood that this method of coming at the abrasive effects of emery does not furnish the mineralogical hardness of this substance, by which we understand the hardness of any individual particle, as evinced by its effect on a substance of less hardness, without regard to the molecular structure of the mineral. Two minerals possessing the same hardness but differing in structure, one being friable and the other resisting, will be found very different in their abrasive effects; for instance, break a piece of quartz in two, subject one of the pieces to a white heat, and after cooling compare the two by rubbing the point against some hard substance; both will be found to scratch equally well. Then try the two in a state of powder by rubbing them between two pieces of glass that have been weighed, and the difference of their abrasive effects will be found very great, because the one subjected to the fire is exceedingly friable and becomes readily crushed to an impalpable powder. This fact is eminently true with reference to emery, many specimens of which containing the same amount of corundum differ somewhat in their *effective hardness*, owing to the more or less compact structure of the corundum.

By the method with the agate and glass I have found the best emery capable of wearing away about one half its weight of the glass (that used was the common French window-glass). The sapphire under the same circumstances wears away more than four fifths of its weight. A tabular view of the results will be given a little farther on.

CHEMICAL COMPOSITION OF EMERY.

This substance consisting of a mixture of corundum and oxide of iron in various proportions, it is easy to see what its composition must be. Yet the chemical examination of this mineral, taken in connection with other properties, is not devoid of interest.

For the purpose of analysis the emery was reduced to a state of powder, in the manner alluded to in speaking of its hardness, with a diamond mortar and sieve. This powder was dried for twenty-four hours over sulphuric acid; a gramme was then weighed in a small platinum crucible of about one fourth of a cubic inch in capacity, fitted with a cover that adapted itself well to it. This small crucible was placed in another of earth, and the space between the two filled with pulverized quartz which also covered the smaller one to the depth of half an inch. Common sand was not used, because during the heating some particles might adhere to the platinum crucible by a semifusion; nor was powdered charcoal employed, because it protected the mineral no better than the pulverized quartz from contact with the air, at the same time a little risk was run in decomposing a small amount of the iron.

Thus arranged, the crucibles were heated to a bright-red for from thirty minutes to one hour. After cooling the platinum crucible was carefully withdrawn and weighed. The loss furnished me with the amount of water in the emery.

It requires a continued red heat to drive out all the water, a circumstance which is true for a number of minerals, particularly for those containing a large amount of alumina, as diaspore and the micas, which will be spoken of in this paper.

The powder, of which the water has been estimated, was next submitted to levigation in a large agate mortar placed on a surface of glazed paper; and when completed it was carefully detached from the mortar, placed in a platinum capsule, heated gently to drive off any hygrometric moisture, and weighed. The increase of weight furnished the amount of silica taken from the mortar.

The levigation of one gramme was accomplished in two operations, each requiring about twenty minutes; and by using a mortar of convenient size and the extremity of a feather or a small brush it is possible to lose but an insensible quantity of the mineral, and to estimate with sufficient precision the amount of silica abraded from the mortar.

Another method by which I accomplished the levigation in some of the analyses was in a steel mortar of the same form as the agate mortar; and when completed the powder was placed in a glass with nitric acid diluted with thirty times its weight of water, and left in it for one hour, agitating it occasionally. The iron taken from the mortar was dissolved, and no part of the mineral attached. The next thing was to filter, and continue the analysis with the substance thus freed from the iron of the mortar, without any second weighing.

Of these two methods I preferred to employ the first for the emery, as it is more expeditious and almost if not quite as exact as the second. There are, however, occasions in which the steel mortar should be resorted to.

The substance once reduced to an impalpable powder, it was necessary to render it *completely* soluble, and my researches to arrive at this were long and tedious. In trying the various known methods the most successful was found to be that with a mixture of carbonate of soda and caustic soda heated to whiteness for one hour; nevertheless I could not obtain a complete decomposition. The decomposition might probably be completed if the levigation was made more thoroughly; but it is easy to understand that with a large number of analyses of the same substance to make, it was a desideratum on my part not to consume the best part of a day in the levigation of a single gramme, particularly as I did not wish to confide this operation to another, as much care was required to lose nothing during the levigation. Mixed with carbonate of baryta and heated in a forge, the decomposition of the mineral was far from being complete; the same may be said for the treatment with the caustic alkalies in a silver crucible.

The bisulphate of potash decomposes it almost entirely by a single operation, but unfortunately a double salt of potash and alumina is formed which is almost insoluble in water or in the acids, and it is only by a solution of potash that it is first decomposed and afterward redissolved. I will not stop to detail all the disadvantages attending this method, but will at once speak of the method which gave me very easily the most accurate results.

It is by means of the bisulphate of soda that all my analyses of emery, of corundum, and of several aluminates were made. I believe that I am the first who has shown the great advantage of using this double salt in the decomposition of certain substances insoluble in the acids; and very probably it will replace in most cases the use of the bisulphate of potash in analytical chemistry. At present all the advantages that may arise from the substitution of the soda for the potash salt can not be mentioned; all that I will say is that the former, in giving a decomposition at least as complete as the latter, furnishes a melted mass quite soluble in water, and in the future operations of the analyses there is no embarrassment from a deposit of alum.

The bisulphate of soda was prepared by adding an excess of pure sulphuric acid to the pure carbonate or neutral sulphate of soda, and heating it in a capsule until all the water had been expelled and sufficient of the acid to allow of the mass becoming solid on cooling. That obtained in commerce is not sufficiently pure.

The pulverized emery is placed in a large platinum crucible with six or eight times its weight of bisulphate of soda, and the mixture is heated over a lamp in the same manner and with the same precautions as are employed when using the bisulphate of potash. From fifteen to thirty minutes suffice for the operation. The mass is allowed to cool, and water with a few drops of sulphuric acid are added to it, and the whole heated, when it soon dissolves, with the exception of a little silica that renders the solution milky, and a small quantity of undecomposed mineral that is readily detected by rubbing a glass rod against the bottom of the capsule. The liquid is now filtered, and the filter is washed once with a little water; then with its contents it is placed in a platinum crucible, burnt completely, and the residue is heated with a little bisulphate of soda, which completes the decomposition; and when treated with water and a drop or two of sulphuric acid all except the silica is dissolved. The liquid which passes the filter in this case is added to the first and the analysis continued. The silica obtained is diminished by the quantity taken up from the mortar in order to arrive at what is actually contained in the mineral. The filtered solution is heated with a little nitric acid to convert all the protoxide of iron into peroxide; then treated with an excess of caustic soda and a little carbonate of the same alkali; this redissolves the alumina first precipitated, and thus separates it from the oxide of iron and a trace of lime. The iron and lime are separated in the ordinary way; the alkaline solution of alumina was acidulated, and the alumina precipitated with carbonate of ammonia.

Thus analyzed, the emery from different places gave the following results:

No.	LOCALITIES.	Effective hardness.—Sapphire 100.	Specific gravity........	CHEMICAL COMPOSITION.					
				Water....	Alumina.	Oxide of Iron.....	Lime.....	Silica.....	Total.....
1	Kulah..........................	57	4.28	1.90	63.50	33.25	0.92	1.61	101.18
2	Samos..........................	56	3.98	2.10	70.10	22.21	0.62	4.00	99.03
3	Nicaria	56	3.75	2.53	71.06	20.32	1.40	4.12	99.43
4	Kulah	53	4.02	2.36	63.00	30.12	0.50	2.36	98.34
5	Gumuch........................	47	3.82	3.11	77.82	8.62	1.80	8.13	99.48
6	Naxos...........................	46	3.75	4.72	68.53	24.10	0.86	3.10	101.31
7	Nicaria	46	3.74	3.10	75.12	13.06	0.72	6.88	98.88
8	Naxos...........................	44	3.87	5.47	69.46	19.08	2.81	2.41	99.23
9	Gumuch........................	42	4.31	5.62	60.10	33.20	0.48	1.80	101.20
10	Kulah..........................	40	3.89	2.00	61.05	27.15	1.30	9.63	101.13

I ought to mention that the analysis afforded other substances in small quanties in some of the emeries, as titanic acid, oxide of manganese, oxide of zirconium, and sulphur (existing in pyrites); but these substances are unimportant in the composition of emery, and are in such minute quantities that it is necessary to operate on a considerable quantity of the mineral to obtain satisfactory results concerning them.

The analyses marked 6 and 8 were made by decomposing the emery as it came from the sieve, without pulverization in the agate mortar. It was by accident that it occurred, and I was not aware of the neglect until it was fused with the bisulphate of soda; but, not wishing to lose the analysis, the operations were continued as in the other cases, only using a little more of the bisulphate in the second decomposition; and somewhat to my surprise the decomposition was quite as perfect as in the other cases. I had nearly completed all my analyses in the manner detailed when this fact became known, so that I have but these two cases to report. It will simplify the analysis of corundum if pulverization in a diamond mortar be found sufficient, and I propose examining specially into this question.

The water which was found in the emery comes from the corundum, a fact which will be shown when the analysis of pure corundum is given, which will be in the second part of the memoir. A very minute quantity of what has been estimated as water might be a little oxygen lost by the oligist which is sometimes found in emery. Those emeries which contain the least water, every thing else alike, are the hardest, as instanced by that from Kulah, notwithstanding the quantity of iron it contains. The silica existing in emery is most often in combination with alumina or the oxide of iron, or with both. For this reason we must not always regard the quantity of alumina as an indication of the quantity of corundum in the emery.

ANALOGIES.

Emery at first sight may be confounded with several ores of iron, as magnetic iron, certain varieties of oligist, and sometimes with chromate of iron; but the fracture of emery is stony, which differs from these ores of iron, and besides the surface exposed is of a lighter color. From the numerous observations made I may set it down as a general rule that any blackish or dark-blue rock of a strong argillaceous smell, that scratches agate well, with a specific gravity in the neighborhood of 4, is sure to be emery.

THE MINING OF EMERY.

The mining of this substance is of the simplest character. The natural decomposition of the rock in which it occurs facilitates its extraction. As has already been mentioned, the rock decomposes into an earth in which the emery is found imbedded. The quantity found under these favorable circumstances is so great that it is rarely necessary to explore the rock. The earth in the neighborhood of the blocks of emery is almost always of a red color, and serves as an indication to those who are in search of the mineral. Sometimes, before beginning to excavate, the spots are sounded by an iron rod with a steel point; and when any resistance is met with the rod is rubbed in contact with the resisting body, and the effect produced on the point enables a practiced eye to decide whether it has been done by emery or not.

The blocks which are of a convenient size are transported in their natural state, but most frequently they are required to be broken by means of large hammers. When they resist the hammer they are subjected to the action of fire for several hours, and on cooling they most commonly yield to blows. It, however, happens sometimes that large masses are abandoned from the impossibility of breaking them into pieces of a convenient size, as the transportation either on camels or horses requires that the pieces do not exceed one hundred pounds.

At Kulah the quantity of emery detached from the rock was not very considerable, as it had been protected from decomposition by the beds of lava that cover it. Here the marble was quarried to get at the emery, which was done in the early part of 1847 with profit, although the transportation from Kulah to Smyrna is over a distance of one hundred and ten miles on the backs of camels. Since the diminution of the price of emery this mine has been abandoned; for the quarrying into the marble is attended with the greatest difficulty, as the tools used for boring, etc., are thrown out of use in a very short time by the pieces of emery which are encountered at every instant. At present all the emery sent from Asia Minor comes from the mine at Gumuch-dagh, twelve miles from the ruins of Ephesus.

COMMERCIAL CONSIDERATION OF EMERY.

The use of emery in the arts is of very ancient date, a fact proved by works on hard stones that could not have been executed except by emery or minerals of that nature. It is very probable that emery coming from the localities which have been mentioned was used in former ages by the Greeks and Romans. For example, the locality of Gumuch-dagh is immediately by the ancient Magnesia on the Miandre, and between Ephesus and Tralles, twelve miles from each of these cities and the same distance from Tyria. In all of these cities the arts flourished, and none more than that of cutting hard stones, if we are allowed to judge from the specimens of their skill in this art that have come down to us.

Nevertheless, the quantity of emery formerly employed was insignificant in comparison to the quantity now required, more particularly within the last twenty years, since the use of plate-glass has been extended. The annual consumption at the present time is about *fifteen hundred tons.*

For various reasons the island of Naxos furnished for several centuries almost exclusively the emery used in the arts, as much from the facility with which it was obtained as for the uniformity of its quality. The emery exists in very great abundance on this island, and notwithstanding the quantity already extracted there still remain immense deposits of it.

The price of this substance at the end of the last century was from forty to fifty dollars per ton, and between 1820 and 1835 it was at times even less. About this period the monopoly of the Naxos emery was purchased from the Greek Government by an English merchant, who so regulated the quantity given to commerce that the price gradually rose from forty to one hundred and forty dollars per ton, a price at which it was sold in 1846 and 1847. It was at this time that I commenced examining and developing the emery formations of Asia Minor, until then unknown. And after making a report to the Turkish Government the monopoly of the emery of Turkey was sold to a mercantile house in Smyrna, and since then the price of this article has diminished to fifty and seventy dollars per ton, according to the quality. I speak of the prices in the English market.

The different mines explored are those of Naxos, of an ancient date; of Kulah, commenced in 1847, and now abandoned for those nearer the sea; of Gumuch-dagh, commenced in 1847 and worked largely; and of Nicaria, commenced in 1850. From all these different places the emery goes to Smyrna, and from there principally to England, the vessels taking it at a very low price, as it serves for ballast.

The various mines belong to the Turkish and to the Greek governments. The Greek Government now sells its emery in lots of several tons. The Turkish Government sells the entire monopoly of its mines, and consequently its operations are controlled by a single interest; but in all probability this monopoly will be done away with in virtue of a commercial treaty existing between Turkey and the other powers. If this takes place, the price of emery will be still further diminished.

Of the different varieties of emery employed in the arts that of Naxos is still preferred, and with reason, as it is more uniform in its quality than that coming from Kulah and Gumuch; nevertheless, if the best qualities of that from the island of Nicaria are found in abundance, and that only sent into market, it will prove at least equal if not superior to that of Naxos.

PART SECOND.

ON THE MINERALS ASSOCIATED WITH EMERY: CORUNDUM, HYDRARGILLITE, DIASPORE, ZINC SPINEL, PHOLERITE, EPHESITE (A NEW SPECIES), EMERYLITE (A NEW SPECIES), MUSCOVITE, CHLORITOID (A NEW VARIETY), BLACK TOURMALINE, CHLORITE, MAGNETIC OXIDE OF IRON, OLIGIST IRON, HYDRATED OXIDE OF IRON, IRON PYRITES, RUTILE, ILMENITE, AND TITANIFEROUS IRON.

Now that it has been shown that emery is found in considerable abundance in certain parts of the world, occupying almost the position of a rock, it is useful to mention the different accidental minerals, or *minerals of elimination*, that are found with emery, and what new facts have been observed with relation to them. Corundum may be first mentioned.

CORUNDUM.

Although emery is constituted principally of corundum, the examination of this substance in its pure state, or rather in the form of those prismatic crystals which I have sometimes found in contact with emery, has brought to light several new and well-established facts that could not have been satisfactorily ascertained from a mixed mineral like emery

At Gumuch-dagh it is not difficult to find large pieces of this mineral, pure or mixed with a little diaspore and emerylite; sometimes the crystals are very distinct under the form of six-sided prisms. The small crystals found in the cavities are sometimes terminated by a summit of six faces. The color of the corundum found in the different places alluded to in this memoir is blue, except that of Kulah and of Adula, which is of a greenish-gray. All that I have to add to what is already known of this mineral relates to its *composition* and *effective hardness;* the latter was ascertained in the way already described in speaking of the emery, and it has been found to vary with the composition of the mineral. The analyses were made in the same manner as those of the emery, and the results which I have obtained are as follows:

Localities.	Corundum.		Composition.					
	Effective hardness.—Sapphire 100.	Specific gravity........	Water........	Alumina......	Magnetic Oxide Iron.....	Lime........	Silica........	Manganese...
Sapphire of India..................	100	4.06	......	97.51	1.89	......	0.80	....
Ruby of India......................	90	......	......	97.32	1.09	......	1.21	....
Corundum of Asia Minor.......	77	3.88	1.60	92.39	1.67	1.12	2.05	trace.
Corundum of Island of Nicaria	65	3.92	0.68	87.52	7.50	0.82	2.01	....
Corundum of Asia..	60	3.60	1.66	86.62	8.21	0.70	3.85	....
Corundum of India...............	58	3.89	2.86	93.12	0.91	1.02	0.96	....
Corundum of Asia................	57	3.80	3.74	87.32	3.12	1.00	2.61	....
Corundum of India...............	55	3.91	3.10	84.56	7.06	1.20	4.00	0.25

The most remarkable fact ascertained by these analyses is the presence of water in variable quantity in all varieties of the corundum except the sapphire and ruby. To me this fact has a certain value in proving that the corundum and the sapphire are formed under different circumstances and do not belong to the same geological formation. The different structure of these two species of corundum might make one suspect a difference in the condition of their formation; and this is somewhat confirmed by the results of the beautiful experiments of M. Ebelmen in making artificial corundum by subjecting alumina and borax to the heat of a porcelain furnace for many hours—circumstances under which he always obtained crystals under some of the modifications of hyaline corundum, and never as prismatic corundum. In addition to this I remark that in my most thorough examination of the localities of emery not the slightest trace of sapphire or ruby was found.

The quantity of water found to exist in corundum coming from different localities is variable, and it would appear that, all other things being equal, those containing the least water are the hardest. I will not insist on the slight difference between the hardness of the sapphire and ruby, having made only one experiment upon each of these minerals.

The two varieties of corundum are so evidently united by their system of crystallization that I would not undertake to separate them on account of the presence of water in one of them, and that in variable quantity; nevertheless, the fact is important, as it explains to a certain extent their differences in structure and hardness. I would remark that great pains were used to ascertain whether the water might not be due to the presence of diaspore or some other hydrate of alumina; but after the most careful and repeated examinations this has been decided in the negative.

HYDRARGILLITE.

Hydrargillite is rarely met with. I have one specimen with this mineral forming the external coating of a crystal of corundum, and also a hexagonal prism of the same mineral. It was not analyzed, but its physical properties and its reactions under the blowpipe served to prove its identity with this mineral. The specimen in my possession comes from Gumuch-dagh.

DIASPORE.

This mineral up to the present time has not occupied a very important position in mineralogy, and has been found only in two or three localities. In the course of this article I hope to show that it plays a somewhat important part in the emery and corundum formations. Before my attention was drawn to the minerals, first discovered by M. Lelievre, it was studied by M. Dufrenoy on that coming from Siberia, and by M. Haidinger on the diaspore of Schemnitz. Before going farther I would remark that the gangue of the latter, which has been described as analogous to steatite, was found by me not to be such, but a hydrated silicate of alumina, similar to one found with the emery of Naxos.*

To the localities of diaspore already known I have to add

* The gangue of the Schemnitz diaspore has been examined by Hutzelmann (see Pogg. Ann., LXXVIII, 575), who makes it to contain three distinct hydrates of alumina; but this fact can not be considered as sufficiently established. One of these hydrates is named *Dillnite*, and another is near Agalmatolite.

those of Gumuch-dagh and Manser in Asia Minor, and the islands of Naxos, Samos, and Nicaria in the Grecian Archipelago; and there is reason to believe that this mineral will be found in almost every corundum locality. I have already found it on crystals of corundum from China.

In examining the emery formations one of the first things that struck my attention was the existence of diaspore and corundum together, then observed for the first time. The same year M. Marignac discovered it in the limestone of St. Gothard, along with the well-known crystals of corundum that exist there. Having found the diaspore under these new circumstances, it has been examined with much attention.

At Gumuch-dagh the diaspore is found in flattened and rounded prisms, with the surface streaked with lines that afford by reflected light an iridescence. Crystals with perfect summits are rarely found, and during two or three days' examination on the place I found only five small crystals with one of the summits perfect; they were, however, very beautiful, and finer probably than any yet known. Not wishing to lose so favorable an occasion to verify the crystallography of diaspore, I requested M. Dufrenoy to undertake the measurement of the angles, and it is to this able professor that we are indebted for the crystallographic results here given.*

The crystals are elongated needles crossing each other in all directions, like an acicular variety of aragonite from the Vosges. They resemble small crystals of topaz in luster and in the disposition of the vertical striæ on the faces *g*. Their color is yellowish-white. They are strongly dichroitic; the summits under certain inclinations appear black, as if the light was completely polarized. The cleavage is very easy parallel to the face g^1, and it is this cleavage that gives a lamellar structure to that diaspore which is not in the form of needles. This cleavage, notwithstanding its facility, does not expose surfaces that reflect with great accuracy; it is the only angle which offers the difference of a half degree; repeated measurements of the other angles never varied more than four minutes. The pearly luster of the cleavage in connection with its striated

* Three of the crystals measured are in the Cabinet of the School of Mines and Garden of Plants at Paris. The second crystal above is nearly as thin as the first, although represented thicker, in order to show well all the planes.

character are the causes of this difficulty, which at first sight would not appear to exist, only becoming evident when the angle is examined.

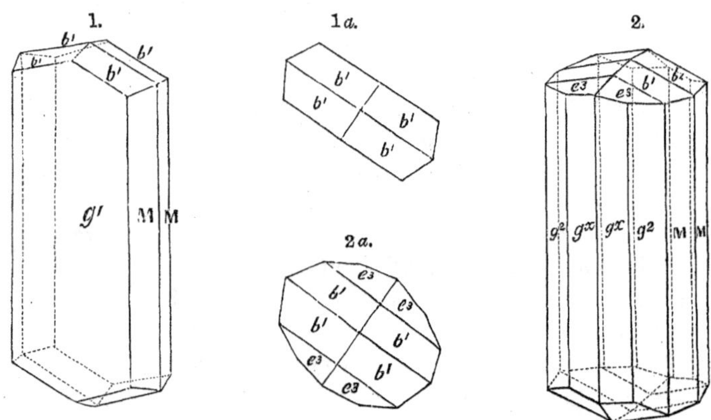

The crystals, very much flattened, parallel to the face g^1 are represented by figures 2 and 3; the face g^1 does not exist, being replaced by three series of faces g, the angles of which could not be measured; but the almost absolute identity of these crystals with those of St. Gothard, which M. Marignac first described, authorizes one to suppose that they are represented by the crystallographic signs g^2 and $g^{\frac{5}{3}}$. The faces M and those of the summit have a very bright luster. The primitive form of the diaspore is undoubtedly a right rhombic prism of 130° 2′; the fact that the base is horizontal is shown by the identity of the angles of the faces b^1 on the anterior faces M and the faces b^1 on the posterior faces of the same. This position is verified in seeking for the angle of the edge b^1 on M, which ought to be a right-angle; in fact the calculation of a spherical triangle composed of the faces M, b^1, and g^1, of which all the angles of incidence were measured, gave for this edge 90° 2′ 30″, which differs from a right-angle by only two minutes and a half.

The following table is made up of the measurements of the angles of the diaspore of Gumuch-dagh (near Ephesus) by M. Dufrenoy, of that of St. Gothard by M. Marignac, and of that of Schemnitz by M. Haidinger; also the measurement of some angles of the hydrated peroxide of iron of Cornwall by M. Dufrenoy, which are here given to show an interesting connection, first pointed out by M. de Senarmont, and which consists in the isomorphism of diaspore and the hydrated oxide of iron. Thus, while the peroxide of iron or oligist iron is isomorphous with alumina or the corundum, the hydrates of the same oxides are isomorphous.

	Diaspore of St Gothard—Marignac..........	Diaspore of Schemnitz — Haidinger.....	Diaspore of Gumuch-dagh, near Ephesus —Dufrenoy*...	Hydrated Oxide of Iron of Cornwall—Dufrenoy.....
M : M	130°	129° 54′	130° 2′	130° 57′
M : b^1	....	....	125° 17′	....
M : b^1 (posterior faces)....................	....	....	125° 18′	....
M : g^1	115°	....	114° 58′	....
b^1 : g^1	104° 12′	....	104°	....
b^1 : b^1	151° 36′	151° 36′	151° 35′	....
b^1 : b^1 (posterior faces)....................	....	....	151° 33′	151° 34′
b^1 : b^1 (opposite faces).....................	116° 38′	....	116° 18′	....
b^1 : c^3	....	....	167° 6′	....
g^1 : g^2	144° 40′	....	....	....
g^2 : g^2	145° 40′	....	....	....
i : i	126° 12′	....	....	126° 20′
g^1 : i	116° 56′	....	....	116° 55′
$e^{1/2}$: $e^{1/2}$	117° 46′	....	....	117° 10′

I have found crystals of diaspore in hydrated oxide of iron, the needles traversing the oxide in all directions. There is a specimen in my possession composed of a small crystal of diaspore, surrounded by a kind of scabbard of crystallized göthite. One of the summits of the crystal is exposed. In breaking the oxide of iron which contains these crystals they become detached, leaving on the oxide an impression with a very brilliant surface.

The diaspore of Gumuch-dagh is also found, of a lamellar structure, but very rarely; that of Naxos, Nicaria, and Samos are all lamellar. Yet of all the specimens that I have collected none offer so much interest as those composed of diaspore imbedded in corundum. Here we see the two minerals passing one into the other, without being able in many places to distinguish the line of separation, so imperceptible is the gradation. After what has been said in respect to corundum it is

* The values of these angles are just as given by the goniometer, without any correction by calculation.

not astonishing to see this connection of alumina more or less hydrated with a hydrate of alumina of definite composition.

After a knowledge of this fact one might seek to explain the existence of water in corundum by the intimate mixture of diaspore with this mineral. If this be the case, the crystal of corundum from the Carnatic, which gave me three per cent. of water, must contain twenty-three per cent. of diaspore, although neither the eye nor the microscope could detect its presence.

As to the properties of diaspore, I have nothing to add to what has already been published on the subject, except that the specimens I examined do not decrepitate to the same extent as that of Siberia. Its specific gravity is 3.45, and hardness above 7. The following analyses were made, the mineral being attacked with the bisulphate of soda. They afford the formula $\dddot{Al}\,\dot{H}$.

LOCALITIES.	Silica.....	Alumina.	Lime.....	Oxide of Iron....	Magnesia	Water....
Gumuch-dagh	0.67	82.20	0.41	1.20	trace.	14.52
Gumuch-dagh	0.82	83.12	trace.	0.66	trace.	14.28
Naxos	0.26	82.94	0.35	1.06	....	14.81

ZINC SPINEL.

I possess a single specimen of this spinel in chloritoid on a piece of emery from Gumuch-dagh. It is in octahedral crystals agglomerated, of a dark emerald-green color. The quantity being small, I have been prevented from making an exact analysis. The quantity of oxide of zinc appears to be from thirty to forty per cent.

PHOLERITE.

A mineral resembling pholerite in composition has been found with the emery of Naxos associated with emerylite. It is white, lamellar, and somewhat crystalline, sometimes gray. It is soft to the touch like steatite, infusible before the blowpipe, and when heated with nitrate of cobalt becomes strongly colored blue. It is scratched with the nail, and has a specific gravity of 2.564. Its composition is identical with the pholerite of Guillemin, also with the mineral forming the gangue of the diaspore of Schemnitz. For analysis it was decomposed with carbonate of soda. It afforded:

		Pholerite of Guillemin.	Gangue of diaspore of Schemnitz—Smith.
Silica	44.41	42.93	42.45
Alumina	41.20	42.07	42.81
Lime	1.21		trace and mag. trace.
Water	13.14	15.00	12.92

This corresponds to the following formula, $\dddot{Si}\,\dddot{Al}+2\dot{H}$; but it is a question whether or not we should consider the water as existing in any definite proportion, and whether or not they did not all contain more water when first taken from their localities. These hydrated silicates of alumina are numerous, and bear various names; but it is doubtful if many of them are entitled to much consideration as distinct species.

EPHESITE (A NEW SPECIES).

This silicate is found with the emery of Gumuch-dagh, and occurs on specimens of magnetic oxide of iron. It is of a pearly-white color, and lamellar in structure; cleavage difficult. It scratches glass easily, and has a sp. grav. of from 3.15 to 3.20. Heated before the blowpipe it becomes milk-white, but does not fuse. At first sight it might be taken for white disthene. It is decomposed with great difficulty by carbonate of soda, even with the addition of a little caustic soda. I used also very successfully in the analysis the bisulphate of soda, either in attacking the mineral from the commencement or in operating first with carbonate of soda, and then acting on the part not decomposed with bisulphate of soda. The alkalies were separated by means of hydrofluoric acid.

Silica	31.54	30.04
Alumina	57.89	56.45
Lime	1.89	2.11 }
Protoxide of iron	1.34	1.00 }
Soda with a little potash	......	4.41 }
Water	3.12	3.06

This corresponds very nearly to the formula $\dot{R}^2\,\dddot{Si}+\dddot{Al}^2\,\dddot{Si}+4\dot{H}$.

	ATOMS.	AT. WEIGHT.	PER CENT.	OXYGEN RATIO.
Soda	2	781.6	7.08	1
Silica	6	3400 2	30.77	9
Alumina	10	6416.2	58.08	15
Water	4	450.0	4.07	2

This mineral has been designated ephesite because of its occurrence at the emery locality near the ancient city of Ephesus.

EMERYLITE (A NEW SPECIES).

This mineral, which I have designated by the name of emerylite, is a new species belonging to the family of micas. I have already published a note indicating its existence,* but have reserved for the present time a complete description of it.

I first discovered this mineral with the emery of Gumuch-dagh in Asia Minor, and subsequently in that of Manser, the islands of Naxos and Nicaria, and also with the emery of Siberia. Its connection with all the emeries that have come under my observation, except that of Kulah, induced me to call it emerylite. When I announced this discovery to Prof. Silliman, jr., he hastened to examine the minerals coming from the corundum localities of the United States, and has succeeded in finding the emerylite with the corundum of several localities. The specimen from Siberia on which I found this mineral is in the collection at the Garden of Plants at Paris, and I have also reason to think that I have found it with the corundum of China.

The emerylite is lamellar, like mica; the plates are easily separated, and possess a little elasticity. Sometimes it is in the form of a mass composed of very small pearly scales, which are very friable, resembling some species of talc. The plates are commonly convex and concave, grouped in such a manner as to form a triangular prism. I have also found it massive, with a micaceous structure, but with an irregular fracture. The aspect of this variety is waxy; it comes from Gumuch-dagh. The crystalline form of this mineral is difficult to determine; but if we are permitted to judge from the streaks on the surface, and the imperfect cleavage in two directions, it would appear to belong to an oblique rhombic prism.

Its color is white and luster silvery. The hardness taken on a specimen from the island of Nicaria is from 4 to 4.5. The sp. grav. taken on ten specimens varies from 2.80 to 3.09. This difference is not remarkable in a lamellated mineral.

*See Amer. Jour. of Science and Art, 2d ser., VII, p. 285, 1849.

That which gave me the greatest specific gravity contained some small specks of titaniferous iron visible to the eye. Its optical properties have not been examined, for the want of a transparent piece of sufficient size and thickness. This mineral is not attacked by the acids. Heated before the blowpipe, it emits a bright light, and melts with great difficulty on the edges, which assume a blue color if touched with the nitrate of cobalt and reheated. Heated in a tube, it furnishes water frequently having an acid reaction due to fluoric acid.

The composition of several specimens subjected to analysis is as follows:

LOCALITIES.	Silica.....	Alumina.	Lime.....	Oxide of Iron.....	Magnesia	Potash & Soda....	Water...	Manganese....
Gumuch-dagh.........	29.66	50.88	13.56	1.78	0.50	1.50	3.41	....
Island of Nicaria....	30.22	49.67	11.57	1.33	trace.	2.31	5.12	....
Island of Nicaria....	29.87	48.68	10.84	1.63	trace.	2.86	4.32	....
Island of Naxos......	30.02	49.52	10.82	1.65	0.48	1.25	5.55	....
Island of Naxos......	28.90	48.53	11.92	0.87	not estimated.	not estimated.	5.08	....
Island of Naxos......	30.10	50.08	10.80	not estimated.	"	"	4.52	....
Gumuch-dagh.........	30.90	48.21	9.53	2.81	"	"	4.61	....
Gumuch-dagh.........	31.93	48.80	9.41	1.50	"	2.31	3.62	trace.
Siberia..................	28.50	51.02	12.05	1.78	"	not estimated.	5.04	....

The oxide of iron may be regarded as an impurity which exists between the plates of the mineral. The composition of the emerylite is represented by

	ATOMS.	AT. WEIGHT.	PER CENT.	OXYGEN RATIO.
Lime.............	2	700.0	13.48	2
Silica.............	3	1700.1	32.74	9
Alumina.........	4	2566.5	49.44	12
Water............	2	225.0	4.34	2
		5191.6		

Formula: $\dot{R}^2 \ddot{Si} + 2\dddot{Al}^2 \ddot{Si} + 2\dot{H}$.

As is seen, the specimens examined came from four distinct localities, and were all taken under different circumstances; yet their analyses accord perfectly, and also agree with those of the United States coming from Village Green and Unionville of Pennsylvania and Buncombe County, North Carolina.

Localities.	Silica.....	Alumina..	Lime.....	Magnesia	Potash & Soda....	Water....		
Village Green..............	32.31	49.24	10.66	0.30	2.21	5.27	=100	Craw.
Village Green..............	31.06	51.20	9.24	0.28	2.97	5.27	=100	Craw.
Village Green..............	31.26	51.60	10.15	0.50	1.22	4.27	=100	Craw.
Village Green..............	30.18	51.40	10.87	0.92	2.77	4.52	=100.46	Craw.
Unionville..................	29.99	50.57	11.31	0.72	2.47	5.14	=100.10	
Unionville	32.15	54.28	11.36	0.05	not estimated.	0.50	F̈e *trace.*	Hartshorne.
Buncombe County.......	29.17	48.40	9.87	1.24	6.15	3.99	HF 2.03 =100.80	Silliman, Jr.

My analyses were made in the ordinary way, only with more carbonate of soda than is usually employed. The alkalies were separated either by means of hydrofluoric acid or by carbonate of lime, which is preferable to the carbonate of baryta for the decomposition of the silicates.

It is seen that potash and soda are present in small quantities in all the specimens. The composition of this mineral is remarkable for the large proportion of alumina present; but when we look at its origin it is not astonishing to find a silicate of alumina with a small amount of silex.

I regard emerylite as a mineral of elimination from emery, the result of an effort by which the corundum in its formation purifies itself. It is not remarkable that from the mass in which the corundum crystallizes the silica, finding itself in presence with an excess of bases, combines with as large a quantity as its affinity admits of. In speaking of the formation of emery I have already alluded to a nodule in my possession that exemplifies this in a very exact manner.

Notwithstanding the recent discovery of emerylite, there is no other species of mica that can be considered so well established as this mineral or so constant in its composition. Up to the present time this mineral has not been found except with emery or corundum, which frequently contain it in the interior of the mass as well as on the surface. Some emeries contain it in such quantity that it has the aspect of gneiss, as I have already said with reference to certain specimens from Nicaria.

The most beautiful specimens of emerylite come from Naxos, and as the blocks of emery from this island frequently contain it there will be no difficulty in procuring specimens for cabinets. It is often mixed with diaspore.

MICA (MUSCOVITE?)

This mica is found on all the emeries which I have examined, but especially on that coming from Kulah. It is always in small plates on the surface of the emery. The analyses of four specimens are as follows:

LOCALITIES.	Silica......	Alumina..	Lime.....	Oxide of iron.....	Magnesia	Potash with little soda	Water....	Manganese.....
Gumuch-dagh......	42.80	40.61	3.01	1.30	trace.	not estimated.	5.62	trace.
Kulah................	43.62	38.10	0.52	3.50	0.25	7.83	5.51	trace.
Kulah................	42.71	37.52	1.41	2.32	trace.	not estimated.	5.95	trace.
Island of Nicaria..	42.60	37.45	0.68	1.70	trace.	9.76	5.20	trace.

The composition is very nearly that of the muscovite or Muscovy glass, and until further examination I shall retain it under that species, as particular care should be exercised in making new species among the micas.

CHLORITOID (A NEW VARIETY OF THIS MINERAL).

It is found with the emery of Gumuch-dagh in considerable abundance. Its structure is lamellar, cleaving without much difficulty, and the surfaces exposed are always very brilliant. In thin fragments it transmits the light and appears of a dark-green color. The powder is greenish-gray. Its hardness is 6 and specific gravity 3.52. Heated in the flame of the blowpipe, it loses water and becomes brown from the absorption of oxygen, but does not melt. When heated without being in contact with the air it loses its brilliancy and acquires the aspect of scales from the blacksmith's forge.

This mineral is attacked by the strong acids, but is only completely decomposed by sulphuric acid. Melted with four or five times its weight of carbonate of soda, it is rendered easily soluble in hydrochloric acid. Great precaution was taken to see that nothing but perfectly pure chloritoid was submitted to analysis, and the possession of well-crystallized specimens enabled me to do this without much difficulty.

The method of analysis was to break the mineral in small fragments, to place it in a small platinum crucible, which was introduced into an earthen crucible and surrounded by pulverized quartz; in one word, I pursued the same method as that for estimating the water in emery. For the other ingredients a new portion was taken, pulverized finely, and attacked either by concentrated sulphuric acid or melted with carbonate of soda, and afterward dissolved in hydrochloric acid with the addition of a little nitric acid evaporated to dryness, and treated with dilute hydrochloric acid. The liquid separated from the silica is treated with an excess of caustic soda, and the filtered liquid is neutralized by hydrochloric acid and the alumina precipitated by carbonate of ammonia.

The contents of the filter, which are essentially peroxide of iron, are placed in a capsule, dissolved by hydrochloric acid, heated and precipitated by ammonia, and thrown on a filter. From the filtered solution the lime and magnesia are separated in the ordinary way. The peroxide of iron remaining on the filter, after being well washed and dried, is weighed and decomposed in a current of hydrogen gas. To the oxide thus reduced nitric acid diluted with thirty times its weight of water is added, and digested at 100° to 120° C. for about an hour, stirring frequently, when, if the iron has been thoroughly reduced, it will be taken up by the acid, and a little alumina left, which is weighed and added to the first portion. Ordinarily I never have found more than from one to two per cent. of alumina with the oxide of iron. Care must be taken to decompose the iron completely, as otherwise the iron will not be entirely taken up by the acid. The mineral thus analyzed afforded as follows:

	Silica.....	Alumina.	Protoxide of Iron.....	Water.....	Lime.....	Magnesia	Titanic Acid.....	Manganese.....	Potash & Soda.....
Decomposed by sulphuric acid	24.10	39.8	27.55	6.50	not estimated.	not estimated.	not estimated.	not estimated.	0.30
Decomposed by carb. soda......	23.94	39.52	28.05	7.08	0.45	0.80	trace.	0.52	......
Decomposed by carb. soda......	23.20	40.21	27.25	6.97	0.83	0.95	trace.	not estimated.	......

These analyses correspond to the following composition:

	ATOMS.	AT. WEIGHT.	PER CENT.
Silica..........................	2	1133.40	23.87
Alumina.......................	3	1925.88	40.57
Protoxide of iron............	3	1350.00	28.44
Water..........................	3	337.50	7.12

The most probable formula is $\ddot{Al}^3\ \ddot{Si}+\dot{Fe}^3\ \ddot{Si}+3\dot{H}$.

The minerals which are brought under this species are the chloritspath or chloritoid of the Ural, the Sismondine of St. Marcel, and the Masonite of Rhode Island; their analyses and formulas are as follows:

	I.	Oxygen..	II.	Oxygen..	III.	Oxygen..	IV.	Oxygen..	V.	Oxygen..	VI.	Oxygen..
Silica...............	27.48	6	24.40	2	24.1	9	28.27	6	25.18	5	23.91	2
Alumina...........	35.37	6	45.17	3	44.2	15	32.16	6	33.61	6	39.52	3
Protoxide of iron	27.05 }	3	30.29	1	23.8	4	33.72	3	35.31	3	28.05	1
Magnesia..........	4.29 }											
Water	6.95	3	....	...	7.6	5	5.00	2	5.88	2	7.08	1

I. Chlorite spar or Chloritoid of the Ural by Bonsdorff. $(\dot{Fe}, \dot{Mg})^3\dddot{Si}+\ddot{\overline{Al}}^2\,\dddot{Si}+3\dot{H}$.

II. Chlorite spar of the Ural, Erdmann. $\dot{Fe}^3\,\ddot{\overline{Al}}+2\ddot{\overline{Al}}\,\dddot{Si}$.

III. Sismondine of St. Marcel, Delesse. $\dot{Fe}^4\,\dddot{Si}^3+5\ddot{\overline{Al}}\,\dot{H}$.

IV. Masonite of Rhode Island, Whitney. $\dot{Fe}^3\,\dddot{Si}+\ddot{\overline{Al}}^2\,\dddot{Si}+2\dot{H}$.

V. Chlorite spar according to Rammelsberg requires $3\dot{R}^3\,\dddot{Si}+2\ddot{\overline{Al}}^3\,\dddot{Si}+6\dot{H}$.

VI. Chrilotoid of Asia Minor, J. L. Smith. $\ddot{\overline{Al}}^3\,\dddot{Si}+\dot{Fe}^3\,\dddot{Si}+3\dot{H}$.

This mineral is found very abundantly with the emery of Gumuch-dagh; it covers the surface of the blocks, and sometimes enters largely into the substance of the emery. It is easy to see from the composition of this mineral that it is formed by elimination from the mass of emery at the time of its consolidation, which by this means tends to purify itself. The nodule of which I have already spoken under the head of emery and of emerylite goes to sustain this view of the question.

On the emery of the other localities I have not found this chloritoid. Its composition is not in perfect accordance with the known varieties of chloritoid, and differs from Sismondine (which it approaches most in composition) by its imperfect solubility in hydrochloric acid.

BLACK TOURMALINE.

This mineral is found abundantly with the emery of Naxos, and also in small quantities with that of other localities. It appears to have replaced the chloritoid that is found so abundantly with the emery of Gumuch-dagh.

The crystals are found agglomerated on the surface, and also disseminated in the interior of the emery. This mineral, like the last, is strongly basic, containing a little more than thirty per cent. of silica.

CHLORITE.

With the emery of Gumuch-dagh we find a chlorite. It is in compact masses, composed of an agglomeration of small crystalline plates, and contains octahedral crystals of magnetic oxide of iron. Analysis gives as its composition:

Silica	27.20
Alumina	18.62
Protoxide of iron	23.21
Magnesia	17.64
Water	10.61

It is identical with the chlorite of *Mont des sept-Lacs* which, gave M. Marignac

Silica	27.14
Alumina	19.19
Protoxide of iron	24.76
Magnesia	16.78
Water	11.50

It is the same as the chlorite of St. Christophe and the ripidolite of Rauris and of St. Gothard. The formula given by von Kobell is $2\text{Mg Al}+3(\text{Mg Fe})^2\,\text{Si}+6\text{H}$.

MAGNETIC OXIDE OF IRON.

This is found with the emery of every locality. It enters into the composition of the emery itself, and is also found on the surface in regular octahedral crystals. We find it frequently massive and of strong polarity. That of Gumuch-dagh contains a trace of titanic acid.

OLIGIST IRON.

It is associated with all the emeries, and sometimes enters into their composition. It is also found in detached masses, either amorphous or as crystallized specular iron.

HYDRATED OXIDE OF IRON.

This oxide of iron is not unfrequently found with emery, covering the surface. It is found with pyrites having resulted from the decomposition of this mineral.

IRON PYRITES.

Pyrites is found principally with the emery of Gumuch and Nicaria. At the latter locality it is in small crystals in the interior of the mass. At Gumuch it is principally on the surface, but much less abundant than at Nicaria.

RUTILE.

This oxide of titanium is found with the emery of Gumuch-dagh and of Kulah, where I obtained some large detached crystals. I have also a specimen with it in small crystals on diaspore attached to emery from Gumuch-dagh.

ILMENITE.

It has been found on the gangue of the emery of Kulah in minute crystals of the usual form of this mineral.

TITANIFEROUS IRON.

Titaniferous iron is found with almost all the varieties of emery that I have examined, but I have analyzed none but that associated with the emery of Nicaria. Care being first taken to see that it was anhydrous, one gramme of it was calcined in a current of oxygen and it augmented .019, which indicated the presence of .171 gramme of protoxide of iron, and corresponds to .190 gramme of peroxide of iron; the same portion then decomposed by a current of hydrogen gas, and the loss sustained was equal to .222 gramme of oxygen, which corresponds to .740 gramme of peroxide of iron; deducting from this the quantity of peroxide equal to the protoxide (.171) contained in the mineral, we have .550 gramme for the quantity of peroxide present. The mass reduced by hydrogen was treated with hydrochloric acid, and the part not dissolved (.230 gramme) was titanic acid with a little alumina. The acid solution contained .010 lime and a trace of alumina. The titanic acid was examined as to its purity and was found to contain no silica and only a trace of alumina. The result of the analysis is

Protoxide of iron	17.10
Peroxide of iron	55.00
Titanic acid	23.01
Lime	1.00
Alumina	a little, not estimated.

This titaniferous iron corresponds in composition to the Washingtonite of Prof. Shepard analyzed by M. Marignac, and to the titaniferous iron of Arendal analyzed by M. Mosander. Its sp. grav. is 4.78.

There are still two or three minerals that I have found associated with emery, but their specific characters have not been well established on account of the difficulty of obtaining enough in a state of sufficient purity for analysis.

The study of these accidental minerals in contact with emery has led to several general conclusions which have been mentioned under the description of the different species; and now I risk little in saying that the hydrates of alumina, as diaspore—as well as the silicates, as emerylite, chloritoid, and tourmaline—and the minerals of iron, as magnetic, titaniferous iron, etc., will be found almost every where with the emery and corundum.

My labors on this subject are thus terminated, and it is to be hoped that the examination of the emery of Asia Minor has served to elucidate the geology and mineralogy of this substance, until now but little known except in its uses.

RAPPORT

Sur un Mémoire de M. Laurence Smith, *ayant pour objet l' étude du gisement de l' émeri de l' Asie Mineure, et des minéraux qui y sont associés.*

Commissaires, MM. Cordier, Élie de Beaumont, Dufrénoy rapporteur.

CONCLUSIONS.

"Il résulte de l'exposé sommaire que nous venons de donner du travail de M. Smith, que ce géologue a fait connaître:

"1°. La nature précise du gisement de l' émeri dans l'Asie Mineure et dans l'Archipel grec.

"2°. Qu'il a décrit la manière d'être et les propriétés des minéraux qui lui sont associés, notamment du diaspore et de l' émerilite; ce dernier minéral forme, par l' identité de sa composition dans les divers gisements où l' auteur l' a étudié, un mica constituant une espèce nouvelle et bien déterminée.

"3°. Enfin, qu'il a donné un moyen pour déterminer les qualités de l' émeri, ainsi que leur valeur commerciale; ce procédé, éminemment pratique, offre en outre de l'intérêt sous le point de vue scientifique, en ce qu'il permet d'apprécier la différence de ténacité de minéraux de dureté égale.

"Ces recherches de géologie, de minéralogie et de chimie analytique constituent, par leur ensemble, ainsi que par les faits nouveaux qu'elles fournissent à la science, un travail du plus haut intérêt. Vos Commissaires vous proposent, en conséquence, de remercier M. Smith de la communication qu'il en a a faite à l'Académie, et, vu l' importance de ce travail, d'en ordonner l' insertion dans le *Recueil des Mémoires des Savants étrangers.*"

Les conclusions de ce Rapport ont été adoptées.

For the Report in full see *Comptes rendus des séances de l'Académie des Sciences*, October 28, 1850.

EMERY MINE OF CHESTER,

HAMPDEN COUNTY, MASS.,

WITH REMARKS ON THE NATURE OF EMERY AND ITS ASSOCIATE MINERALS.

Considerable interest is attached to the recent developments of an extensive deposit of emery in Chester, Hampden County, Mass., by Prof. C. T. Jackson; and my name has been associated in various ways with it, without my having had any thing directly to do with it. Sundry communications have also been received by me from various parties These communications are best answered by the facts embraced in this article, some portion of which it has always been my intention to publish, without reference to the special interest of any one in the matter.

Prior to 1846 emery was simply known as a mineral, coming to us from a few remote localities, and was used in the arts without our having any knowledge of its true geological position or its mineralogical relations. About that period circumstances favored my commencing those geological and mineralogical discoveries in relation to emery that were afterward embodied in two papers presented to the Academy of Sciences of Paris in 1850, in which the subject was thoroughly discussed, and I might say almost exhausted. The light in which those discoveries were considered will be seen by the conclusions of the report of the committee of the Academy, consisting of Messrs. Dufrenoy, Elie de Beaumont, and Cordier, viz.:

"It results from the review just given of the labors of Dr. Smith, that he has made known—1. The precise nature of the geology of emery in Asia Minor and the Grecian Archipelago. 2. That he has described the properties of the principal minerals associated with it, and the manner in which they occur, especially diaspore and emerylite; this last mineral forms, by the identity of its composition in the different formations that the author had occasion to study, a mica constituting a new species, and one well determined. 3. Finally that he has given a means for determining the qualities of emery, and consequently their commercial value. This process, eminently practical, offers besides an interest in a scientific point of view, inasmuch as it permits of determining the difference in the tenacity of minerals of equal hardness. These researches of geology, mineralogy, and of analytical chemistry constitute a work of the highest interest, both as a whole as well as from the new facts they promise to science. Your committee consequently propose to thank Dr. Smith for having communicated them to the Academy, and in consideration of the importance of the work, to order the insertion of his paper in the *Receuil des Mémoires des Savants étrangers.*"

At that time I had discovered six new localities of emery in Asia Minor and the Grecian Archipelago. Those localities were far removed from each other, and furnished so many different places for the study of emery and its associate minerals in addition to the old locality of Naxos; and consequently many points of general interest were brought out, besides others connected with the line of study. Those who may feel interested in the subject will find the investigation and results there arrived at in the American Journal of Science and Arts, Vols. X. and XI., 1850 and 1851; they embrace the geology, mineralogy, chemical composition, manner of mining, commercial considerations, associate minerals, etc.

The study of the associate minerals I considered of great importance, as they would be guides in future explorations in other parts of the world; and even prior to completing the researches on the subject I wrote to Professor Silliman and asked him to examine the American corundum localities for these minerals, one of them in particular, which he immediately did. With the corundum from the locality in Chester County, Pennsylvania, and Buncombe County, North Carolina, he "soon found the mineral indicated," and communicated the same to the American Journal of Science and Arts, November, 1849, pp. 379 and 383.

Nothing further came to my notice in relation to emery until I received from Prof. C. T. Jackson a letter dated October 9, 1864, containing what follows: "You discovered emerylite or margarite in Asia Minor as an associate mineral with emery. On the 22d of October last, 1863, I discovered, while surveying an iron-mine in Chester, Mass., some beautiful veins of the margarite, from half an inch to two inches wide, and of a fine, delicate rose color, or light-pink. The nature of this mineral I did not discover until my return to Boston, but at first supposed it was lepidolite; on analysis it proved to be margarite, and from that I ventured to predict the occurrence of emery; but no attention was paid to this prediction by the owners of the mine, who were more intent on the iron-ore. A few weeks since I saw Dr. Lucas, one of the owners, resident in Chester, and called him into my office, and explained to him the great value of emery, and told him how to detect it, and he promised to make the search I required, and took exact directions from me. The next day after his return to Chester he found the emery, a big vein nearly six feet wide, which had been mistaken by him for iron-ore, it being very magnetic. I write you this to show you the importance of your discovery of the emerylite or margarite (for this appears to be identical) as an associate of emery, and also as an interesting case of deduction from scientific memoirs."

Accompanying the letter he sent me a paper giving me a summary of a communication he had made to the Boston Society of Natural History on the subject, concluding by remarking that "had not the occurrence of emerylite and chloritoid called his attention to the probable existence of emery at this locality, it would have been overlooked to this day, and no one knows how much longer. The fact was mentioned as an example of the real uses of supposed useless minerals; and the Doctor took occasion to express his obligations to Dr. Smith of Louisville for his valuable contributions to our knowledge of the associate emery minerals of the Grecian Archipelago and Asia Minor."

These statements are sufficient to show how far my geological observations served as a guide to Prof. C. T. Jackson in his deductions with reference to the existence of emery in Chester, and with what diligence Dr. H. S. Lucas followed up the latter's directions, resulting in the valuable development of emery.

I have since visited the locality, having done so in the month of March last. The geological character and position of the rocks was not as well made out by me as might have been done in a more favorable season; but as my observations accord, as far as they go, with those of Dr. Jackson and Prof. Shepard, I prefer inserting their observations, rather than my own, in describing the geology of the emery locality.

"The mine is situated nearly in the center of the Green Mountain chain as it traverses the western border of the state, at a point not far from half way between the Connecticut and Hudson rivers. It is included in the metamorphic series of rocks, here consisting of vast breadths of gneiss and mica-slate, with considerable interpolations of talcose slate and serpentine. The general direction of the stratification is N. 20° E. and S. 20° W., the relation to the horizon varying from vertical to a dip of from 75° to 80°, sometimes east, sometimes west.

"The immediate vicinity of the mine presents a succession of lengthened rocky swells with rather precipitous sides, having summits between seven hundred and fifty and one thousand feet above the level of the principal streams by which the hills are traversed. The longer axis of the elevations generally coincides with the directions of the strata.

"The emery-vein traverses in an unbroken line the crests of two of these adjoining mountains, and scarcely deviates as a whole from the magnetic meridian. Each mountain is estimated to have a length of two miles, thus giving four miles extent to the metalliferous stratum, for such it may be truly called, consisting as it does so largely of the metals iron and aluminum. The Westfield River, here a small stream about four rods in width, flows directly across the northern end of the vein, dividing it into two equal portions. The height of each mountain is estimated at seven hundred and fifty feet.

"The emery-vein, whose average width may be taken at four feet, is situated at the junction of the great gneiss formation constituting the western flank of the mountains, with the mica-slate forming the eastern slope. To speak more exactly, however, it lies just within the gneiss, having throughout a layer of this rock of from four to ten feet in thickness for its eastern wall. Nor does the mica-slate advance quite up to this outside layer of the gneiss; but in place thereof an extensive intrusion of talcose slate occurs, having an average thickness of twenty feet on the south mountain, and widening out at the north mountain to a breadth of nearly two hundred feet as it reaches the terminus of the vein, in the bed of the Westfield River.

"The gneiss, more especially in the vicinity of the vein, is a very peculiar rock. It abounds in thick seams of a coarse-grained, very black and shining hornblende, and where this is not found, it is much veined and penetrated by epidote. The stratification is much contorted also; and when the surface of the formation happens to be weathered or water-worn, its basseting edges strikingly resemble in color some of the serpentine marbles. It is also noticeable that in it quartz is every where singularly deficient. Traces of a white calcareous spar (calcite) are now and then visible upon the joints of the gneiss, with occasional specks of yellow copper, together with malachite stains; but no corundum, emery, or magnetite particles have thus far been detected as constituents of the gneiss. It is quite otherwise, however, with the talcy rock exterior to the wall of gneiss; for that formation in all its different varieties of talcose slate, soapstone, chloritic aggregates (with included seam of indianite), talcy dolomite, etc., which together constitute the stratum separating the gneiss from the mica-slate, contain here and there disseminated grains of either emery, corundum, or magnetite; but, like the gneiss again, are strikingly free from quartz or uncombined silica in any of its forms. Indeed, this generally abundant substance is altogether wanting, not only in the emery vein, but in the talcose formations constituting its eastern boundary.

"It makes its appearance, however, in abundance in the mica-slate as soon as the talcose rocks are passed—showing itself not only as the usual constituent of the slate, but in more or less continuous seams, from a few inches thick up to above six inches, and sometimes a foot, in width. Where the seams are thin and discontinuous, the included masses thin out at each end before disappearing, the sharp edges being curved in opposite directions, so as to form frequent white patches upon the surface of the rocks in the shape of the letter S."

MINERALOGICAL CHARACTER AND COMPOSITION OF THE CHESTER EMERY.

It resembles more nearly that from Gumuch-dagh (near Ephesus) than any other that I know of. It is of a fine grain, and dark-blue bordering on black, not unlike certain varieties of magnetic iron-ore; with it there are frequently found pieces of corundum of some size. The interior of the mass is free from micaceous specks, such as are found in the emery of Naxos. Its powder examined under the microscope shows the distinct existence of more than one mineral, which are often so inseparably connected that the smallest fragments contain them together. The two predominating are *corundum* and *magnetic oxide of iron.*

Several specimens were submitted to chemical examination from those most largely impregnated with magnetic oxide of iron to those that appeared to contain the least. They all consisted essentially of alumina and oxide of iron; but I invariably found a little titanic acid and silica, and most commonly a minute quantity of magnesia. No. 1 was an inferior specimen; No. 2, the better quality of rock; No. 3, the emery rock crushed and prepared for market in the form of emery; No. 4, the same, and called emery crystals.

	1	2	3	4
Alumina..................	44.01	50.02	51.92	74.22
Magnetic oxide of iron	50.21	44.11	42.25	19.31
Silica	3.13	3.25	5.46	5.48

I examined a specimen of No. 2: grain fine, and treated repeatedly with hydrochloric acid and water over a water-bath; a great deal of oxide of iron and a little alumina were dissolved; the residue on analysis proved to be nearly pure corundum, giving

Alumina	84.02
Magnetic oxide of iron..	9.63
Silica*	4.81

All the chemical and physical examinations made go to show that the emery of Chester is, like all other emeries, a mixture of corundum and oxide of iron—a fact that will be reverted to again a little further on.

Prof. Jackson analyzed two specimens, after digesting them with nitro-muriatic acid, and has given as the composition

	1	2
Alumina..	60.40	39.05
Protoxide of iron............................	39.60	40.95

* No attempt was made to estimate the water.

and then goes on to state, "from which it would appear that protoxide of iron is an essential chemical ingredient in emery, and not an accidental admixture.* Dr. J. Lawrence Smith's experiments lead to the same result, but he considers the oxide of iron to be an irregular mixture with the alumina and not a regular chemical constituent. In either case I think emery ought to rank as a separate species, and not as a granular variety of corundum, from which it differs so in physical characters."

I would here remark that Dr. Jackson's conclusion would be correct in the first state of the case, were the iron an essential chemical ingredient; but in the latter it would be erroneous, and introduce inextricable confusion into the science of mineralogy by admitting mere mechanical mixture as a specific distinction.

Prof. C. U. Shepard, writing on the same point, says: "His conclusions (Dr. Jackson's) would obviously be acquiesced in were it not for the strong resemblance in striæ and cleavage between the emery and common corundum, making it impossible for us to separate the substances crystallographically from one another. Nothing like a perfect crystal of emery has yet been found at the mine; but it is quite remarkable that the mineral is here generally coarsely massive, or in large separate individuals often of the size of kernels of Indian corn, whose cleavages are perfect, and which present on their planes the delicate striæ so characteristic of corundum from the Carnatic." Yet Prof. Shepard is for making emery a new mineral species and calling it *Emerite*, with the formula $\dot{\mathrm{F}}\mathrm{e}\,\dddot{\bar{\mathrm{A}}}\mathrm{l}$.

If the views of Profs. Jackson and Shepard are to be taken as correct, the question as to the mineralogical position of emery is easily settled without resorting to any new mineral species. It is simply a massive iron-spinel (hercynite) with the anomaly of having a hardness equal to corundum.

	IRON-SPINEL.	EMERIES.	
		Jackson.	Shepard.
Alumina	58.75	60.49	$\dddot{\bar{\mathrm{A}}}\mathrm{l}\,\dot{\mathrm{F}}\mathrm{e}$
Protoxide iron	41.25	39.60	

I would say at this point that if the mineral of Chester is to

* An examination of my analyses in 1850, which it is supposed are the ones referred to here, most certainly do not sustain the conclusion.—J. L. S.

be regarded as an aluminate of iron, the rock called emery coming from Naxos and other well-known localities is not that compound, *and that if one is emery the other is not.* But as I do not take their view of the matter, *I consider the Chester mineral as true an emery as that of Naxos.*

As there seems to be some mistake and incorrect quotation in regard to my analyses of emery and corundum, I reproduce the tabular statement of the analyses and effective hardness, referring the reader to the original paper for a correct view of what is understood by the effective hardness.

No.	LOCALITY.	Effective Hardness.—Sapphire 100.	Specific Gravity	Composition.				
				Water	Alumina.	Magnetic Oxide of Iron	Lime	Silica
	EMERY.							
1	Kulah	57	4.28	1.90	63.50	32.25	0.92	1.61
2	Samos	56	3.98	2.10	70.10	22.21	0.62	4.00
3	Nicaria	56	3.75	2.53	71.06	20.32	1.40	4.12
4	Kulah	53	4.02	2.36	63.00	30.12	0.50	2.36
5	Naxos	46	3.75	4.73	58.53	24.10	0.86	3.10
6	Nicaria	46	3.74	3.10	75.12	12.06	0.72	6.88
7	Naxos	44	3.87	5.47	69.46	19.08	2.81	2.41
8	Ephesus	42	4.31	5.62	60.10	33.20	0.48	1.80
9	Kulah	40	3.89	2.00	61.05	27.15	1.30	9.63
	CORUNDUM.							
1	Sapphire of India	100	4.06	......	97.51	1.89	......	0.80
2	Ruby of India	90	......	......	97.32	1.09	......	1.21
3	Corundum, Nicaria	77	3.88	1.60	92.39	1.67	1.12	2.05
4	Corundum, Asia Minor	65	3.92	0.68	87.52	7.50	0.82	2.01
5	Corundum, Asia	60	3.60	1.66	86.62	8.21	0.70	3.85
6	Corundum, India	58	3.89	2.86	93.12	0.91	1.02	0.96
7	Corundum, Asia	57	3.80	3.74	87.32	3.12	1.00	2.61
8	Corundum, India	55	3.91	3.10	84.56	7.06	1.20	4.00
	EMERY.							
1	Chester, Massachusetts	33	......	......	44.01	50.21	......	3.13
2	Chester, Massachusetts	40	......	......	50.02	44.11	......	3.25
3	Chester, Massachusetts	39	......	......	51.92	42.25	......	5.46
4	Chester, Massachusetts	45	......	......	74.22	19.31	......	5.48
5	Chester, Massachusetts	......	......	......	84.02	9.63	......	4.81

By the above it will be seen that the magnetic oxide of iron in the emery of Naxos, Ephesus, etc., varies from 13 to 33 per cent., the water from 1.9 to 5.0 per cent., the silica from 1.6 to 9.6 per cent. All of these ingredients form minerals apart from the corundum, which is represented by the principal portion of alumina. Some of the alumina found in the analysis is associated with the above ingredients to form associate minerals which have been fully studied. This last will serve to explain why it is that emeries having the same amount of alumina may have different degrees of effective hardness. Thus: Nos. 9 and 4, both Kula emeries, containing about the same amount of alumina, have effective hardness in the proportion of 40 to 53; but it will be seen that No. 9 contains 9.6 per cent. of silica, which doubtless appropriates a portion of the alumina, thus reducing the alumina attributable to corundum; so that were it possible to ascertain the exact amount of corundum present in 9 and 4, it would doubtless be in proportion to their effective hardness. So again, if we compare Nos. 8 and 1, the effective hardness will be found in the proportion of 42 to 57, while their amounts of alumina vary only as 60 to 63; but if we regard the amount of water in the two, it is as 5.6 to 1.9, much of this water coming from diaspore that is intimately mixed with the corundum; and in several specimens I possess the two minerals shade into each other so completely that it is impossible to tell where one begins and the other ends. The above facts were all well examined when my first memoirs appeared on this subject, which accounts for the following remark then made: " Those emeries which contain the least water, every thing else alike, are the hardest, as instanced by that from Kulah, notwithstanding the quantity of iron it contains. The silica existing in emery is most often in combination with alumina or the oxide of iron, or both; for this reason we must not always regard the quantity of alumina as an indication of the quantity of corundum in emery."

In concluding this part of the subject I would state that while I do not consider my opinions infallible in this matter, still all my experience and research gathered from such varied sources point to the conclusion that emery is a mixture of several minerals, principally corundum and magnetic oxide of iron, the former being the effective agent in the mechanical abrasion to which it is applied; the oxide of iron is not to be considered as an unimportant ingredient, it serving by its presence to destroy to some extent the harsh cutting action of the corundum.

MINERALS ASSOCIATED WITH THE EMERY OF CHESTER.

Corundum.—This mineral, as might naturally be expected, is found with the emery, sufficiently distinct and separate to be at once recognized, sometimes in thin seams, massive in its character, but more commonly in flattened crystals of small dimensions.

Diaspore.—Very excellent and beautiful specimens of this hydrate of alumina have been found at this emery locality; it is often in distinct and separate prismatic or bladed crystals, quite colorless and transparent.

Emerylite or Margarite.—Some of the finest specimens of this mineral that are known have been found at this locality. It will be seen by referring to my former papers on emery that I first discovered this mineral associated with emery; its composition showed it to differ from any other then known mineral. I compared it subsequently with margarite, which had been discovered before, and suspected the identity of the two minerals; but as the analysis made out and accepted as the composition of margarite did not accord with that of emerylite, I undertook to re-examine margarite, when I found that its composition had been erroneously determined, and that it was in fact the same mineral with emerylite, which last name has had to yield to the priority of date of the other.

I have analyzed the margarite from Chester and find its composition as follows:

Silica	32.21
Alumina	48.87
Lime	10.02
Oxide of iron	2.50
Manganese	.20
Magnesia	.32
Soda and little potash	1.91
Lithia	.32
Water	4.61

There is a little titanic acid with the oxide of iron that I did not estimate.

Chlorite.—This mineral as found with the emery is the so-called *corundophilite* of Shepard. On examination it proves to be, both chemically and physically, a chlorite of the variety ripidolite.

Biotite.—In examining a specimen of dark-green micaceous mineral which I took to be chlorite (the corundophilite of

Shepard) and from its purity expected to get a very accurate idea of its composition, but in the very commencement of the examination it was discovered to be well-characterized *biotite.*

This mineral occurs on the surface of a white rock that Professor Shepard calls indianite, but which I have not had time to examine. It is in small thin micaceous crystals perpendicular to the surface of the indianite; in the mass it is of a dark-green color, so dark that at a little distance it looks like lamellar plumbago. A careful analysis gave the following composition:

Silica	39.08
Alumina	15.38
Magnesia	23.58
Peroxide of iron	7.12
Manganese	.31
Potash	7.50
Soda	2.63
Water	2.24
Fluorine	.76
	98.60

This corresponds with the composition of the biotite from Monroe County, New York, as made out by Prof. Brush and myself in our re-examination of American minerals several years ago.

Corundophilite proved to be a Chlorite.—About the time I published my memoirs on emery in 1850 and 1851, Prof. Shepard made the announcement of a new mineral (American Journal of Science and Arts, 1851, XII, p. 211), stating that it "occurs with corundum near Ashville, in Buncombe County, North Carolina, in imperfect stellate groups, and also spreading out in laminæ between the layers of corundum; color leek-green, etc." An analysis afforded silica 34.76, protoxide iron 31.25, alumina 8.55, water 5.47, *making a loss of nearly* 20 *per cent.*, *a portion of which he attributes to alkalies;* neither lime nor magnesia were detected. He operated on one hundred and forty milligrammes. This mineral was considered a new one, and Prof. Shepard called it corundophilite. Supposing that I had observed the same mineral in certain specimens of emery and emerylite from Chester, Massachusetts, I inclosed a fragment of the specimen to Prof. Shepard to ascertain if this was the mineral he called *corundophilite;* he returned the specimen announcing that it was. I then analyzed the same and found it to be, both chemically and physically, a chlorite, identical no doubt with the chlorite I found associated with the emery of Asia Minor. Both the Asia Minor and Chester varieties occur in compact mass, composed of an agglomeration of small crystalline plates—identical with the chlorites of Mont des Sept-Lacs and of St. Christophe, and the ripidolites of Rauris and St. Gothard. In the following analysis I do not pretend to furnish that of the pure mineral, as from the thinness of the layers in the specimens at my disposal it can not be separated in that state of purity I am in the habit of seeking for in all minerals that I examine:

Silica	25.06
Alumina	30.70
Protoxide of iron	16.50
Magnesia	16.41
Water	10.62
	99.29

The optical characters were not examined, there being no means at hand.

I may remark that the alumina and magnesia were separated by resolution and reprecipitation three times.

Tourmaline.—This mineral is also found with the emery of Chester in the same manner as with the emery of Naxos.

Titaniferous iron (ilmenite).—This is found principally in flattened crystals in the margarite.

Oxide of titanium (brookite or rutile.)—With the diaspore we found some beautiful flattened hair-brown crystals. The specimen in my possession does not furnish the face of the crystals so as to enable me to make out what form of titanium oxide it is. Prof. Shepard thinks he has sufficient evidence to pronounce it to be brookite.

Magnetic oxide of iron—This ore of iron is found in great abundance associated with the emery, and is worked for the manufacture of iron; it contains a little oxide of titanium.

The above, as well as some other associated minerals of less importance, justify the concluding remarks of my paper on emery fifteen years ago, viz.: "I do not risk much in saying that the hydrate of alumina or diaspore, as well as the silicates, as emerylite, chlorite, and tourmaline, and the minerals of iron, as magnetic, titaniferous iron, etc., will be found almost every where with the emery and corundum."

5

MINERALS OF CHILE.

The minerals collected by the United States naval astronomical expedition were almost exclusively those of silver and copper. The specimens of the ores of these two metals, taken in connection with all authentic accounts, would lead one to believe that Chile hardly has a parallel in any region in the globe for the abundance as well as purity of these ores. Were it not for the physical difficulties connected with the surface of the country, and the scarcity of water and fuel, the wealth accruing to Chile from the working of these mines would be far greater than it is now.

Although the expedition furnishes no geological report of the country, it is thought proper, before describing the minerals in detail, to give some general idea of the geology of the country, more especially as connected with the minerals collected; and for this purpose recourse is had to the labors of M. Domeyko and M. L. Crosnier, as published in the "*Annales des Mines.*"

A general idea of the geological structure of Chile is readily formed, although we might be led to suppose otherwise from the great disturbing forces that have operated in that part of the world in the form of injected masses of igneous rock, as well as from the present changes produced by existing volcanic action, and the gradual elevation of the whole country, with daily recurrence of earthquake action. These disturbing forces do not, however, in any way interfere with our study of the general geology of the country, while of course it renders the investigation of the geology of any particular region exceedingly embarrassing.

The great chain of the Andes extends parallel to the coast of Chile at a distance of from ninety to one hundred miles. On the eastern side it descends by gradual slopes toward the immense plains of the Argentine Republic. On the western side, where the upheaving force appears to have concentrated all its energy, the slopes are abrupt, and transformed frequently into vertical precipices of considerable height. The mountains appear heaped confusedly one on top of the other, and the first impression is that in the midst of so much confusion it is vain to seek for the primitive condition of the surface of Chile. Stratified rocks disappear entirely from north to south for the mean width of forty-five miles—from the desert of Atacama to Valdivia. These rocks, although they once existed, are now profoundly altered or entirely melted by contact with the enormous masses of granite. The clay-shales, which doubtless constituted the mass of the original stratified rocks, are now transformed into porphyries of every shade and of the most varied composition, alternating in some parts with beds of compact quartz. Even when the rocks are seen stratified, far removed from the masses of granite, and in beds sensibly horizontal or little inclined, still the numerous injected veins which traverse them and ramify in all directions prove that hardly any where have the rocks escaped the modifying force of igneous action.

Two immense granite elevations appear to have disturbed Chile in its entire length parallel to the coast. One is immediately on the coast, with an average breadth of forty-five miles, while the other is one hundred miles east in the midst of stratified rocks. The first range plunges into the sea, having valleys in various parts of it filled with tertiary deposits. As regards the respective ages of these two ranges there appears to be a difference of opinion; some supposing that the range on the coast was first upheaved, and at a subsequent period the inner range, while others suppose them to have originated at the same time. But whichever one of these suppositions is true, the general characters of the rock of the two ranges are the same, as well as the metalliferous veins and accompanying vein rocks. Associated with the granite of these ranges are hornblende rocks of the greatest variety, porphyries of all shades, containing crystals of feldspar sometimes of considerable size. Besides these, there are other compact rocks which can not be properly classified.

The principal masses of secondary rocks that lay between the two ranges of mountains are composed of metamorphic porphyry of a great variety of shades of color. Sometimes the porphyry is entirely altered; it then contains well-formed crystals of feldspar, and appears to have been melted where it now rests; and at other times it is earthy, as if the transformation had been incomplete. Large masses of reddish, yellow, and violet quartz alternate with the porphyry in certain points; also calcareous beds, sometimes fossiliferous. These stratified rocks are elevated on the flanks of the Andes, and form some of the most prominent peaks of this range. These strata are so completely pierced and elevated in every direction by the masses of granite as to modify in every possible manner their direction, inclination, and mineralogical character.

Besides the secondary stratified rocks just made mention of, there are other stratified rocks which are horizontal, having been deposited since the elevation of the mountain-chains. They are all, however, of recent origin and of small extent, disseminated along the coast, with the exception of the sandy plain that extends between Huasco and Copiapó, having a length of from one hundred and twenty to one hundred and thirty miles, with a variable width. This plain has, however, been elevated since its formation; in fact, M. Domeyko has determined three distinct terraces of successive and gentle elevation. There are also alluvial deposits now going on in some of the valleys of the elevated portions of the mountains, consisting of a fine clay, transported there by the mountain streams.

According to the observation of M. Crosnier, he has encountered but one formation that appears to be of lacustrine origin, and this is situated in the cordilleras of Chillan, forty-five miles north of Lavederos.

The tertiary deposits subsequent to the elevation of the Andes contain in many parts lignite. Some of these places are worked. The principal mines are situated to the south of Biobio, some twenty miles distant from the mouth of this river, on the sea-shore. The mines are called Lota and Lotilla.

Some of the departments of Chile have been examined with minuteness by M. Domeyko, more especially that of Copiapó; which, although little else than a vast desert, is the richest department of Chile in mines of every description, there not being a single mountain where the veins are not of sufficient importance to be worked. And it is worthy of remark that no mines are found higher than four thousand five hundred feet above the level of the sea; and this peculiarity I believe pertains to all parts of Chile.

Taking the Bay of Copiapó as a starting-point and going east, we find the underlying rock of the country granite, the surface being covered with tertiary deposits of very modern origin, the same that is found at the mouth of all the Chilean rivers. These deposits form two and three terraces, and consist principally of sand mixed with shell and gravel. At about six miles from the sea solid calcareous beds show themselves, containing species of crustaceæ now found living on the shore. The granite of this coast is fine-grained, having the same aspect as that in the neighborhood of Coquimbo, and is the same as that of the mountains of Carrisal, San Juan, and La Higuera, celebrated for their copper-mines. Granite-hills project frequently above the tertiary planes that extend to and rest on the first chain of granite-rocks which are low and rounded. It is in these rocks wherever seen, whether on the coast or projecting above the tertiary planes, or, when still further east, projecting through secondary strata, that the copper and gold are found. A good example of this is the Cerro del Cobre mountain, which elevates itself at the bottom of the valley of Copiapó. This mountain is composed of an elevated mass of porphyritic diorite, traversed by veins of iron and copper ores, containing considerable quantities of magnetic iron and ferruginous oxide of copper, copper pyrites, etc. It forms a species of granitic island in the midst of stratified porphyritic and other compact rocks more or less calcareous, and preserves all the characters of the coast rocks, even to the nature of the veins that it contains.

Further east, overlying the granite and dioritic rocks, are stratified porphyries; and here, at a height of two thousand two hundred and fifty feet above the level of the sea, as at Ladrillos, commence the indications of silver, disseminated in extremely fine particles of chloro-bromide; but on excavating this indication soon disappears, and it is not until we reach a more elevated point that silver is found very abundantly, and where the stratification becomes more perfect. Above the stratified porphyries there are calcareous and schistose rocks, more or less disturbed from their original position.

What is here said of the geological structure of the country east of Copiapó is true of many other parts of Chile from the coast eastward. From these general views of the geology of Chile I next pass to the consideration of the minerals collected by the expedition, accompanying the mineralogical discription of them with an account of the manner of their occurrence. For the latter I am also indebted to the geologists already made mention of.

NATIVE GOLD.

The specimens of this metal were contained in quartz-rock, exhibiting all the usual characteristics of auriferous quartz. The gold contains silver with but a trace of copper. In Chile this metal is found in veins as well as in the drift; the whole granite of the country is traversed by quartz containing more or less gold associated with the peroxide of iron, and at some depth from the surface with iron pyrites; sometimes with cupreous pyrites, arsenical pyrites, blende, galena, and sulphuret of anatomy. These veins by their decomposition furnish auriferous deposits of considerable extent, that are now worked.

Mention is made by M. Crosnier of a number of gold deposits irregularly disseminated in the midst of decomposed granite and red clay, which contains a large quantity of peroxide of iron, and which appears not to have originated from the decomposition of regularly-formed veins. This fact is apparent in the neighborhood of Valparaiso. It is also stated that gold is found in clay more or less ferruginous, arising from the decomposition of the granite in the most elevated portions of certain mountains; consequently in a situation where it should not have been carried by water.

It is supposed that the gold came up with the mass of granite at the time of the elevation of the latter, and not by subsequent injections of veins; and in most instances iron pyrites is regarded as its original associate. This character of auriferous formation is of course the exception, as in most instances the gold is traceable to regular veins, or to the decomposition of these veins. Although gold seems to be quite generally distributed through Chile, but few of the deposits remunerate exploration; the most extensive are on the flanks of the Andes, about forty miles east of Chillan, where it exists to the depth of thirty-five feet, in a very fine yellow clay mixed with black sand. The yield of gold is not very great.

NATIVE COPPER.

This is very commonly found in all the copper-mines of Chile. In one specimen from Andacollo (Coquimbo) it was found crystallized in modified octahedrons; it is very commonly associated with the red oxide of copper, as beautifully shown by a specimen from Illapel (Coquimbo). It is also found in quartz at Andacollo (Coquimbo). Others of the specimens came from San José, San Pedro Nolasco, Hinchado, Higuera, and Aconcagua.

RED COPPER.

This mineral is found beautifully crystallized in octahedrons more or less modified. The most beautiful specimens of this description are from Coquimbo; other specimens are massive and granular.

Its hardness is 3.5; specific gravity 5.9. Its color is various shades of bright-red, and the crystals are transparent, although from the exceeding intensity of their color they must be examined by a strong light.

This mineral is quite brittle, and is composed of

Copper	88.88
Oxygen	11.12
	100.00

Formula is Cu^2 O.

It sometimes forms veins coated with green and blue silicates of copper in the mines of Camarona and Cortadera, in the province of Coquimbo. In the Andacollo mine it is found pure and abundant below the oxysulphuret, resting on metallic copper, with which it is very commonly mixed. Aconcagua also afforded specimens. At Illapel it is found containing native silver.

CAPILLARY RED COPPER.

This beautiful form of the oxide of copper is found in fine, delicate rhombohedral crystals. It was found in the cavities of massive specimens of the red copper from Aconcagua. The crystals are as small as the finest hair, and sometimes half an inch in length. Its color is crimson-red; specific gravity 5.8. Its composition is the same as the last described mineral.

TENORITE OR BLACK OXIDE OF COPPER.

This is found massive, almost always mixed with other minerals of copper. It has a black metallic luster, and when pure contains

Copper	79.86
Oxygen	20.14
	100.00

Its formula is $Cu\,O$.

ATACAMITE.

This mineral was first discovered in the sands of the desert of Atacama, and hence its name. It is crystallized in modified rectangular prisms and rectangular octahedrons. Its color is of a dark emerald-green, almost black at times. It is translucent; has a hardness of from 3 to 3.5, and a specific gravity of about 4. It consists of water, chloride, and oxide of copper, and contains, according to analysis of Ulex—

Chlorine	16.12
Oxide of copper	56.23
Water	11.99
Copper	14.56
Silica	1.20
	100.00

Corresponding to the formula $Cu\,Cl+3\,\dot{Cu}+3\,\dot{H}$.

This mineral is also found in the district of Tarapaca. It is ground up in Chile, and is used as powder for letters under the name of *arsenillo.*

COPPER GLANCE.

The specimens of this mineral examined were all massive, of a black metallic luster, soft, and easily cut with a knife, having a specific gravity of 5.7. It commonly has green and blue carbonate disseminated through the mass. It is composed of

Copper	79.8
Sulphur	20.2
	100.0

Having for its formula $Cu^2\,S$.

It is most abundant in those mines furthest from the coast, existing in secondary stratific porphyry, and sometimes containing a notable amount of silver. It is also found abundantly in the mines of Chile that are near the coast, and are in dioritic and porphyritic rocks; but in them it is rarely found pure, being almost always mixed with the black oxide of copper or the oxychloride. The specimens examined were from Copiapó, although there are numerous localities. It is remarkable that at San Antonio this mineral is associated with native silver, and yet often contains hardly more than one thousandth of this latter metal. Specimens of pure sulphuret of copper are found, in which metallic silver is imbedded in the form of grains or little plates; and the same sulphuret contains grains and plates of native copper entirely separate from the silver.

ERUBESCITE OR PURPLE COPPER.

This is one of the most abundant of the minerals of copper found in Chile. It is procured in large quantities at the mines of Tamaya, in Coquimbo, Los Sapos, and Higuera. No crystals were seen. It is massive, of a purplish, variegated color, with a metallic luster. It is brittle, and not very hard. When the surface is freshly broken it is of a brass color that very often tarnishes, acquiring a purplish hue. The massive varieties of this mineral always vary more or less in their composition. The specimens examined contained from fifty-five to sixty-five per cent. of copper. Three specimens, that have been thoroughly analyzed by M. Domeyko, gave

	Tamaya.	Los Sapos.	Higuera.
Copper	66.7	56.1	59.5
Iron	8.9	17.7	18.2
Sulphur	22.8	23.1	20.5
Quartz	1.6	3.1	1.8
	99.8	100.0	100.0

The formula is $Fe\,S+2\,Cu^2\,S$.

This mineral furnishes a great deal of the copper produced in Chile.

COPPER PYRITES.

This is the most abundant copper-ore of Chile, and is found in immense quantities in the province of Coquimbo. Some of it, as that from Tamaya, contains .0025 per cent. of silver, while that of another mine contains gold. All the specimens were massive, of a brass-yellow color, metallic luster, fresh fractured surfaces tarnishing readily. In fact it possesses all the known characteristics of this mineral as found elsewhere. Its composition when perfectly pure is

Sulphur	35.05
Copper	34.47
Iron	30.48
	100.00

Several specimens examined gave

	1	2	3	4
Sulphur	33.05	37.22	......	......
Copper	36.60	33.67	31.02	35.01
Iron	29.33	28.56	......	......
	98.98	99.45		

Its formula is $Cu^2 S + Fe^2 S^3$.

This mineral is rarely found in granite, but often in hornblendic and porphyritic transition rocks, accompanied by iron pyrites, magnetic iron, asbestos, quartz, and various species of clay; very rarely with carbonate of lime. The most important mines yielding the copper pyrites are Carrisal, Atacama, and Higuera, Brillador, Tambillos, etc., in Coquimbo.

ARSENICAL GRAY COPPER.

Gray copper appears not to be found very abundantly in Chile. There are, however, three varieties of it; one of which contains quite an amount of mercury, another having the composition of ordinary gray copper, while a third abounds in arsenic. They all three possess the ordinary physical characters of gray copper; namely, a steel-gray and iron-black color, with metallic luster, rather brittle; hardness, 3 to 4, with specific gravity varying from 4.5 to 5. No specimen of this variety was obtained. It is found at San Pedro Nolasco, and its composition, as made out by M. Domeyko, is

Copper	48.5
Iron	4.8
Zinc	2.3
Silver	0.3
Arsenic	11.4
Antimony	6.4
Sulphur	26.1
	99.8

MERCURIAL GRAY COPPER.

This is found in some of the mercurial mines of Chile in small amorphous masses, disseminated in a quarter gangue, accompanied by the blue carbonate of copper and a red earthy substance of deep-red color, apparently an antimoniate of mercury. This also has been analyzed by Domeyko, with the following result:

Antimony	20.7
Iron	1.5
Zinc	trace.
Copper	33.6
Mercury	24.0
Sulphur	20.2
	100.0

ANTIMONIAL GRAY COPPER.

This is the common form of gray copper, and several specimens were brought home by the expedition. It contained but a small amount of silver, as seen by the following analysis:

Sulphur	26.83
Antimony	23.21
Arsenic	3.05
Copper	36.02
Iron	2.36
Zinc	4.52
Silver	3.41
	99.40

The formula of gray copper is represented by—

$$(4\ Cu^2,\ Ag,\ Fe,\ Zn)\ S+(Sb\ As)\ S^3.$$

Besides the above species of gray copper, others are found which, whether arsenical or antimonial, contain only a few thousandths of mercury. These varieties are almost invariably destitute of silver.

DOMEYKITE, ARSENICAL COPPER.

This mineral is massive, of a tin-white color, with a metallic luster, and specific gravity of 4.5. It is about the hardness of copper pyrites. The specimen examined was not a pure one; it furnished

Arsenic	22.08
Copper	72.41
Iron	3.22
Sulphur	2.01
	99.72

Perfectly pure specimens, according to Domeyko, contain

Arsenic	28.36
Copper	71.64
Which give the formula $Cu^3 As$.	100.00

It is found pure, without any admixture of sulphuret, near Illapel, in the same veins which near the surface yield red copper with native silver. It is also found in some of the silver-mines of Atacama, particularly in those of San Antonio.

It is almost always mixed with copper pyrites in varying proportions, and sometimes with the oxide and amorphous green arseniate of copper.

Besides this species, there is found in the cordilleras a kind of white native copper, containing from 3 to 5 per cent. of arseniuret of copper, and resembling native silver.

OLIVENITE, ARSENIATE OF COPPER.

It always accompanies the arseniurets, and is amorphous, with a compact earthy structure, green color, with varying shades, and is always mixed with carbonate and silicates of copper. This mineral, it appears, is never found perfectly pure in Chile; but when pure, as found elsewhere, it contains

Arsenic acid	31.78
Phosphoric acid	6.57
Oxide of copper	58.34
Water	3.31
	100.00

And the formula is $\dot{C}u^4 (\dddot{A}s, \dddot{P}) + \dot{H}$.

CHRYSOCOLLA, SILICATE OF COPPER.

This is very commonly found in all the copper-veins of Chile, always massive, sometimes in the form of mamillary coatings and concretions. It is of various shades of green and blue, and sometimes of a dark and almost black color. Its specific gravity is 2.2. It is easily crushed. It is not an easy matter to find the chrysocolla perfectly pure. The specimen that furnished the material analyzed was a mass of copper pyrites, covered with a mamillary coating of the silicate, which was detached with much care. It furnished

Oxide of copper	42.51
Silica	31.35
Water	21.62
Oxide of iron	1.97
Alumina	2.82
	100.00

Corresponding very nearly to the formula, $\dot{C}u^3\dddot{S}i^2+6\dot{H}$. Other specimens were found to contain oxide of copper varying from twenty to fifty per cent.

The name Llanca is given by miners to a silicate of different shades of green and blue, which very often accompanies the copper minerals, especially the oxysulphurets, forming the envelope of some veins, constituting masses in which native copper, red oxide, carbonate, and at times sulphurets of copper, are found. Most of the copper-veins in Chile abound in these silicates near the surface. The basic silicate found in many of the copper-mines of Coquimbo are always in the upper parts of the veins, forming narrow seams, between red oxide and green and blue Llanca. It is frequently mixed with the black silicate. La Higuera and San Lorenzo furnished the specimens examined.

AZURITE, BLUE CARBONATE OF COPPER.

This occurs both crystallized and massive. Among the specimens was one, crystallized on copper pyrites, from Andacollo. It possesses all the common characteristics of this mineral as found elsewhere, and is composed of

Oxide of copper	69.09
Carbonic acid	25.69
Water	5.22
	100.00

The formula representing it is $2\dot{C}u\ddot{C}+\dot{C}u\dot{H}$.

It is found in many localities associated with the ores of copper.

MALACHITE, GREEN CARBONATE OF COPPER.

This mineral exists abundantly in Chile, but is not found in those large compact masses (such as are procured from Siberia and other places) out of which ornaments are made. It has no peculiar properties in which it differs from the malachite of other localities. Crystallized specimens were procured from Tortolas and Tamaya. Other specimens came from Tarienta, San José, etc. Its composition is

Carbonic acid	20.00
Oxide of copper	71.82
Water	8.18
	100.00

Formula is $\dot{C}u^2\ddot{C}+\dot{H}$.

BLUE VITRIOL, SULPHATE OF COPPER.

This salt is found, associated with the sulphate of iron and alumina, at Tierra Amarilla, in the valley of Copiapó. It arises from the decomposition of copper pyrites. It is constituted of

Oxide of copper	32.14
Sulphuric acid	31.72
Water	36.14
	100.00

Its formula is $\dot{Cu}\,\dddot{S}+5\,\dot{H}$.

VOLBORTHITE, VANADATE OF COPPER AND LEAD.

This rare mineral was first noticed in Chile by M. Domeyko, in the Mina Grande, about six miles from the silver-mines of Arqueros. It is an amorphous substance, porous, heavy, and of a dark-brown color. It lines the cavities of an arsenio-phosphate of lead. At first view it would be confounded with the hydrated oxide of iron, from which it differs, however, by its great fusibility and ready solubility in nitric acid. There were no specimens sufficiently pure for analysis. Those examined by M. Domeyko gave

	1	2
Oxide of lead	54.9	51.97
Oxide of copper	14.6	16.97
Vanadic acid	13.5	13.33
Arsenic acid	4.6	4.68
Phosphoric acid	.6	.68
Chloride of lead	.3	.37
Silica (?)	1.0	1.33
Lime	.5	.58
Oxide of iron and alumina	3.5	3.42
Earthy residue	1.0	1.52
Loss by heat	2.7	2.70
	97.20	97.55

Giving for its formula $\dot{Pb}^6\,\dddot{V}+\dot{Cu}^6\,\dddot{V}$.

This differs somewhat from the formula furnished by the analysis of the volborthite, as found in the copper-mines between Miash and Katherinenberg, Russia; but as the Chile variety has not yet been found crystallized, the differences may be due to impurities.

REMARKS ON THE COPPER MINERALS.

The minerals of copper have been described after gold, from the fact that the great mass of them occur in Chile in the same geological formation as the gold. It is the granite that is most commonly traversed by copper-veins, sometimes of a considerable size. Along the coast it is found in the form of copper pyrites alone, or associated with two varieties of iron pyrites, and also as peacock or purple copper. Galena and blende are rarely found in them, and scarcely ever gray copper. Native copper, red oxide, oxychloride, oxysulphuret, green carbonate, and hydrous and anhydrous silicates of copper of a great variety of colors are also abundant, especially at the upper part of the veins. The silicates sometimes line the walls of the veins, and penetrate to some distance in the inclosing rock, which becomes unequally colored blue or green. The numerous veins of copper are disseminated very irregularly in the granite, and their value is equally variable; sometimes the veins have a breadth of from six to nine feet, as at Tamaya, near Coquimbo, where at the depth of six hundred feet there is a daily yield of from eight to ten tons of an ore yielding seldom less than fifty and oftentimes as much as seventy-five per cent. of copper.

NATIVE SILVER.

This is found in more or less abundance in the various silver-mines of Chile. Most frequently it is associated with dolomite, calcareous spar, sulphate of baryta, and some of the minerals of cobalt. Much of it is found in the form of thin sheets, as at San Pedro Nolasco; at Calabaço (Illapel) it is in small, irregular grains; and at various mines in Copiapó it exists in the form of threads, along with native arsenic and other arsenical minerals. At Chañarcillo it occurs associated with the chloro-bromides, in dendritic forms; and at San Antonio and some other mines it is found in both small and large grains in arseniuret of copper and arseniuret of cobalt. At Illapel it is found in red oxide of copper.

SILVER GLANCE, SULPHURET OF SILVER.

This mineral occurs in all the mines of silver, although in no considerable quantity, and is rarely if ever crystallized. It is of a black-lead color, of a metallic luster, having a specific gravity of 7.3, and is readily reduced, on a piece of charcoal, by the action of the blowpipe. Its composition is

Silver	85
Sulphur	15
	100

Its formula is Ag S.

SULPHURET OF SILVER AND COPPER.

This compound is made mention of by M. Domeyko as existing in the mines of San Pedro Nolasco and Catemo. His analysis gave the following as its constitution:

	San Pedro Nolasco.		Catemo.	
	1	2	3	4
Silver	28.8	24.1	16.6	12.1
Copper	53.4	53.9	60.6	64.0
Iron	0.0	2.1	2.3	2.5
Sulphur	19.8	19.9	20.5	21.4
	100.0	100.0	100.0	100.0

From the variable nature of its composition I should consider it merely a mixture of silver and copper glance.

RUBY-SILVER.

It occurs both crystallized and massive, possessing a very dark crimson-red color; the color is commonly so intense that the mass appears black except when examined by transmitted light in thin pieces; it is easily cut with the knife, and furnishes silver under the blowpipe when heated on charcoal. Its most constant companions are native arsenic, arseniuret and sulpho-arseniuret of iron, arsenical cobalt, blende, calcareous spar, silver glance. It is sometimes found crystallized in metastatic dodecahedrons; at other times it is in masses disseminated in the midst of different spars and argillaceous gangues. It is found in microscopic crystals in the cavities and crevices of native arsenic and of arseniuret and sulpho-arseniuret of iron. The principal sources of it are at Chañarcillo in the lower part of the veins, and in other mines in the province of Atacama.

There are two distinct compositions to the dark and light ruby-silver; the former being a sulphuret of antimony and silver, and the latter a sulphuret of arsenic and silver.

	Dark Ruby-Silver.
Silver	58.98
Antimony	23.46
Sulphur	17.56
	100.00

The formula of this is $3\ Ag\ S + Sb\ S^3$.

	Light Ruby-silver.
Silver	65.38
Arsenic	19.46
Sulphur	15.16
	100.00

The formula being $3\ Ag\ S + As\ S^3$.

The latter is the most common variety in Chile; one specimen, analyzed by M. Domeyko, furnished

Silver	63.85
Iron	.96
Cobalt	.19
Arsenic	13.85
Antimony	.70
Sulphur	18.00
Gangue	1.60
	99.15

ANTIMONIAL SILVER.

It is found both massive and crystallized near Coquimbo; it does not exist abundantly, is of a tin-white color with metallic luster, having specific gravity of 9.5. This mineral is frequently mixed with arsenical and native silver; when pure it contains

Silver	77
Antimony	23
	100

Having for its formula $Ag^4\ Sb$.

POLYBASITE.

Found in considerable quantity in the province of Atacama, massive, of an iron-black color, and a specific gravity of 6.2; it is composed of

Silver	66.25
Copper	4.08
Arsenic	5.22
Antimony	2.56
Iron	2.34
Sulphur	18.68
	99.13

Its formula is considered to be $9\ (Ag\ Cu^2)\ S + (Sb\ As)\ S^3$.

BISMUTH-SILVER.

In the mines of San Antonio, in the province of Copiapó, an alloy of silver and bismuth is found; its color is tin-white, high metallic luster. The only analysis we have of this variety of bismuth-silver is one by M. Domeyko; the following are the results:

Silver	60.1
Bismuth	10.1
Copper	7.8
Arsenic	2.8
Gangue	19.2
	100.0

HORN-SILVER, CHLORIDE OF SILVER.

This is one of the most abundant silver minerals in Chile, as it is found there in quantities far exceeding any thing that is elsewhere known. It is commonly massive, resembling wax of a grayish color, when the surface is freshly broken; but soon tarnishes on the exposure to light, acquiring a purplish tint. Sometimes it is of a greenish tint. Its luster is resinous; easily cut with a knife; specific gravity 5.4. It possesses all the properties of the artificial chloride. Its composition is

Silver	75.33
Chlorine	24.67
	100.00

Formula, Ag Cl.

Several very fine specimens were brought by the expedition from the Chañarcillo, Valenciana mines, in Atacama, and other localities.

BROMIC SILVER.

This compound of silver is likewise found in Chañarcillo, and in many respects resembles the chloride; its color is greener, and it never occurs in such masses as the chloride. It is equally soft, having a little higher specific gravity—5.8. Composition when pure:

Silver	58
Bromine	42
	100

Formula, Ag Br.

EMBOLITE, CHLORO-BROMIDE OF SILVER.

This mineral is found both crystallized and massive in several of the mines of Chile, in the provinces of Atacama and Coquimbo. It is less abundant than the chloride, although more so than the bromide. Externally it is greenish, internally a sulphur-yellow. It has the same luster as the chloride; it is, however, harder than the latter; its specific gravity is the same as the bromide. The composition of it is

Silver	66.96
Chlorine	13.20
Bromine	19.84
	100.00

Formula is Ag (Cl Br).

IODIC SILVER.

This beautiful and rare mineral has been found in some little quantity in the silver-mines of Algodones, province of Coquimbo. The mineral is of a pale sulphur-yellow color, very fragile and soft, having a specific gravity of 5.5. One specimen that I saw had crystalline faces, indicative of a rhombic dodecahedron. It is commonly lamellar, and M. Domeyko has recognized in some small pieces three rhomboidal cleavages; two of the cleavages appear quite perfect, having a pearly luster. It is more brittle and more fusible than either the chloride or the chloro-bromide. The presence of iodine and silver are readily recognized by the ordinary tests. Its gangue is composed partly of carbonate of lime and partly of a brick-red fine clay. In the Carmen mine a considerable amount of iodide was found in the first part of the vein; at the depth of twelve *varas* (thirty-three feet) it disappeared, and chloro-bromide made its appearance in identically the same gangue; and at a still greater depth the latter mineral disappeared, and was replaced by the chloride, accompanied with the sulphuret of silver. It has also been found in small quantities at one of the mines of the Chañarcillo district.

This interesting mineral has the same atomic constitution as the other natural haloid salts of silver, as originally shown by M. Domeyko; although, in referring to certain works on mineralogy, Domeyko is quoted as giving for its composition one atom of silver and two of iodine, while the chloride and bromide of silver are alluded to as constituted of atom and atom, forgetting that the I^2 used (as is frequently done) corresponds to I commonly used by American and English chemists, making the formula as given by Domeyko Ag. I, which formula is sustained by my analyses, as well as those made by M. Domeyko.

The results I obtained are as follows:

	1	2
Iodine	52.834	53.109
Silver	46.521	46.380
Chlorine	trace.	trace.
Copper	trace.	trace.
	99.455	99.489

The formula Ag. I gives as percentage—

Iodine	53.85
Silver	46.15
	100.00

ARQUERITE.

This mineral is found in great abundance at the mines of Arqueros, near Coquimbo; in fact it is the ore of those mines. It is quite like native silver in appearance, with, however, a little more greasy luster. It is disseminated through a calcareous rock. Several specimens examined furnished different proportions of silver and mercury, the proportions of silver varying from eighty-three to ninety-two per cent. M. Domeyko, who has had opportunity of examining a greater variety of specimens, gives the following fixed composition:

Silver	86.49
Mercury	13.51
	100.00

The formula is $Ag.^6$ Hg.

In all likelihood there is a definitely constituted silver amalgam at Arqueros, but in most instances is altered by admixture with native silver.

REMARKS ON THE GEOLOGY OF THE SILVER-ORES.

In speaking of the copper and gold veins it was remarked that they traversed the granite and other old unstratified rocks. M. Domeyko thinks that he has established a law in the distribution of the metalliferous veins of Chile. It is that gold and copper veins, exempt from arsenic, antimony, and silver, abound in the granite-rock; while all the silver-veins, without reference to the associates of the silver, belong to the stratified rocks; and also that the copper-veins found in stratified rocks are very frequently argentiferous. M. Crosnier, however, points out two exceptions to this rule in the province of Copiapó, viz., the Pampa Larga and Garin mines. The Pampa Larga veins traverse compact feldspar, a portion of which near the surface is transformed into kaolin. The upper portion of the vein contains chloride and sometimes native silver; but at a certain distance from the surface the entire mass of the vein is composed of compact native arsenic, in which we find occasionally sulphuret of antimony, realgar, arsenio-sulphuret of silver (sometimes in very beautiful transparent crystals). Arsenical pyrites and calcareous spar are also found.

The Garin and Pampa Larga mines are the only two exceptions pointed out to the general law first mentioned.

The best method of furnishing a correct idea of the mineralogical and geological relations of the different kinds of silver-ores is to give an account of how they occur in one or two of the principal mines.

Some of the most remarkable mines are those in the Chañarcillo Mountain, which is from twenty-five to thirty miles in a direct line from the coast. This mountain is composed of calcareous rocks, more or less argillaceous. Some of the calcareous rocks are dolomitic, while others are without magnesia. The stratification is regular and almost horizontal. The argillaceous matter in the rocks are of two kinds—a white clay, and another composed of a silicate of alumina and iron.

This locality has been thoroughly examined by M. Domeyko, and he finds no organic remains in those parts of the mountain where the metal veins are found. The same geologist has, however, been informed that an ammonite was found in the rock of Reventon Colorado at some distance beneath the surface. In other parts of this mountain organic remains are abundant in the calcareous rocks, especially the Turritella Andii and Terebratulæ.

From the summit of the Chañarcillo Mountain to the lowest workings of the mines is a little less than one thousand feet, and in that space there can be distinguished something like three distinct divisions in the formation of the rocks.

The plane at the summit of the mountain is composed of a dolomitic rock, having in some places a thickness of one hundred feet; it consists of about one third clay. The rock is split in all directions, and the surface of the fissures covered with small crystals of calcareous spar. In some places it is so much split that it looks more like a mass of broken rocks piled together, the interstices being filled with an earthy matter as pulverulent as chalk, and composed of one third carbonate of lime and two thirds clay. It is in these fissures of the upper layer that very considerable masses of chloro-bromide of silver have been found.

The second division of the rocks differs but little in character from the last, being an argillaceous limestone; it is, however, more regular and not so much fissured; at the same time the metalliferous veins traversing it are much poorer. The thickness of this division is over three hundred and twenty feet; and here commences the third division, where the limestone contains less clay and but a little trace of magnesia. The color of the rock is a bluish-gray mottled with yellow; of a compact structure and conchoidal fracture. This rock contains the principal wealth of the Chañarcillo mines, and in it seems to be the principal deposit of chloro-bromide of silver. The thickness of this bed is estimated at nearly four hundred feet. Below this again lies another bed, where the calcareous rock is again more argillaceous and the veins poorer. In this portion of the mountain porphyritic rocks are found at the lowest depths to which the workings have gone.

Numerous metalliferous veins traverse this mountain in every direction. The materials constituting these veins (and mixed with which the silver-ores are found) are the carbonates of lime, iron, and magnesia, zinc and manganese, and the sulphate of baryta, which, however, exists in less quantity in these mines than in those in other parts of Chile. The metalliferous portions of these veins are composed principally of chloro-bromide of silver mixed with native silver and a small portion of sulphuret and sulpho-arseniuret of silver. The chloro-bromide does not show itself in equal abundance at all depths of the productive calcareous bed already mentioned; it is, particularly in the upper, one or two hundred feet; below this depth the gangue becomes less and less calcareous and the mineral changes its nature. At first it is the pure chloride or little mixed with sulphuret; then the proportions of sulphur, antimony, native arsenic, and ruby-silver commence to increase; so that at three hundred feet depth hardly a trace of chloro-bromide is found, the silver being associated with sulphur, arsenic, and antimony.

These are the general features of these famous silver-mines, and as here described some general idea can doubtless be formed of their geological character. Although the general character of the mines resembles those just described, still the minerals and the containing rock frequently differ. Thus in the San Antonio mine, in the valley of Potrero Grande, the rock of the country is porphyry, regularly stratified, and the gangue rock of the veins a dark ashy-gray argillaceous rock of an earthy fracture. It is oftener found impregnated with calcareous and pearl spars, which form veins and nodules in the midst of the gangue. The iron found in these veins is in the form of protoxide, while that at Chañarcillo is in the form of hydrated peroxide. Again, the mines of this latter locality abound in chloride and chloro-bromide of silver, while on the sulphuret of the San Antonio mine there is arseniuret and native silver. Taking the chloride and chloro-bromide as a distinguishing mark between the mines, they may be divided into two classes; those like Chañarcillo and Agua Amarga abounding in these two minerals, and those like San Antonio, San Lorenzo, San Pedro Nolasco, etc., the prominent minerals of which are the sulphuret and arseniuret of silver, with barely traces of the chloride.

CINNABAR.

This mineral of mercury occurs in no great masses in Chile. It is usually found in the granite formation near veins of gold and copper, as in Coquimbo and Aconcagua; also in a vein of quartz, in some stratified porphyry, near the gold mines of Andacollo. The gangue accompanying cinnabar is quartz, with micaceous and hydrated oxide of iron. The composition of the cinnabar is

Mercury	86.2
Sulphur	13.8
	100.0

The formula is Hg S.

GALENA.

It is found in some parts of Chile, commonly associated with the sulphurets of other metals. Composition:

Lead	86.66
Sulphur	13.34

Formula, Pb S.

MIMETENE, CHLORO-ARSIENATE OF LEAD.

This compound of lead has been found in an impure state at Mina Grande, east of Arqueros, mixed with the vanadates of lead and copper. The analysis of a specimen by Domeyko gives

Chloride of lead	9.05
Oxide of lead	58.31
Oxide of copper	.92
Arsenic acid	11.55
Phosphoric acid	5.13
Vanadic acid	1.86
Lime	7.96
Alumina and peroxide of iron	1.10
Clay	2.00
Ignition	1.12
	99.00

Mimetene when pure has for its formula

$$3\,(\dot{P}b,\ \dot{C}a)\ 3\,(\dddot{A}s\ \dddot{P})+Pb\ Cl.$$

VANADINITE.

This is found at the same locality as the last mineral, and mixed with it and vanadate of copper and lead. It has not been discovered crystallized, nor has it been separated in a state of purity from the accompanying minerals.

WULFENITE, MOLYBDENATE OF LEAD.

It is found in the province of Coquimbo, in orange-colored octahedral crystals; also in lemon-yellow plates, with the usual composition:

Oxide of lead	60.81
Molybdic acid	39.19

Having for its formula $\dot{P}b\ \dddot{M}o$.

Domeyko gives the analysis of a specimen where lime appears to replace part of the lead. It is as follows:

Oxide of lead	43.0
Molybdic acid	42.2
Lime	6.3
Peroxide of iron	8.5
	100.0

METEORIC IRON.

This is found scattered in some parts of the desert of Atacama, in pieces from the size of a small nut to lumps weighing fifty pounds and more. It is of a porous nature, the pores being filled by a yellowish and greenish olivine, sometimes the olivine constituting one fifth the mass. We have no account of the falling of these meteoric masses. One specimen that was examined gave

Iron	90.08
Nickel	9.12
Cobalt	.39
Copper	.03
Phosphorus	.13
	99.75

The olivine accompanying was also analyzed :

	Pulverulent Olivine.	Compact Olivine.
Silica	40.50	39.51
Peroxide of iron	11.54	13.38
Magnesia	46.41	47.37
Manganese	.35	.16
Lime	trace.	trace.
	98.80	100.42

MAGNETIC OXIDE OF IRON.

Found in veins of copper at Higuera and various other parts of the provinces of Coquimbo, Copiapó, and Chillan. Its constitution is

Iron	72.40
Oxygen	27.60
	100.00

Formula, $\dot{Fe}\,\dddot{\bar{F}}$.

MICACEOUS OXIDE OF IRON.

It is abundant in Higuera and Punitaque, where it accompanies minerals of copper, gold, and mercury. Its most constant companion is gold. Small veins of carbonate or silicate of copper are frequently contained between the scales, and occasionally red oxide of copper. Its composition is

Iron	70
Oxygen	30
	100

The formula is $\dddot{\bar{F}}$.

GOTHITE.

Commonly found in scales or plates, disseminated or grouped, and is sometimes mistaken for cinnabar. It is also found in the form of geodes, particularly in Topocalma and Valdivia; in the geodes marine shells (turritella) are frequently found of very modern alluvial formation, like that in which the lignites of Concepcion and Colcura are found. Breithaupt called a prismatic crystalline variety of this mineral from Chile *Chileite*, without, however, any just grounds of separating it from the gothite proper. The analysis of the Chileite, as given by Breithaupt, is

Peroxide of iron	83.5
Water	10.3
Copper	1.9
Silica	4.3
	100.0

Formula, $\not{F}e\ \dot{H}$.

PYRITES.

The different varieties of iron pyrites are found in all parts of Chile. They sometimes contain an appreciable amount of gold.

COQUIMBITE—WHITE COPPERAS.

The Tierra Amarilla, near Copiapó, is a seam of pyrites that crosses compact feldspathic rocks, and from its decomposition several minerals result. The one in question occurs in regular hexagonal plates of a yellowish-white color and pearly luster. It has a strong astringent taste, and is quite soluble in water. It is a neutral sulphate of iron, as shown by Rose's analysis :

Peroxide of iron	24.11
Sulphuric acid	43.55
Alumina	.92
Lime	.73
Magnesia	.32
Silica	.31
Water	30.10
	100.04

Its formula is $\not{F}\ \ddot{S}^3+9\dot{H}$.

COPIAPITE—YELLOW COPPERAS.

This occurs associated with the last, and is most commonly found in fibrous masses, of a beautiful silky luster when the fracture is fresh; it, however, soon becomes of a rusty color. It is not so soluble as the last, and is a basic salt. Its specific gravity is 1.84. The analysis of a fine specimen furnished me

	1	2
Sulphuric acid	30.25	30.42
Peroxide of iron	31.75	30.98
Water	38.20	not estimated.
Undissolved	.54	
	100.74	

The analyses correspond to the formula $\not{F}e\ \ddot{S}^2+11\dot{H}$.

ARSENIURET OF IRON.

This mineral is of metallic luster, of a silver-white color. Specific gravity, 7.3. It is found in several of the silver-mines of Chile, especially those of Carriso, where it is accompanied by mispickel, iron pyrites, blende, native antimony, ruby-silver, and native silver. A specimen analyzed by M. Domeyko furnished

Arsenic	70.3
Iron	27.6
Sulphur	1.1
Silver	.2
	99.2

The formula is Fe As.

MISPICKEL.

Is found with copper and cobalt minerals near Coquimbo, with copper and tungsten near Illapel, and with ruby-silver, antimonial silver, and native silver in the mines of Chañarcillo, in the lower part of the veins; also near to Carriso. A specimen examined gave

Arsenic	44.30
Sulphur	20.25
Iron	30.21
Cobalt	5.84
	100.60

The formula of mispickel is $Fe\ As+Fe\ S^2$, with cobalt replacing the iron to a greater or less extent.

CARBONATE OF IRON AND MANGANESE.

This is described as a distinct mineral by M. Domeyko, but in all likelihood it is merely a mixture. It accompanies the sulphuret of copper and gray copper in the silver-mines of San Pedro Nolasco, in a formation of secondary stratified porphyry. This species is of a dark blackish-gray and semi-metallic luster; its structure is foliated in their laminæ, diverging and grouped together in such a manner that the whole forms globular concretions, covered with small crystals of pearl-spar. The mineral is soft; the powder is attracted by the magnet. It dissolves readily in cold acids, and according to M. Domeyko's analysis consists of

Oxide of iron	32.10
Oxide of manganese	30.50
Lime	2.75
Magnesia	trace.
Carbonic acid	32.80
Not dissolved	.35
	98.50

OXIDE OF MANGANESE.

This is found at Arqueros, near the silver-veins in secondary porphyry. The varieties that appear to exist there are psilomelane and pyrolusite.

SMALTINE—ARSENICAL COBALT.

This mineral of cobalt is found in Atacama, in transition and secondary formation, often accompanying ruby-silver, native arsenic, and arsenical nickel. It occurs both crystallized and massive, possessing all the properties peculiar to this mineral. The composition of the specimen examined was

Arsenic	70.85
Cobalt	24.13
Iron	4.05
Copper	.41
Nickel	1.23
Sulphur	.08
	100.75

The formula of the mineral is Co As, part of the cobalt being frequently replaced by other metals.

COBALTINE—SULPHO-ARSENICAL COBALT.

This is found in Coquimbo, in small, brilliant octahedral crystals, with truncated corners. It is also found granular and massive, in pieces of considerable size. The specimens from the mines of Volcan and San Simon are of a steel-gray color, imperfect foliated structure, metallic luster, hard, amorphous, accompanied with arseniuret of copper. It is also found associated with copper pyrites; and there is one vein of it running parallel to a vein of copper pyrites. Its composition is

Arsenic	44.23
Sulphur	19.82
Cobalt	34.12
Iron	3.01
	101.18

The formula is $Co^1\ S^2 + Co\ As$.

COBALT BLOOM—ARSENIATE OF COBALT.

It is found in all the veins containing the arseniurets of cobalt, and also in most of the silver veins, but never in any considerable quantity. At Arqueros it is found with the native amalgam, and with native and horn silver, in the mines of Argua, Amarga, Chañarcillo, Punta Brava, Tunas, etc. It is crystallized in radiating crystals of a peach-blossom color, and consists of

Arsenic acid	38.21
Oxide of cobalt	35.92
Oxide of nickel	.08
Oxide of iron	2.13
Lime	.32
Water	23.16
	99.82

The formula is $\dot{C}o^3\ \ddot{\ddot{A}}s+8\ \dot{H}$.

NICKEL GLANCE—ARSENICAL NICKEL.

This is found in Atacama. It is of a steel-gray color; freshly broken surfaces soon tarnish. No analysis was made of this mineral from the above locality; and we know of none that has been made. When pure its constitution should be

Arsenic	45.16
Sulphur	19.33
Nickel	35.51
	100.00

Its formula is $Ni\ S^2+Ni\ As$. Other metals, especially iron, frequently replace the nickel to some extent.

NATIVE BISMUTH.

This is found, alloyed with silver, in the San Antonio mine, Atacama. The mineral has already been described, under the head of the silver minerals. It commonly contains from fourteen to fifteen per cent. of bismuth.

NATIVE ANTIMONY.

This is found in considerable quantity in the silver-veins in the mines of Carriso. It is disseminated in small irregular veins, and in laminæ, like galena. The most constant companions of it are native silver, ruby-silver, gray antimony, gray copper, etc. The gangue is carbonate of lime and heavy spar.

WHITE ANTIMONY

Accompanies the last-mentioned mineral in several of its localities. It has been found massive; is of a snow-white color, with sometimes a reddish hue. We have no analysis of this mineral from any of the localities in Chile. It is an oxide of antimony, and when pure should consist of

Antimony	84.32
Oxygen	15.68
Its formula is Sb O^3.	100.00

ANTIMONY GLANCE.

This is also found in the localities furnishing native antimony, with all the ordinary properties of this well-known mineral. Its composition is

Antimony	72.89
Sulphur	27.12
Its formula is Sb S^3.	100.00

NATIVE ARSENIC.

This substance occurs abundantly in the provinces of Atacama and Coquimbo. It is of a tin-white color that soon tarnishes; it is volatilized completely by the action of heat, and possesses all the other peculiarities of this metal. It often contains a little antimony and iron. It accompanies ores of silver, particularly ruby-silver, antimonial and sulphuret of silver, native silver, arsenical cobalt, arseniuret and sulpho-arseniuret of iron. I am not informed of the existence of any other arsenical minerals in Chile, but presume the oxide and sulphuret must also be found.

BLENDE—SULPHURET OF ZINC.

This ore of zinc is found near the Leona mine in Rancagua. Specimens examined by M. Domeyko contained a notable amount of iron; one of his analyses is as follows:

Zinc	43.0
Iron	12.4
Sulphur	28.6
Gangue	14.7
	97.7

Its formula is Zn S, with iron sometimes replacing a portion of the zinc.

Besides these minerals described, there were a few others of a non-metallic character collected by the expedition, which will be simply enumerated.

LAPIS LAZULI.

This beautiful mineral occurs in no inconsiderable quantities in the province of Coquimbo. Carbonate of lime runs through the mass, in small veins, and iron pyrites is intimately mixed with it in small crystals. It being impossible to separate the two last-mentioned minerals from the lapis lazuli, no analysis was made of it. A specimen of the mineral from the Andes was analyzed by Mr. T. Field with the following results:

Silica	37.60
Alumina	11.21
Sulphur	1.65
Iron	.08
Magnesia	.36
Soda	9.66
Lime	24.10
Carbonic acid	15.05
	99.71

Although this analysis differs somewhat from the mineral procured from other localities, still the difference may be accounted for by the unavoidable impurities.

CALCAREOUS SPAR.

This is found in all parts of Chile, and is one of the most common gangue-rocks of the silver-ores.

DOLOMITE.

This is also a common mineral in Chile, forming in many places beds of immense thickness.

HEAVY SPAR

Exists in the silver veins forming ore of the gangue-rocks.

ASBESTOS (GREEN).

A specimen was brought from the copper-mines of Coquimbo, and another from Tambillos.

TUNGSTATE OF LIME.

This mineral is found in the copper-mines of Llamaco, near to Chuapá, and contains about three per cent. of oxide of copper in its constitution.

LIGNITE.

This variety of coal has been found in some little abundance at Concepcion, and is worked to some extent. These lignites ordinarily form but one seam that is thick enough to repay exploration; it is often accompanied by a second thin seam and one more irregular. It is seldom that the seams are found more than six or nine feet above the level of the sea, and most always dip to the west beneath the ocean. It has been found on the shores of Concepcion, of Valdivia, and on the shores of the island of Chilóe. The mines that have been worked are one near Penco, another near Lirquen, the mines of Talcahuana, of Las Tierras Colorados, of Lota and of Lotilla; the two last mines are considered those of most importance.

M. Crosnier gives the analysis of several of these lignites as follows:

	Lota.	Lotilla.	Penco.
Coke	52.3	42.7	39.9
Volatile matter	44.6	54.3	51.8
Ash	3.1	3.0	8.3
	100.0	100.0	100.0

The coke is light and porous; it is sufficiently solid when well burnt.

MINERAL WATERS.

Five specimens of mineral waters were submitted to examination; but as there was only about one pint of each, the analyses can not be considered as satisfactory as it is desirable that they should be.

No. 1. From the baths of Apoquindo, east of and about five hundred feet above Santiago, in the first range of the Andes. When the water was collected its temperature was 74°, the air being 57°. The specific gravity of it is 1.00226.

Solid contents in one litre 2.743 grammes, composed of

Chloride of calcium	1.665
Chloride of sodium	1.008
Chloride of magnesium	trace.
Sulphate of lime	.032
Oxide of iron	.018
Organic matter	trace.
Silica	.020

No. 2. From the baths of Colina. The temperature of the water at the source is 89½° Fahr.; specific gravity 1.00053. The amount of solid contents in one litre are 0.428 gramme, composed of

	GRAM.
Sulphate of lime	.120
Sulphate of soda	.089
Chloride of calcium	.077
Chloride of sodium	.142
Oxide of iron	trace.
Organic matter	trace.
Silica	trace.

No. 3. This is also from the baths of Colina, and when collected was 79° Fahr.; sp. grav. 1.00045. The composition of the water is the same as the last. Solid contents in one litre 0.435 gramme, composed of

Sulphate of lime	.118
Sulphate of soda	.094
Chloride of calcium	.087
Chloride of sodium	.136
Oxide of iron	trace.
Organic matter	trace.
Silica	trace.

No. 4. From Cauquenes *Tibia* bath; sp. grav. 1.00270; solid contents in one litre 3.3032 grammes, composed of

Sulphate of lime	.0600
Sulphate of soda	.0320
Chloride of calcium	2.1682
Chloride of sodium	1.0310
Chloride of magnesium	trace.
Oxide of iron	.0020
Organic matter	trace.
Silica	.0100

No. 5. Cauquenes *Pelambre* bath; sp. grav. 1.00283. It is constituted the same as the last. Solid contents in one litre 3.3923 grammes, composed of

Sulphate of lime	.0630
Sulphate of soda	.0410
Chloride of calcium	2.1751
Chloride of sodium	1.1012
Chloride of magnesium	trace.
Oxide of iron	trace.
Organic matter	trace.
Silica	.0120

ANALYSIS OF WATER BROUGHT FROM THE RIO DE MENDOZA BY LIEUT. MACRAE.

The bottle contained a large amount of mud sediment. The clear water, on evaporation, gave 540 grammes of solid matter to the litre, composed of

Carbonate of lime	.110
Carbonate of magnesia	.072
Sulphate of lime	.792
Sulphate of magnesia	.108
Sulphate of soda	.192
Sulphate of iron	.036
Chloride of sodium	.228
Silica	.112
Organic matter	.150

THERMAL WATERS OF ASIA MINOR.

THE THERMAL WATERS OF BROOSA.

There are few countries where thermal waters are so numerous and cover so extensive a surface as in Western Asia Minor; many of them still bear marks of the estimation in which they were held by the ancient Romans and Greeks for the purpose of supplying their baths.

Owing to the difficulty of obtaining proper vessels or corks at or near the springs, coupled with the risk of breakage by the necessary transportation on the backs of horses over rough and mountainous roads, travelers have been deterred from collecting these waters for the purpose of analysis. In my travels through certain parts of this country I took along with me bottles and corks, and collected between twenty and thirty specimens of different localities, some of them in considerable quantity; and of that number fifteen or sixteen have arrived safely at my laboratory, where most of them have been already examined.

In my remarks upon them I will first allude to the thermal waters of Broosa or Prusia, which are the most important at the present day, and the most accessible from Constantinople. The spot itself is hallowed by many interesting historical associations. The city was founded by Hannibal during a friendly visit which this great Carthaginian general made to Prusias, the king of Bythinia, whose name was given to it. Like all other cities of so ancient date, it has gone through many changes, passing successively into the hands of the Greeks, Romans, and Turks. Since 1326 the Turks have continued masters of this part of Asia Minor, it having been conquered by Osman just prior to his death, for many years after which event it remained the capital of the Ottoman Empire.

Broosa is readily reached from Constantinople by a steamer that goes from this latter place to Modania, on the gulf of the same name, about seventy miles from Constantinople. From Modania a ride of about twenty miles on horseback brings you to Broosa, at the foot of the Bythinian Olympus. The warm baths of this place have been celebrated from the earliest epochs, and the visit of Constantine with his wife in 797 is recorded in history as having resulted favorably in restoring the latter to health. And at a still later period Sultan Soleman the Great visited these baths on account of an attack of gout, and to commemorate his cure he had a large dome constructed over the source to which he attributed the beneficial effects derived by him; the dome still stands.

As it is not my object to enter here into the details of baths well known to all travelers in this part of Asia Minor, I shall at once proceed to the description of the sources. The sources of thermal waters near Broosa are seven in number, all situated in a little valley which separates Mount Olympus from Mount Katairli, and they are comprised within the distance of a mile and a half. In the immediate neighborhood of some of these sources, and sometimes in direct proximity, are sources of cool and delightful water that serve to regulate the temperature of the water used in the baths, of which there are as many as twenty private and public. These sources furnish waters of two descriptions, the sulphurous and the non-sulphurous, and I shall commence with a description of the former.

Thermal Sulphur Waters.

There are two sources of this class of water near Broosa, or rather two places near to each other where it flows out of the mountain, for my examination goes to prove that they are the same water. Their names are Kukurtlu and Bademli-Baghtsche.

KUKURTLU SOURCE.

The name of the source signifies sulphur. It flows rapidly from the side of the mountain near to its base, through a bed of calcareous tufa, furnishing upward of twenty gallons a minute, which, along with the water from a cold spring near by, is made to flow through the baths. There is a very sensible odor of sulphureted hydrogen proceeding from the water of this source, more especially as it issues forth from the mountain; for there is a large amount of gas bubbling through the small reservoir into which the water rises, accompanied with a larger amount of vapor. As the water flows it leaves an incrustation of carbonate of lime, more or less colored with some organic matter. This source is held in particular veneration by the Greeks of the country, who usually assemble here twice a year to commemorate the martyrdom of St. Patrice, which was ordered by the proconsul of Broosa, and executed by his being thrown into this almost boiling spring.

The country is *geologically* made up of the older rocks, as granite, gneiss, limestone, etc., a siliceous variety of the latter overlying the other two; in some parts, however, the limestone is remarkably pure, and has doubtless furnished to these waters that carbonate of lime so extensively deposited at the base of this part of the mountain in the form of tufa, which, for a mile or two of extent, rises several hundred feet above the plain at the foot of the mountain.

Physical Properties.—The water as taken from the source is perfectly clear and transparent, and remains so when kept in well-corked bottles, but otherwise a yellow deposit is soon formed, which is probably crenate of lime. A slight odor of sulphureted hydrogen, not perceptible when the water is cold. The taste of the water when cold is in no way peculiar, and it is very pleasant to drink. Specific gravity 1.00118. Temperature (atmosphere at 66° Fah.) 182° Fah., which varies but a few degrees with the seasons.

Chemical Composition.—The *gas* which escapes from the source was collected in inverted bottles, well corked and sealed, and in one thousand parts was found to contain

Carbonic acid	886
Nitrogen	99
Oxygen	11
Sulphureted hydrogen	4

Solid contents in one litre of the water, 0.970 gramme. The water is alkaline, and when concentrated to one third its bulk gives a very sensible alkaline reaction with reddened litmus paper. It is found to contain the following ingredients in one litre:

	Grammes.		Grammes.
Carbonic acid, free*	.3420	Lime	.1415
Carbonic acid, fixed	.1820	Magnesia	.0142
Hydro-sulphuric acid	.0012	Alumina	.0012
Sulphuric acid	.2140	Silica	.1100
Chlorine	.0103	Iron	trace
Soda	.2600	Organic matter (crenic acid)	.0350
Potash	.0110		

These acids and bases may be represented as combining in the following manner:

Bicarbonate of soda	.4100	Sulphate of alumina	.0043
Bicarbonate of lime	.1830	Chloride of sodium	.0170
Bicarbonate of magnesia	.0460	Hydro-sulphate of soda	.0033
Sulphate of soda	.1950	Carbonate of iron	trace
Sulphate of potash	.0202	Silica	.1100
Sulphate of lime	.1710	Organic matter	.0342

The incrustation from this spring was next examined. One gramme of a beautiful crystalline portion was analyzed and found to contain

Carbonate of lime	.970	Silica	.003
Carbonate of magnesia	.016	Organic matter	trace
Sulphate of lime	.008	Fluoride of calcium	trace
Peroxide of iron	.011		

There are some portions of the incrustation richer in organic matter than this, but then the mixture is sensible to the eye, and does not represent the pure crystalline deposit of the spring.

The Kukurtlu source supplies two baths with water, one called the Buyuk Kukurtlu and the other the Kutschuk Kukurtlu.

The other source of sulphur thermal water is called

BADEMLI-BAGHTSCHE.

This source is about three hundred feet from the latter, and flows from three or four openings in the tufa. On my visit to it the entrance to the sources was closed up with masonry, and the door could not be opened by the Turks from some superstitious motive. I was, however, enabled to procure the water a few feet from the source as it flowed through an open gutter.

* What is here meant is such of the carbonic acid as can be expelled in boiling the water.

Gas is said to escape abundantly from the source, just as in the Kukurtlu source.

Physical Properties.—It is clear and transparent, remaining so in well-corked bottles; exposed to the air, it gradually becomes cloudy, and deposits a yellowish sediment. Has a slight odor of sulphureted hydrogen when warm. Specific gravity 1.00116. Temperature (atmosphere at 67°) 184° Fah.

Chemical Composition.—Solid contents 0.978 gramme in one litre. The water when concentrated reacts strongly alkaline. In one litre there are the following ingredients in grammes:

Carbonic acid, free	.2920	Lime	.1378
Carbonic acid, fixed	.1875	Magnesia	.0160
Hydro-sulphuric acid	.0010	Alumina	.0005
Sulphuric acid	.2160	Silica	.1100
Chlorine	.0112	Iron	trace
Soda	.2650	Organic matter (crenic acid?)	.0402
Potash	.0130		

The combination of the acids and bases may be represented in the following manner:

Bicarbonate of soda	.4070	Sulphate of alumina	.0020
Bicarbonate of lime	.1790	Chloride of sodium	.0192
Bicarbonate of magnesia	.0520	Hydro-sulphate of soda	.0019
Sulphate of soda	.2000	Carbonate of iron	trace
Sulphate of potash	.0225	Silica	.1100
Sulphate of lime	.1660	Organic matter	.0402

Two baths are also supplied from this source, the one called Yeni-Kaplidja and the other Kainardja.

It will be seen that in physical properties and chemical composition the water of this source is identical with that of Kukurtlu; at which fact I was at first somewhat surprised, as an approximate analysis, made some years ago by Dr. Bernard, led me to look for a difference in the composition of these waters; and it was not until my analysis was completed that I became convinced that the waters of the Kukurtlu and Bademli-Baghtsche sources were the same, making its way through different openings in the tufa. I would merely remark here that the analysis made by Dr. Bernard must have been quite crude, as among other things he gives to a litre of the Kukurtlu water 0.332 gramme of sulphureted hydrogen, water which when cold has no hepatic odor, and has hardly a sensible effect on lead-water.

None of the other sources near Broosa evolve a trace of sulphureted hydrogen, and contain less solid matter; they are all alkaline, and give an alkaline reaction when concentrated.

Thermal Alkaline Waters.

Of the alkaline waters I have examined three sources, situated at some distance from each other.

The Kara Mustapha source is about two hundred yards from the Kukurtlu, and almost on the border of the plain of Broosa; it supplies a bath bearing the same name.

Physical Properties.—Clear when taken from the source and kept in well-stopped bottles. As the opening in the mountain from which it escapes is bricked over, it was impossible for me to ascertain if there were an abundant escape of gas. Temperature 127° Fah. Specific gravity 1.00094.

Chemical Composition. — Solid contents in one litre 0.541 gramme, and the same quantity of the water contains

Carbonic acid, free	.104	Lime	.115
Carbonic acid, fixed	.150	Magnesia	trace
Sulphuric acid	.068	Iron	trace
Chlorine	.005	Silica	.066
Soda	.132	Organic matter not estimated.	

The combinations of the acids and bases may be represented as follows in grammes:

Bicarbonate of soda	.2600	Chloride of sodium	.0084
Bicarbonate of lime	.2380	Carbonate of iron	trace
Sulphate of soda	.0452	Silica	.0660
Sulphate of lime	.0670	Organic matter not estimated.	
Carbonate of magnesia	trace		

Incrustations of carbonate of lime are deposited from this source, but not so abundantly as from the two first mentioned.

TSCHEKIRGHE SOURCE.

The Tschekirghe source is about a mile and a half from Broosa, and supplies four baths: those of Boigusel, Vani, Tschekirghe, and Yeni-Han.

Physical Properties.—Clear, and does not readily deposit a sediment; the incrustation much less than at the other sources. No gas escapes from it as it flows from its source. Temperature (air at 72° Fah.) 113° Fah. Specific gravity 1.00068.

Chemical Composition.—Solid contents in one litre 0.550 gramme; the same amount in the water contains

Carbonic acid, free	.040	Lime	.168
Carbonic acid, fixed	.094	Magnesia	trace
Sulphuric acid	.152	Iron	trace
Chlorine	trace	Silica	.040
Soda	.039	Organic matter not estimated.	

Combined as

Bicarbonate of lime	.2336	Sulphate of lime	.2190
Carbonate of soda	.0480	Chloride of sodium	trace
Carbonate of magnesia	traces	Silica	.040
Carbonate of iron	traces	Organic matter not estimated.	
Sulphate of soda	.0250		

The last source that I shall allude to is a very small one, near to the Kukurtlu, and not connected with any bath; it is, however, used by the natives for the treatment of diseased eyes.

GUEUZAYASMA SOURCE.

The Gueuzayasma source rises slowly in an excavation in the side of a rock, no gas whatsoever escaping from its surface; an incrustation is formed from it, that is in some places covered with a thin green coat resembling some of the salt of nickel or copper; it is, however, entirely of a vegetable character, and exhibits under the microscope a beautiful lace structure.

Physical Properties.—Clear and transparent. Temperature 113° Fah. Specific gravity 1.00122.

Chemical Composition.—Solid contents in one litre 0.901 gramme. One litre of the water contains

Carbonic acid, free	.220	Lime	.175
Carbonic acid, fixed	.150	Magnesia	trace
Sulphuric acid	.215	Iron and alumina	trace
Soda	.151	Silica	.114
Potash	.006	Organic matter not estimated.	

Combined as follows:

Bicarbonate of soda	.2405	Sulphate of lime	.2370
Bicarbonate of lime	.2249	Sulphate of alumina	trace
Bicarbonate of magnesia	strong trace	Carbonate of iron	trace
		Silica	.1140
Sulphate of soda	.1160	Organic matter not estimated.	
Sulphate of potash	.0110		

The incrustation of the spring contains ninety-seven per cent. of carbonate of lime; the remainder is composed of carbonates of iron and magnesia and the sulphate of lime, without a trace of fluorine.

These various springs, it will be seen, supply nine public baths, which vary in their size and magnificence, that of *Yeni-Kaplidja* being the largest and most beautiful. They are constructed on the usual plan of the Eastern baths, and consist of three parts.

First, a large hall, with an elevated platform all around, two feet high, and sometimes galleries attached. It is on the platform that one disrobes himself prior to entering the bath, and it is also here that the bather reposes on a couch in retiring from the bath. This apartment is frequently ornamented with considerable luxury; it is well lighted, and there is sometimes in the middle a fountain, the falling of whose waters in the basin produces a freshness, and at the same time invites to slumber. This apartment is called by the Turks *Djamekian* (Vestiarium).

The next division in the bath is the *Soouklouk*, where one begins to experience the temperature of the inner bath, and where he reclines on a marble slab, and is either shampooed or places himself in the hands of the barber to be shaved, cupped, or bled.

The third division is the *Hammam*, or bath properly speaking, where there is an atmosphere of 105° to 110° Fah., filled with the vapor of water arising from the heated marble floor. Here there are various recesses, with small marble basins, in which streams of hot and cold water are allowed to flow; and once seated by one of them, an attendant of the bath takes possession of you and puts you through a series of operations that can be better felt than described. The baths at Broosa have usually in the *Hammam* a large basin of hot water, into which the bathers can plunge; the one in the *Yeni-Kaplidja* is about five feet deep by thirty in diameter.

There is in some of these baths a small room called the *Boghoulouk* (Sudatorium), where the temperature is from 120° to 130° Fah. Once through the various operations of the bath, one returns to the first room, reclines on a bed, and indulges for a half hour or more in the Eastern luxuries of smoking and drinking coffee or sherbet.

This is a hasty sketch of the operations that the bather

usually undergoes at these baths; but as numbers of invalids visit them, arrangements are made by which they can bath in whatsoever way they may think best or the physician prescribe, for there are private apartments attached.

These thermal waters are in great repute in Turkey, and their effects are said to be most marked on chronic irritation of the abdominal organs, chronic rheumatism, gouts, chronic irritation of the mucous membrane of the intestines, diseases of the bladder, of the skin, and of the eyes, etc. These waters are also recommended to be taken internally when cold.

In the calcareous incrustation of three of these springs that were examined, I found the remains of two or three varieties of siliceous infusoria after the lime had been dissolved out by an acid.

THERMAL WATERS OF YALOVA.*

The shortest way of reaching the springs of Yalova is by landing on the south side of the Gulf Nicomedia, near to Angori (three hours distant from Constantinople by steamer), and proceeding along a beautiful plain, that gradually narrows until terminating in a valley closely shut in by hills. The springs in question are situated in this valley, about six miles from the sea; they are at the foot of a hill, which on the south-west closes in the valley of Yalova, and are known in the country by various names, as *Couri-Hamam*, *Dagh-Hamam*, etc.

On the road that approaches the springs there are extensive remains of the foundations of old Roman and Grecian buildings, and still nearer the remains are more perfect in the form of arches, aqueducts, baths, etc. Their extent gives evidence of the celebrity they enjoyed in former times. The styles of their architecture belong to different periods. The remains of the brick edifices are evidently of the period of the Lower Empire, for on many of the bricks are to be found an impress of the cross and Latin words written in Greek letters. To judge from the form of these letters, particularly the epsilons, sigmas, and omegas, one is led to believe that they date from the Justinian age. The massive stone arches which support the

* The locality of these waters is described very fully, as it is little known, being seldom visited by travelers.

vault under which the waters rise seem to have been constructed by the Romans. Their structure presents nothing which opposes the idea received by the inhabitants of the surrounding villages—namely, that they were constructed during the reign of Constantine the Great. And what seems to sustain this hypothesis is the popular legend that the mother of Constantine was indebted to these waters, at one period of her life, for her restoration to health; and from this fact (according to the authority of the celebrated archæologist, the Patriarch of Constantius) Yalova was formerly called Helenapolis.

In want of more exact data we cite as sustaining this supposition the custom of the Greek villagers of the neighborhood, kept up for many centuries, of assembling at these baths on the anniversary of the *fêtes* of St. Constantine and St. Helena to celebrate the virtue of these waters. Von Hammer, in his history of the Turkish Empire, alludes to this place in the following words: "Some leagues from Cara-Mursal, on the south side of the Gulf of Nicomedia, there exist the baths of Yalova (ancient Sergla or Trepanon). This place was adorned with a great number of palaces and hospitals by the Empress Helen, whose father had kept an inn there. This place was afterward raised to the rank of a city by Constantine, the founder of the Byzantine Empire, and called Helenapolis in honor of his mother. It was to this place that the first army of the crusaders, conducted by Peter the Hermit and Gautier *sans-avoir*, took refuge after being routed near Nice. It was here also that the Saracens constructed pyramids and towers with human bones. Helenapolis has been at all times celebrated for its thermal waters. Near their source is to be seen the tomb of an *Abdal*—that is, an enthusiastic dervish—who, armed with a wooden sword, undertook at the head of a body of Mussulmans to conquer this city."

There are several ancient authors who allude to these springs, among whom are Ammianus Marcellinus, Mela, and Anna Comnena.

Yalova, which is now but a small village, was formerly the place of debarkation for the inhabitants of the celebrated cities of Nicomedia, of Nicea, and of the numerous cities of Bithynia, who visited these springs. The port of Couri, whose antiquity is indicated by several Greek inscriptions, was probably, as now, frequented by those coming from Constantinople and other cities of the Propontide.

After the fall of the Roman Empire these baths went to ruin, and were almost forgotten; nevertheless the reservoirs and aqueducts remain as in the time of the Lower Empire. It is only a few years since an Armenian banker purchased the place and constructed houses for the reception of the sick.

These waters have at least *nine* sources. They flow from the sides and bottom of a hill, rising through a sandy bottom accompanied with bubbles of gas, and differ but little in their temperature and composition. The character of the surrounding rocks is not easily made out; I am inclined to refer them, from my observations higher up the gulf, to the older tertiary. The waters in their course leave not the slightest deposit, so that the ancient aqueducts have never become obstructed.

According to the accepted classification, the mineral waters of Yalova belong to the *hot sulphurous waters*. They have at their source a very slight odor of sulphureted hydrogen, but the quantity is so small, either in the water or the gas, that it could not be estimated. The temperature of the waters is from 151° to 156° Fah., and varies but little with the changes of the atmospheric temperature. The water is limpid and transparent, and has the specific gravity 1.00115. The gas which escapes at the source gave, on analysis,

Nitrogen	97 per cent.
Oxygen	3

One litre of the water gave 1.461 gramme of solid matter. The same quantity of water contains in grammes

Sulphuric acid	.690	Magnesia	.002
Chlorine	.086	Alumina	trace
Soda	.393	Silica	.035
Lime	.208		

Combined in the following manner:

Sulphate of soda	.807	Sulphate of magnesia	.005
Sulphate of lime	.414	Sulphate of alumina	trace
Chloride of sodium	.072	Silica	.035
Chloride of calcium	.068		

The composition of these waters resembles that of the Bath waters of England; the latter, however, not being of so high a temperature. They act powerfully on the nervous system and on the secretions and excretions, particularly those of the skin, which renders them so efficient in rheumatism, gout, etc.

THERMAL WATERS OF HIERAPOLIS.

The ruins of the ancient city of Hierapolis are among the most interesting in the south-western part of Asia Minor. The place is situated about six miles from Laodicea (one of the seven churches), and about one hundred and ten miles from Smyrna. The site is seen for many miles before it is reached, as it rises abruptly from the north side of an extensive plain, and the sides of the hill are covered with an incrustation of dazzling whiteness for upward of a mile in length, and from this it has received its present name, *Pambuk-kelesey* (cotton castle).

This place was much admired in former times, if we are to judge from the inscriptions still to be seen on different parts of the ruins, to the following effect: "Hail, golden city, Hierapolis; the spot to be preferred before any in wide Asia; revered for the rill of the nymphs, adorned with splendor." The people, in some of these inscriptions, are styled "the most splendid" and the senate "the most powerful."

It is a place well known to travelers in this part of the world, and therefore I shall confine myself strictly to what concerns its thermal waters, which have ever been its principal object of note, as evinced by the extensive ruins of baths. In fact, the very hill upon which the city stands owes its formation to the deposition of carbonate of lime from these waters, and it now rises upward of a hundred feet above the plain, with a width of about six hundred feet. Immediately behind the city rises another set of hills of calcareous rock, from which flow the waters in question; they first enter a large pool in front of the theater, and from this the water flows in numerous little streams that course in channels made by incrustations from the water. The amount of water is very great, and it is so highly charged with carbonate of lime as to incrust all bodies that it comes in contact with, and it takes place so rapidly that the concretion does not possess great solidity, and frequently has a granular form, resembling driven snow.

It is this incrustation, as I before said, which gives to the immediate site of the city its remarkable character. In some places, as the waters flow over the steeply-inclined sides of the hill, it forms a succession of terraces at regular distances, that require but little effort of the imagination to liken to an amphitheater with its marble seats. At other places it flows over the precipitous sides sixty or seventy feet high, and one or two hundred feet wide, incrusting the precipice with a snow-white sheet, which might be likened to a consolidated cataract; and what adds to the delusion, at the base the incrustations have accumulated an irregular mass not unlike foam. This petrified stream extends several hundred feet into the plain. It has formed walls and dikes, and incrusts the grass and vegetation that it flows over, and many of the tufts of grass, in perfect verdure, are thickly incrusted near the roots with this white carbonate of lime.

The channels that conduct the water through the city are made by deposits from these waters; many of them are very deep and almost arched over. The incrustations in and about the city have elevated the level of it some fifteen or twenty feet since its destruction. Strabo tells us that in his day the people of this city conducted these waters into their gardens or other places where they desired to form a wall, and in a short time they obtained fences of a single stone, so rapid was the deposition. The road which leads from the plain to the city is a causeway which is formed entirely from the water.

Physical Properties.—The water is of remarkable transparency, and remains so after being kept for any length of time. Having lost my thermometer the day before arriving at this source, I was unable to ascertain its exact temperature. I judged it to be about 130° Fah. Specific gravity 1.00143.

Chemical Properties.—Solid contents in one litre 1.934 gramme. One litre contains in grammes:

Carbonic acid, free	.352	Potash	trace
Carbonic acid, fixed	.462	Lime	.589
Sulphuric acid	.541	Magnesia	.164
Chlorine	.012	Silica	.008
Soda	.182	Phosphoric acid	.005

The combination of acids and bases may be represented in the following manner:

Bicarbonate of soda	.078	Sulphate of lime	.119
Bicarbonate of lime	1.368	Sulphate of magnesia	.431
Bicarbonate of magnesia	.041	Phosphate of lime	.012
Chloride of calcium	.020	Silica	.008
Sulphate of soda	.341		

The *incrustation* of the spring was analyzed; it is remarkably white, and almost pure carbonate of lime. The composition is as follows:

Carbonate of lime	98.2
Silica	00.6
Magnesia, Phosphate of lime, Fluoride of calcium	1.2=100.0

At the present day these waters are not used, and the neighboring country is quite deserted, with the exception of a miserable village of some half dozen huts. In former times, however, besides the use of this water for the baths, it was greatly in repute among the dyers in a purple color made from a kind of root; and owing to the remarkable adaptation of this water for that purpose, the tint obtained is said to have rivaled the more costly purple, and to have constituted the principal source of riches to the city. The company of dyers is alluded to in the inscription on a square building among the sepulchers. These waters also seem to have possessed medical virtues, if we are to judge from some of their medals, on which you find Apollo (the tutelar deity of Hierapolis), with Esculapius and Hygeia.

Strabo alludes to a circumstance connected with these waters that I inquired into while there, but without success. It is the existence of what that author calls a *plutonium*, described as an opening about the size of a man in the side of the hill, with a kind of inclosure of half an acre in front of it; from the opening there issued constantly a dark vapor that filled the inclosure in front of it. He states that all animals entering this inclosure became suffocated, but that the sacred eunuchs who attended in the temples could enter with impunity.

I sought to discover this *plutonium*, but without success; it was doubtless an opening in the rock, from which issued a mixture of carbonic acid and vapor of water, that had subsequently become obstructed.

THERMAL WATERS OF ESKI-SHEHR.

Eski-Shehr is the ancient Dorylæum, the plains of which are very extensive. It is mentioned by the Byzantine writers as the field of the first battles between the soldiers of the Eastern Empire and the Turks. It is situated on the river Pursceck or Thymbius, which empties into the Sangarius, that flows into the Black Sea, and is equidistant from that sea and the sea of Marmora, being a little over one hundred miles from each. Eski-Shehr is a city of some importance to the Turks, and it is from here that Europe derives the greater portion of that mineral called meerschaum, used in making pipes.

In a certain quarter of this city, by excavating to the depth of a few feet, hot water is obtained, which is a matter of great annoyance to the inhabitants, as they can have no wells of drinking-water. It is in this quarter that are situated the celebrated hot baths, doubtless used for more than two thousand years, with such change in structure as time and the habits of the people required. There is here a large excavation sixty or eighty feet square, closed in with stone and roofed over; its depth I did not measure, but am told that it is twelve or fifteen feet. The water arrives from many sources at the bottom of this reservoir.

The reservoir was made by the Greeks and repaired some years ago by the Turks. The amount of water furnished is very great, and forms half the water used in turning a mill in the neighborhood.

The water is allowed to flow from the great reservoir into a large Turkish bath, as well as from different hydrants for the purpose of washing dyed stuffs, etc.

Physical Properties.—This water is clear and transparent, and when cold it is very agreeable to the taste; no gas escapes from it, nor is there any deposit, even after very long repose. Temperature 119° Fah. Specific gravity 1.00017.

Chemical Composition.—The solid contents in one litre 0.260 gramme. One litre of the water contains in grammes

Carbonic acid, free	.118	Soda with a little potash	.119
Carbonic acid, fixed	.195	Lime	.040
Sulphuric acid	.030	Silica	.008
Chlorine	trace		

Combined as follows:

Bicarbonate of soda	.219	Sulphate of lime	.029
Bi-carbonate of lime	.078	Chloride of calcium	trace
Sulphate of soda	.021	Silica	.008

As is seen by the analysis, this water is remarkably free from solid matter, nor is it supposed by the inhabitants of the country in its neighborhood to possess any other than the ordinary properties of water.

The geological character of the contiguous country has nothing in it that would induce one to suspect the existence of such abundant sources of warm water. The plain of *Eski-Shehr* appears to be one of those extensive lacustrine regions so common in the western portion of Asia Minor; the deposits consist of a consolidated breccia. Imbedded we find the rocks of the neighboring mountains, as well as the meerschaum or silicate of magnesia, so extensively worked for exportation. Thermal waters are obtained in numerous parts of the plain as well as at Eski-Shehr.

THERMAL WATERS OF TROY.

Near the plains, in which are supposed to have been situated the ancient city of Troy, are numerous sources of thermal waters, of several of which I procured specimens; only two, however, have been analyzed, the others not having arrived at my laboratory. These springs are those alluded to by Homer, and they have enjoyed more or less reputation from the time of the Trojans to the present date. The two that I have examined are saline and their sources near each other. Analysis shows them to be identical. The physical properties will be alluded to when the other waters from this locality have been examined.

Chemical Composition.—One litre contains of solid matter 21.301 grammes. The same quantity of water has in its composition

Carbonic acid, fixed	.0595	Lime	1.4000
Sulphuric acid	.0680	Magnesia	.3012
Chlorine	12.8000	Oxide of iron	trace
Bromine	trace	Silica	.0600
Soda	9.2110		

Combined as follows:

Carbonate of lime	.1225	Chloride of magnesium	.7031
Sulphate of Soda	.0607	Bromide of magnesium	trace
Sulphate of lime	.0540	Chloride of iron	trace
Chloride of sodium	17.4450	Silica	.0600
Chloride of calcium	2.5078		

THERMAL WATERS OF MITYLENE.

On this island, the ancient Lesbos, there are several warm springs, and much of the geological structure of the place is volcanic. I visited two of the springs; the first is near to the village of Mitylene, and immediately on the shores of the gulf of Olives; it is called *Kelemyeh Oulinjah*, and there are two baths attached to it.

KELEMYEH OULINJAH SOURCE.

The water is clear and flows without leaving a deposit. Its temperature is 102° Fah. (atmosphere at 77°), and when cold there is nothing marked in its taste.

Chemical Composition.—There are 1.250 grammes of solid matter in a litre of the water, which contains the following ingredients in grammes:

Carbonic acid, free	.155	Soda	.278
Carbonic acid, fixed	.075	Lime	.152
Sulphuric acid	.040	Magnesia	.070
Chlorine	.570	Silica	.015

Combined as follows:

Bicarbonate of lime	.2450	Chloride of calcium	.0865
Sulphate of soda	.0357	Chloride of magnesium	.1628
Sulphate of lime	.0330	Silica	.0150
Chloride of sodium	.6510		

The other source visited on the island of Mitylene is about six miles north of the village, and is called *Touzla;* there are baths attached to it, and the waters are strongly saline.

TOUZLA SOURCE (SALINE).

Physical Properties.—The water does not flow clear, being more or less tinged with yellow produced by some organic acid in combination with lime. This deposit is seen to mark the course of the water as it flows down the beach into the sea, which is very near to it. Temperature of the water 117° Fah. (atmosphere at 76°). Specific gravity 1.0263.

Chemical Composition.—There are 34.520 grammes of solid matter in a litre of the water, which contains the following ingredients in grammes:

Carbonic acid, fixed	.050	Lime	2.534
Sulphuric acid	1.648	Magnesia	.110
Chlorine	18.440	Alumina	.012
Bromine, minute quantity, not estimated.		Iron	.003
		Soda	14.858

Combined as follows:

Carbonate of lime	.0912	Chloride of calcium	2.0040
Sulphate of soda	1.4625	Chloride of magnesium	.2023
Sulphate of lime	1.3000	Carbonate of iron	.0038
Sulphate of alumina	0.0221	Bromide of magnesium, minute quantity.	
Chloride of sodium	28.0260		

There are several other sources of thermal water in various parts of this island; the one reputed to have the greatest temperature is about eighteen miles from the latter, and called *Fezilkeh;* this source I could not visit, and can therefore say nothing of it from personal examination. There is yet one other source that I will allude to—the

TIBERIAD THERMAL WATERS.

The source of these waters is on the border of the sea of Galilee and within a mile of the city of Tiberias, of the solid structure of which repeated earthquakes have left but little. The surrounding country shows marked evidence of extensive volcanic action.

There are several sources at the place I visited, but they seem to vary little from each other. They flow into Turkish baths, and from them pass into the sea, on their way leaving a slight yellow deposit, which is doubtless, as in many of these waters, a crenate of lime.

Their temperature was not accurately ascertained for want of a thermometer, but I should consider it about 120° Fah.

Chemical Composition.—In one litre 23.540 grammes of solid matter. The same quantity of water furnished in grammes

Carbonic acid, fixed	.006	Lime	.443
Sulphuric acid	.197	Magnesia	.119
Chlorine	13.989	Silica	.006
Soda	8.751		

Combined as follows:

Sulphate of soda	.0620	Chloride of calcium	.7800
Sulphate of lime	.0386	Chloride of magnesium	.1850
Sulphate of magnesia	.0151	Silica	.0060
Chloride of sodium	22.2330	Carbonate of lime	.0106

The quantity of water brought away was too small to examine for the presence of bromine.

This is the last of the thermal waters of Asia Minor which have been examined; there are a few others that may yet reach me, when the composition will be made known as soon as examined.

CAUSE OF THE THERMAL WATERS IN WESTERN ASIA MINOR.

The cause of the abundance of warm springs in this quarter of the globe (in all formations from the alluvial to the oldest rocks) is doubtless owing to the extensive igneous action within no great depth beneath the surface of the country; a fact evinced by the frequency of earthquakes, but more especially by their extent; for they almost invariably extend from one end of it to the other, as well as to the neighboring islands.

Neither time nor change of government has contributed so much to the destruction of the hundreds of magnificent cities which once covered this country as the desolating influence of the earthquake, and many are the cities that now exist which have been prostrated over and over again and rebuilt, each time in diminished splendor, until at last they are little better than collections of huts when contrasted with their original condition. All the country at the present day seems to be as much subject to them as formerly.

The only part of Western Asia Minor where phenomena are seen strictly analogous to those of active volcanoes is in the *Catacecaumene*, or burnt district, situated in Lydia, about one hundred miles east of Smyrna. Numbers of volcanic cones exist in the neighborhood of Koola, the craters of many of which are quite distinct, especially the one called *Kaplar Alan*, which has a perfect crater about half a mile in circumference, and two or three hundred feet deep. The extent of this region is some twenty miles long by eight broad. We have no record of any activity in these volcanoes, and Strabo described them in his day quite as they are now, and the Turks give to Satan the full credit of having created such a black, parched-up district. My object at the present time is merely to mention this district, as a full description of it enters into a paper on the earthquakes and volcanoes of Asia Minor, that I propose publishing at some future time; it is brought forward now merely to show what this volcanic center has to do with the thermal springs just described.

REMARKS ON THE OCCURRENCE OF NITROGEN IN THERMAL WATERS.

The only substance connected with these waters that I shall allude to is the nitrogen contained in the gas accompanying many of them, and in some instances constituting almost the entire gaseous product, as in the case of the springs of Yalova. This singular fact attracted my attention several years ago while examining into the gaseous products of various springs, and I then ascertained that the gas was found especially with warm springs; the nitrogen, when found accompanied with oxygen, existed in proportions much greater than in the atmosphere, and in numerous instances it was almost pure. The question naturally arises, whence comes this nitrogen? and as we know of no other natural source of nitrogen than the atmosphere, it occurs to the mind that there is a source of the gas in the thermal waters, which before they pass to the heated substrata absorb a certain amount of air; the oxygen of the air contained in the water combines with the rocks and minerals, or is taken up by some deoxydizing agent in the waters, which as they return to the surface naturally bring the nitrogen of the air freed of all or most of its oxygen.

This explanation, which appears so natural, does not, however, account for the fact, and I have been obliged to abandon it. Did the nitrogen in these waters occur in such small quantities as we might suppose to have been absorbed by water, the explanation would hold good; but the fact in the case of the springs at Yalova and many other sources is that the gas, which is nearly pure, bubbles up in *great abundance.* Again, if the nitrogen evolved by springs be simply such as the water absorbs before penetrating the surface of the earth, how does it happen that this gas escapes from springs of ordinary temperature? For it is reasonable to suppose that the water having once taken into solution a gas, will not give it out except by heat or the presence of a large amount of saline matter, neither of which occur to explain the evolution of nitrogen gas from certain springs.

Feeling thus satisfied that the nitrogen in the gaseous products of springs is not owing to its absorption from the atmosphere, its origin has been sought for elsewhere, but without success, and I am constrained to believe that nitrogen is one of those elements stored up in the interior of the earth, in more or less abundance, either pure or combined, and frequently finds its way to the surface through those fissures by which mineral waters are conducted. Its more frequent occurrence with thermal waters is doubtless owing to the greater depth from which the latter come.

After all, however, that has been said, we must acknowledge the explanation imperfect, and as only furnishing another evidence of the difficulty of learning any thing of the origin or uses of this singular substance, nitrogen, in its elementary state.

ON THE ANALYSIS OF THESE WATERS, PARTICULARLY WITH REFERENCE TO THE SILICA AND ALKALIES.

The general method of analysis adopted differed but little from that usually employed, and the construction of the salts out of the acids and bases has been made entirely from the dictates of my judgment in the matter.

The examination of the silica attracted considerable attention, from the fact that we are in the habit of always estimating it as uncombined silica, even when found in alkaline waters. Although my researches are sufficient to prove to my mind the inaccuracy of this, still I have not thought proper in this paper to deviate from the rule generally adopted, leaving it to more extended research to decide the point.

In the analysis of the waters of Broosa, nearly all of which are alkaline, the following fact has been observed: that on concentrating a considerable quantity of the water to a small bulk all the carbonate of lime is precipitated and a portion of the silica (whether in combination with lime or not is not yet decided); but a large portion of the same still remains in solution, as well as some lime, although the water is alkaline with an excess of carbonate of soda. The silica is in such quantity that it could remain only in solution in combination with an alkali; in fact, there is a silicate of soda and lime present.

The question here arises whether the silica was in a state of combination before the water was concentrated, or is it a result that has taken place during the evaporation; this question can only be decided by more extended investigation.

The observation of the above fact has led me to adopt the following method of estimating the silica in mineral waters: Take a certain quantity of the water, evaporate almost to dryness, add hydrochloric acid, a little more than is required to saturate the carbonates present; continue to evaporate to complete dryness, and then add water acidulated with a little hydrochloric acid; filter and wash the silica that remains on the filter; in this way we are sure to have the silica perfectly free from any silicate.

The method adopted for estimating the alkalies will be mentioned in a few words, as more details of it will be given in a paper devoted especially to that subject; the method has particular reference to the separation of the alkalies from magnesia.

SEPARATION OF THE ALKALIES FROM MAGNESIA.

Take the solution filtered from the silica, evaporate to dryness to drive off the excess of acid, add a little water to redissolve, then add pure lime-water and filter, when the chlorides of the alkaline metals and calcium with excess of lime will pass through, the magnesia, alumina, oxide of iron, etc., remaining on the filter. Separate the lime with carbonate of ammonia, or still better with oxalate of ammonia, evaporate to dryness, and heat to drive off the ammoniacal salts, when nothing but the chlorides of alkaline metals will be left, which can be separated in the ordinary way.

This completes the description of all the thermal waters of Asia Minor which have as yet come under my notice, with the observations that the investigation have given rise to.

RE-EXAMINATION OF AMERICAN MINERALS.*

In the investigations which will be detailed we have endeavored to clear up the doubts due to imperfect analyses and descriptions that exist respecting several American minerals. Every care has been taken to procure the best specimens, and our results have been tested by several analyses of each. We are under obligations to several mineralogists of this country who have placed their cabinets at our disposal for this investigation, and among those to whom we are more specially indebted we take pleasure in mentioning Messrs. L. White Williams and W. W. Jeffries, of Westchester, Pa.; T. F. Seal, of Philadelphia; and Mr. Theo. S. Gold, of West Cornwall, Ct.

1. Emerylite identical with Margarite.

Emerylite was originally found by one of us on the emery of Asia Minor, and also on the same mineral coming from the Grecian Archipelago, Siberia, and China; it was subsequently traced by Prof. Silliman, jr., in connection with the corundum occurring in various parts of this country. Its constant occurrence with emery and corundum (forming one of the minerals of elimination in their formation) suggested the name emerylite as most appropriate, its composition having been found to be different from that of any known mineral.

It was justly considered an interesting species, on account of its accompanying the various forms of corundum wherever observed; and it may be safely said that no mineral has been proved to be as widely distributed in so short a time after the first announcement of its connection with the emery of Asia

* In the first half of this re-examination I was ~~assisted by~~ my friend, George J. Brush.

Minor, and the suggestion that it might be found with the corundum of other localities.

The analyses which were immediately made of this mineral by different chemists, on specimens coming from various parts of the world, showed a uniformity of composition most remarkable in a micaceous mineral, and so it was considered by a committee of the Academy of Sciences at Paris, who reported on this subject.* This fact is most clearly seen by reference to the following table of analyses made in 1850:

LOCALITIES.	$\dddot{\text{Si}}$	$\dddot{\text{Al}}$	$\dot{\text{Ca}}$	$\dddot{\text{Fe}}$	$\dot{\text{Mg}}$	$\dot{\text{K}}$ & $\dot{\text{Na}}$	$\dot{\text{H}}$	$\dot{\text{Mn}}$	
Gumuch-dagh.......	29.66	50.88	13.56	1.78	.50	1.50	3.41	....	J. L. Smith.
Island of Nicaria...	30.22	49.67	11.57	1.33	trace	2.31	5.12	....	"
Island of Nicaria...	29.87	48.48	10.84	1.63	trace	2.86	4.32	....	"
Island of Naxos....	30.02	49.52	10.82	1.65	.48	1.25	5.55	....	"
Island of Naxos....	28.90	48.53	11.92	.87	not est.	not est.	5.08	....	"
Island of Naxos ...	30.10	50.08	10.80	not est.	"	"	4.52	....	"
Gumuch-dagh.......	30.90	48.21	9.53	2.81	"	"	4.61	....	"
Gumuch-dagh.......	31.93	48.80	9.41	1.50	"	2.31	3.62	trace	"
Siberia.................	28.50	51.02	12.05	1.78	"	not est.	5.04	....	"
Village Green, Pa..	32.31	49.24	10.66	....	.30	2.21	5.27	....	W. J. Craw.
Village Green, Pa..	31.06	51.20	9.24	....	.28	2.97	5.27	....	"
Village Green, Pa..	31.26	51.60	10.15	....	.50	1.22	4.27	....	"
Village Green, Pa..	30.18	51.40	10.87	....	.72	2.77	4.52	....	"
Buncombe Co., N.C.	29.17	48.40	9.87	....	1.24	6.15	3.99	H F 2.03	Silliman, jr.
Unionville, Pa.......	29.99	50.57	11.31	....	.62	2.47	5.14	....	W. J. Craw.
Unionville, Pa.......	32.15	54.28	11.36	trace	.05	not est.	.50	....	Hartshorne

It was suspected by us, at the time the species was made, that it might prove identical with margarite; but not having the latter mineral at hand, we had to proceed on the known analyses of it, which we here give. The first is by Dumeril; the second by the Göttingen Laboratory, on the authority of Hausmann.

	$\dddot{\text{Si}}$	$\dddot{\text{Al}}$	$\dddot{\text{Fe}}$	$\dot{\text{Ca}}$	$\dot{\text{Na}}$	$\dot{\text{H}}$
1.	37.00	40.50	4.50	8.96	1.24	1.00=93.20
			$\dot{\text{Fe}}$		Mn	$\dot{\text{Mg}}$
2.	33.50	58.00	0.42	7.50	0.03	0.05=99.50

These analyses differing so materially from those of emerylite, fully justified the formation of the species.

As soon as margarite could be procured it was subjected to analysis, and the inaccuracy of former analyses proved; but we had not at that time sufficient of the mineral to complete the investigation as desired. In the mean time Hermann† reanalyzed it, found a different composition from any previous

* Comptes Rendus de l'Académié des Sciences, Oct. 28, 1850.

† J. f. pr. Chem., liii, 1.

one, and concurring with the one that had been made by us, as well as with those more recently made, which are here given.

	$\ddot{Si}$	$\ddot{\bar{Al}}$	$\ddot{\bar{Fe}}$	$\dot{Ca}$	$\dot{Mg}$	$\dot{Na}$	$\dot{K}$	$\dot{H}$	
1.	28.47	50.24	1.65	11.50	0.70	1.87	trace	5.00	= 99.26
2.	28.64	51.66		12.25	0.68	2.01*		4.76	=100.00

These correspond to the formula $\dot{R}^3\ddot{Si}+3\ddot{\bar{Al}}^2\ddot{Si}+3\dot{H}$.

	Atoms.	At. weight.	Per cent.	Oxygen ratio.
Silica	4	2309.24	30.58	4
Alumina	6	3850.8	50.99	6
Lime...............	3	1054.5	13.96	1
Water	3	337.5	4.47	1

The specimen of margarite examined was received from Dr. Krantz, of Bonn, and came from Sterzing in the Tyrol, the original locality.

By these analyses it will be seen that margarite and emerylite are identical, and the former name having priority of date (although the composition of the mineral was not made out until lately), it must doubtless replace the latter, unless its geological appropriateness can sustain it.

2. Euphyllite.

This mineral was first analyzed by Crooke, but the analysis, having been made by a fusion with carbonate of baryta, was found to be incorrect. It was re-analyzed by Erni and Garrett.† Dr. Erni's analyses gave the formula $\dot{R}^3\ddot{Si}+11\ddot{\bar{R}}\ddot{Si}+3\dot{H}$. Garrett found no water; his analyses give the same formula as Erni's, minus the water.

Our results differ essentially from those heretofore obtained, as is seen by the following analyses:

	1	2	3	4
Silica	40.29	39.64	40.21	40.96
Alumina..................	43.00	42.40	41.50	41.40
Peroxide of iron......	1.30	1.60	1.50	1.30
Lime	1.01	1.00	1.88	1.11
Magnesia	.62	.70	.78	.70
Soda	3.94	3.94	3.25	3.25
Potash	5.16	5.16	4.26	4.26
Water	5.00	5.08	5.91	6.23
	100.32	99.52	99.29	99.21

* By the difference.

† Amer. Jour. Science and Arts, 2d series, viii, 382; Dana's Mineralogy, 3d ed., p. 362.

No. 1 was from a specimen in our own collection. No. 2 was from the original specimen in Prof. Silliman's cabinet. Nos. 3 and 4 were specimens received from Messrs. Williams and Jeffries, of Westchester, Pa. Specific gravity Nos. 1 and 2, 2.83. The analyses give the formula $\dot{R}\ddot{S}i+\bar{\bar{R}}^3\ddot{S}i^2+2\dot{H}$.

	Atoms.	At. weight.	Per cent.	Oxygen ratio.
Silica	3	1731.93	39.63	9
Alumina.........	3	1925.40	44.05	9
Soda...............	½	193.60	4.43 }	1
Potash............	½	394.42	6.74 }	
Water	2	225.00	5.15	2

This mineral in its most beautiful form is of rare occurrence (analyses 1 and 2 are of this variety); there is, however, another variety, not differing essentially in physical characters or in chemical composition, which is found in considerable abundance at the locality.

In all probability the mineral alluded to as Muscovite (?) in the memoir on the minerals associated with emery* is this mineral; and when we are able to get at certain specimens from Asia Minor, containing this mica in a pure state, this point will be investigated. It is of much interest toward tracing out its geological connection with corundum formations widely separated, in which respect it may resemble *emerylite.*

3. Mica from Litchfield, Conn.

This mineral is associated with the kyanite of Litchfield. In general appearance it resembles margarodite. Hardness= 3.35; specific gravity 2.76; almost colorless, having a faint tinge of green; transparent; luster pearly. The results of two analyses gave

	$\ddot{S}i$	$\bar{\bar{A}}l$	$\bar{\bar{F}}e$	$\dot{M}g$	$\dot{C}a$	$\dot{M}n$	$\dot{N}a$	$\dot{K}$	Fl	$\dot{H}$
1.	44.60	36.23	1.34	0.37	0.50	trace	4.10	6.20	trace	5.26
2.	44.50	37.10		undet.	undet.	......	4.00	5.90	......	5.16

These correspond very closely with Liebnerite, as well as with Damourite and some analyses of margarodite. Annexed are the analyses for comparison:

	$\ddot{S}i$	$\bar{\bar{A}}l$	$\dot{F}e$ ($\bar{\bar{F}}e$)	$\dot{M}n$	$\dot{M}g$	$\dot{K}$	$\dot{N}a$	$\dot{H}$	
Liebnerite....	44.66	36.51	1.75	......	1.40	9.90	0.92	4.49	Marignac.
Damourite ...	45.22	37.87	trace	......	......	11.20	......	5.25	Delesse.
Margarodite.	46.23	33.08	3.48	trace	2.10	8.87	1.45	4.12	Delesse.

* Amer. Jour. Science and Arts, 2d series, xi, 62.

The silica and peroxides* in these analyses are identical; but, in common with many of the micas, it is extremely difficult to deduce any one formula that would be a correct expression of their chemical constitution, owing to slight differences in the protoxides. This is rendered more obvious by comparing their oxygen ratios.

	$\dot{R}$	$\dddot{R}$	$\dddot{Si}$
1. Litchfield mica	1 :	7.22 :	9.70
2. Liebnerite	1 :	6.33 :	9.43
3. Damourite	1 :	9.35 :	12.00
4. Margarodite	1 :	6.16 :	8.95

The striking similarity of these species would lead us to suspect that if new analyses were made of specimens from the original localities, they might prove identical. In all physical characters, except structure, there is a complete correspondence.

4. Unionite, identical with Oligoclase.

This mineral was described by Prof. Silliman, jr., in the Amer. Jour. Science and Arts, 2d series, viii, 384. The following are its characters: In general appearance it resembles a soda spodumene; it has a very distinct cleavage in one direction; luster vitreous; color white; hardness 6; specific gravity 2.61. It is found with euphyllite at the corundum locality near Unionville, Pa. The results of three analyses are as follows:

	$\dddot{Si}$	$\dddot{Al}$	$\dddot{Fe}$	$\dot{Ca}$	$\dot{Mg}$	$\dot{Na}$	$\dot{K}$	ign.
1.	64.09	21.45	trace	0.86	0.69	10.94	1.36	1.02=100.41
2.	64.45	20.97	trace	0.77	0.46	10.94	1.36	1.14=100.09
3.	......	21.70	trace	0.85	0.49	......	......	1.02

The third analysis, owing to an accident, is incomplete; the constituents determined are given for comparison. The oxygen ratio of these analyses is very nearly 1 : 3 : 9, which gives the formula $\dot{R}\dddot{Si}+\dddot{Al}\dddot{Si}^2$. This is the formula of oligoclase; the analyses correspond with that species, and the physical characters being the same, there can be no doubt as to the identity of unionite and oligoclase.

It is believed that this is the first time that oligoclase has been observed in the United States.

---

* Considering the iron in Liebnerite as peroxide.

5. Kerolite of Unionville, Pa., a hydrated silicate of Alumina.

Associated with euphyllite and unionite, there occurs a peculiar amorphous mineral, which has been circulated among some of our American mineralogists under the name of kerolite. In our examinations of the minerals from this locality we thought it of sufficient importance to ascertain its chemical composition.

In physical characters it resembles kerolite; hardness 2.25; specific gravity 2.22; color yellowish-white; brittle; crumbles to pieces when thrown in water. Analysis gave

$\dddot{\mathrm{Si}}$	$\bar{\ddot{\mathrm{Al}}}$	$\dot{\mathrm{Mg}}$	$\dot{\mathrm{Mn}}$	$\dot{\mathrm{Na}}$ & $\dot{\mathrm{K}}$	$\dot{\mathrm{H}}$
44.50	25.00	7.75	trace	trace	22.39=99.64

Of the water 1.04 per cent. was lost by twenty-four hours' desiccation over sulphuric acid, 8.81 per cent. by heating to 212°, and the remainder at a red heat.

In chemical composition it is near *halloysite.* It is an imperfectly formed mineral, and consequently is not homogeneous; it passes into euphyllite and feldspar.

6. Bowenite, identical with Serpentine.

This mineral occurs at Smithfield, R. I., and was described by Bowen* as a variety of nephrite. His analysis gave

$\dddot{\mathrm{Si}}$	$\dot{\mathrm{Mg}}$	$\dot{\mathrm{Ca}}$	$\dot{\mathrm{Fe}}$	$\bar{\ddot{\mathrm{Al}}}$	$\dot{\mathrm{Mn}}$	$\dot{\mathrm{H}}$
44.69	34.63	4.25	1.75	0.56	trace	13.42

This composition differed so much from nephrite, and corresponded so closely to the formula $2(\dot{\mathrm{Mg}}\,\dot{\mathrm{Ca}})^2\,\dddot{\mathrm{Si}}+3\bar{\dot{\mathrm{H}}}$, that Professor Dana felt himself justified in noticing it as a distinct species.†

The following are the physical characters of the mineral: Hardness 5 (it will scratch glass if rubbed with a little force against its surface; it first gives way, but ultimately scratches the glass); specific gravity 2.57; color, in large masses, bright apple-green; highly translucent; structure granular, and exceedingly tough. We give analyses of three specimens. No.

* Amer. Jour. Science and Arts, 1st series, vi, 346.

† Dana's Mineralogy, 3d edition, p. 265.

1 was from the cabinet of Professor Silliman, jr.; No. 2 from the mineralogical collection of Harvard University, received from Professor Cook; No. 3 from the Lederer collection in Yale College.

	$\ddot{Si}$	$\dddot{Al}$	$\dot{Mg}$	$\dot{Fe}$	$\dot{Ca}$	$\dot{H}$
1.	42.20	trace	42.50	1.56	trace	13.28=99.54
2.	42.56	trace	43.15	0.95	......	12.84=99.50
3.	42.10	trace	41.23	1.11	1.90	12.77=99.11

These analyses give the oxygen ratio 4 : 3 : 2, and the formula $2\dot{Mg}^3\ddot{Si}^2+3\dot{Mg}\dot{H}^2$, which calculated is

$\ddot{Si}$	$\dot{Mg}$	$\dot{H}$
43.5	43.8	12.7

This is the composition and formula of serpentine, and the fact of its identity with that species is also borne out by its physical characters.

The large amount of lime obtained by Bowen was doubtless due to the limestone and tremolite with which it is often very intimately associated; much care is required to separate these substances entirely from the Bowenite, but the mineral so purified contains no lime.

7. Williamsite, identical with Serpentine.

We notice that this species is considered distinct by Prof. Shepard in the last edition of his mineralogy, notwithstanding it has been shown to be serpentine by Hermann,* and previously from an analysis made by one of us, published in Dana's Mineralogy, page 692. In this analysis referred to, 3.35 per cent. of alumina and iron were obtained. We have since examined the relative proportions of these substances, and find that the amount was due to iron with but a trace of alumina. Two analyses made from very pure specimens gave

	$\ddot{Si}$	$\dddot{Al}$	$\dot{Mg}$	$\dot{Fe}$	$\dot{Ni}$	$\dot{H}$
1.	41.60	trace	41.11	3.24	0.50	12.70=99.15
2.	42.60	trace	41.90	1.62	0.40	12.70=99.22

It is evident from these analyses that the mineral is identical with serpentine, and affords the same formula as the mineral last mentioned. It may be well to remark that great care was taken to see that no magnesia accompanied the oxide of iron in its precipitation by ammonia; not satisfied with adding an

* J. f. pr. Chem., liii, 31.

excess of sal ammoniac to the solution before the addition of the ammonia, we redissolved the precipitate, added sal ammoniac, and reprecipitated the oxide of iron; this was done even a third time, before the last traces of magnesia were got rid of, or that we were sure that the amount of iron would not be increased by containing magnesia — a circumstance in which sufficient precaution is not always used. What is here said of oxide of iron is equally true of alumina.

8. Lancasterite, a mechanical mixture of Brucite and Hydro-magnesite.

While on a mineralogical excursion to the localities near Texas, Pa., a few months since, in company with Mr. W. W. Jeffries, we observed at Wood's Mine a peculiar magnesian mineral, somewhat resembling lancasterite; a chemical examination showed it to be *hydro-magnesite*. The composition of it, as well as its strong resemblance to some specimens of lancasterite, led to a re-examination of the latter species.

Lancasterite is described as occurring "foliated like brucite," but sometimes in crystals "resembling somewhat stilbite or gypsum." As we desired to see whether these forms were identical in chemical composition, a portion of the foliated mineral was carefully selected and the amount of carbonic acid determined. It was but a trace; the magnesia and water being estimated gave the same amount as is found in brucite; there was also a trace of manganese and iron.

Some of the small crystals "resembling stilbite or gypsum" were then examined; analysis showed them to have the same composition as the hydro-magnesite of Kobell.

These results go to prove that lancasterite is not a distinct species, but a mechanical mixture of brucite and hydro-magnesite. In Dr. Erni's analyses of this mineral (made in the Yale Laboratory) we are aware he found great difficulty in obtaining a constant composition, and it was only after a series of analyses that he obtained any concordant results. The specimens he examined were both crystallized and foliated, the folia in some cases overlying the crystalline portion. With this explanation the composition he obtained is easily understood.

The following are the results of our analyses. Nos. 1 and 2 were foliated; Nos. 3 and 4 were of the radiated variety:

	1	2	3	4
Magnesia	66.30	66.25	42.30	44.00
Protoxide of iron	.50	1.00	trace	trace
Protoxide of manganese	trace			
Carbonic acid	1.27	trace	36.74	36.60
Water	31.93	32.75	20.96	19.40
Direct determination of water (3 and 4)			20.10	

The foliated variety gives the exact composition of brucite. In two determinations of loss by heat the numbers 34.30 and 35.67 were obtained; great difficulty was found in obtaining the brucite perfectly pure, owing to its intimate association with the hydro-magnesite.

The radiated variety (as before stated) gives the composition of hydro-magnesite, and to show that the original analyses were made from a mixture of these minerals we give Dr. Erni's results* for comparison:

$\dot{Mg}$	$\dot{Fe}$	$\ddot{C}$	$\dot{H}$	Total.
50.01	1.01	27.07	21.60	99.69
50.72	.96	26.85	21.47	100.00

9. Hydro-magnesite found crystallized.

The hydro-magnesite above mentioned is extremely beautiful, and in appearance resembles very much the thomsonite from Kilpatrick in Scotland. Its structure is highly crystalline, and in some instances forms distinct crystals, which have been considered as monoclinic (?) (Dana); the diagonal cleavage is very distinct; hardness 3–3.5 (scratching calcite with ease); specific gravity 2.145–2.18. It occurs at Wood's Mine, Texas, Lancaster County, Pa., in seams which are sometimes half an inch in thickness, and at Low's Mine in veins generally from one tenth to one fifth of an inch wide, having a beautifully radiated structure. The results of two analyses of a specimen from Wood's Mine are as follows:

	1	2
Magnesia	43.20	42.51
Carbonic acid	36.69	35.70
Water	20.11	21.79
Iron and manganese	trace	trace

* From Dana's Mineralogy, p. 213.

A direct determination of the water gave 19.83 per cent. These analyses give the oxygen ratio 2 : 3 : 2, and the formula $3(\dot{M}g\,\ddot{C}+\dot{H})+\dot{M}g\,\dot{H}$ which calculated gives

$\dot{M}g$	$\ddot{C}$	$\dot{H}$
44.68	35.86	19.46

The composition and formula are the same as obtained by Kobell and Wachtmeister for hydro-magnesite from Negroponte and Hoboken. We are not aware that this species has heretofore been observed with a crystalline structure.

10. Supposed Magnesite of Hoboken shown to be Aragonite.

In connection with the foregoing investigations it was thought that an examination of some of the anhydrous carbonates of magnesia might be interesting. For this purpose a specimen of magnesite from Hoboken, N. J., was submitted to analysis (the variety referred to is that which occupies seams and cavities in the Hoboken serpentine, having an aggregated fibrous structure, and not unfrequently occurring in delicate needle crystals). A careful qualitative examination of the needle crystals showed them to be carbonate of lime, with scarcely a trace of magnesia; they have the form of aragonite. Specimens from Staten Island and the vicinity of Westchester, Pa., were examined, with like results. The crystals from the serpentine quarry near Westchester are frequently transparent, and are among the most beautiful specimens of crystallized aragonite that have been observed in this country.

11. Chesterlite, identical with Orthoclase.

This mineral occurs in implanted crystals on dolomite near East Bristol, Chester County, Pa. In physical characters it resembles orthoclase, but it has been considered triclinic, and Erni's analysis* gave soda as the alkali. The crystals occur frequently as twins, and are often very much distorted—in specimens we have examined the angle T on T′ varies from 121° to 127°—rendering it extremely difficult to determine the normal value of the angle; some of the measurements would, however, lead to the conclusion that it is monoclinic, since the angle of cleavage is by our measurements near 90°. So far as

* Dana's Mineralogy, 3d ed., p. 678.

our opinion is concerned—based on both its chemical and physical character—we unhesitatingly pronounce it an orthoclase. Two analyses gave

	1	2
Silica	64.76	65.17
Alumina	17.60	17.70
Peroxide of iron	.50	.50
Lime	.65	.56
Magnesia	.30	.25
Potash	14.18	13.86
Soda	1.75	1.64
Ignition	.65	.65
	100.39	100.33

These correspond to the composition of orthoclase, and chemically the mineral is identical with it; if it shall be proven that the crystalline form is triclinic, it will be a potash albite, and as such an interesting species.

The specimens examined were received from Messrs. T. F. Seal and William S. Vaux, of Philadelphia.

12. Loxoclase, identical with Orthoclase.

The feldspar, associated with pyroxene at Hammond, N. Y., has been named as a distinct species by Breithaupt.* Its crystalline form, hardness, specific gravity, and other physical characters are the same as orthoclase, and the reasons for forming a new species of it are based upon its cleavage and chemical constitution. The latter Plattner found to be

$\dddot{Si}$	$\dddot{\bar{Al}}$	$\dddot{\bar{Fe}}$	$\dot{Ca}$	$\dot{Mg}$	$\dot{Na}$	$\dot{K}$	$\dot{H}$
63.50	20.29	0.67	3.22	trace	8.76	3.03	1.23=100.70

We have examined two varieties of it. Analyses 1 and 2 are from specimens taken from a large crystal, and were not perfectly pure, owing to intimate association with a lime pyroxene; analyses 3 and 4 are from a very pure crystal.

	1	2	3	4
Silica	65.40	65.69	66.09	66.31
Alumina	19.48 }	20.72	19.15	{ 18.23
Peroxide of iron.	1.25 }			{ .67
Lime	2.26	2.36	.94	1.09
Magnesia	.20	.25	.21	.30
Potash	2.76	2.36	4.35	4.35
Soda	7.23	7.98	7.81	7.81
Ignition	.76	.76	.20	.20
	99.34	100.12	98.75	98.96

* Pogg. Ann., lxvii, 419.

It will be seen at a glance that the only difference between this mineral and orthoclase is the large amount of soda, and in analysis 1 and 2 a small amount of lime, this last, most of which is doubtless an impurity, alters somewhat the oxygen rates.

No. 1 gives	1 : 3.10 : 11
No. 2 "	1 : 3 : 10.60
No. 3 "	1 : 2.90 : 11.08
No. 4 "	1 : 2.74 : 10.83

This slight difference in the ratio (produced by the presence of a considerable amount of soda) is not uncommon in orthoclase. In that from Hohenhagen, Schnedermann found 4.15 potash and 7.53 soda; the flesh-red feldspar from Bathurst, Canada, gave Hunt 6.36 potash, 5.37 soda; and Gmelin found in the feldspar from Laurvig 6.55 potash and 6.14 soda, and in that from Fredicksvärn 7.03 potash and 7.08 soda. These numbers affect to some slight extent the oxygen ratio, but the correspondence of the minerals in physical characters denotes their identity with orthoclase. Moreover, the identity of loxoclase with orthoclase is made obvious when we take the ratio between the silica and alumina, which in the purer varieties (analyses 3 and 4) is as 4 : 1, and analysis 4 gives the ratio 12 : 3.04 : 1.11, or $\dot{R}\ddot{S}i + \dddot{R}\ddot{S}i^3$.

The specimens examined were received from Professor Silliman, jr., and Mr. Samuel W. Johnson.

13. Danbury Feldspars: 1. Oligoclase; 2. Orthoclase.

1. *Oligoclase.*—The feldspar in which the danburite occurs has so strong a resemblance to the oligoclase from Sweden that we have been led to analyze it; the results of our examination prove its identity with that species. The analyses gave

	1	2
Silica	64.03	63.50
Alumina	22.37	22.75
Peroxide of iron	trace	trace
Lime	2.91	3.28
Magnesia	trace	trace
Soda	10.06	9.37
Potash	.60	.50
Ignition	.30	.21
	100.27	99.61

These give the oxygen ratio 1 : 3 : 9 and the formula $\dot{R}\ddot{S}i + \dddot{A}l\ddot{S}i^2$, which are the ratio and formula for oligoclase.

2. *Orthoclase.*—There is also associated with danburite a potash feldspar not unlike the soda feldspar just mentioned; in some cases it is so intimately associated with it as to require great care in selecting for analysis. So far as our observation extends, we have been able to identify the oligoclase by its occurrence in masses with a broad cleavage surface, and another less smooth, meeting at the angle 93°–94°; in orthoclase this angle is 90°, and most frequently it presents at this locality small cleavage faces, and is sometimes of a granular structure.

The following analyses are of the latter variety, which doubtless contain a little oligoclase that it is impossible to separate.

	1	2
Silica	63.80	63.95
Alumina	18.90	19.05
Lime	.80	.61
Magnesia	.20	.20
Potash	11.43	10.95
Soda	3.86	3.69
Ignition	.30	.50
	99.29	98.95

The specimens examined were taken from the locality by one of the authors.

14. Haddam Albite, identical with Oligoclase.

Associated with the iolite at Haddam, Conn., there occurs a glassy feldspar which has heretofore been called albite. Its composition is that of an oligoclase, as will be seen by the following analysis:

	1	2
Silica	63.87	64.64
Alumina	21.82	21.98
Lime	2.14	2.17
Magnesia	trace	trace
Soda	10.18	9.80
Potash	.50	.50
Ignition	.29	.29
	98.80	99.38

The specimens examined were received from Professor Dana.

We have examined the *moonstone feldspar*, from Mineral Hill, Delaware County, Pa., which is also oligoclase.

15. Greenwood Mica—Biotite.

The chemical constitution of only a very few American biotites has been examined. In fact, von Kobell's* analysis of a mica from Monroe, N. Y., is the only one that has been published; unfortunately, even in regard to this there is some doubt as to its locality, for biotite is found in more than one place near Monroe.

The specimens we have examined are from Greenwood Furnace, Monroe, N. Y.; the mineral occurs in large crystals of a dark olive-green color, and the results of the analyses are such as to lead to the supposition that the specimens examined by von Kobell were from this locality. He obtained

$\dddot{Si}$	$\dddot{\bar{Al}}$	$\dddot{\bar{Fe}}$	$\dot{Mg}$	$\dot{K}$	HF	$\ddot{Ti}$	$\dot{H}$
40.00	16.16	7.50	21.54	10.83	0.53	0.20	3.00=99.76

The results of our analyses are

	1	2
Silica	39.88	39.51
Alumina	14.99	15.11
Peroxide of iron	7.68	7.99
Magnesia	23.69	23.40
Potash	9.11 }	10.20
Soda	1.12 }	
Water	1.30	1.35
Fluorine	.95	.95
Chlorine	.44	.44
	99.16	98.95

These give the oxygen ratio

	$\dot{R}$	$\bar{R}$	$\dddot{Si}$
1.	11.31 :	9.31 :	20.72
2.	11.20 :	9.45 :	20.53

or very nearly 1 : 1 : 2, which corresponds to the formula $\dot{R}^3 \dddot{Si} + \dddot{\bar{R}} \dddot{Si}$. A small portion of the oxygen in the mineral is replaced by fluorine and chlorine.

The specimens examined were received from Messrs. Jenkins & Horton, of Monroe, N. Y.

16. Biotite of Putnam County, N. Y.

In appearance this mineral resembles talc, having a wavy, lamellar structure and a soapy feel. Its color is brownish-

* J. f. pr. Chem., xxxvi, 309, and Dana's Mineralogy, third edition, p. 360.

green in mass, and pale yellowish-green by transmitted light. Hardness 2–2.5; specific gravity 2.80. The laminæ are entirely devoid of elasticity. It has been called pyrophyllite by some mineral collectors, but upon what grounds we are ignorant, as it does not possess the remarkable property of exfoliating and swelling up by heat, so peculiar to pyrophyllite. Analysis shows its composition to be identical with biotite.

	1	2
Silica	39.62	39.49
Alumina	17.35	17.06
Peroxide of iron	5.40	5.21
Magnesia	23.85	23.65
Potash	8.95	......
Soda	1.01	......
Water	1.41	......
Fluorine	1.20	......
Chlorine	.27	......
	99.06	

Analysis 1 gives oxygen ratio $\dot{R}$ 11.22 : $\dddot{R}$ 9.73 : $\ddot{Si}$ 20.58, or 1 : 1 : 2; and the same formula as for the mineral last mentioned, $\dot{R}^3 \dddot{Si} + \dddot{R} \dddot{Si}$.

The specimens examined were received from Mr. Silas R. Horton, of Craigville, New York.

17. Margarodite.

This mineral occurs at Lane's Mine, Monroe, Conn. It has been analyzed by W. H. Brewer,* but owing to some impurities in his specimens he obtained an excess of silica.

Specimens very carefully selected, to avoid the fluor spar and other minerals with which it is associated, gave

	1	2
Silica	46.50	45.70
Alumina	33.91	33.76
Peroxide of iron	2.69	3.11
Magnesia	.90	1.15
Potash	7.32	7.49
Soda	2.70	2.85
Water	4.63	4.90
Fluorine	.82	.82
Chlorine	.31	.31
	99.78	100.09

* Dana's Mineralogy, third edition, p. 359.

These correspond to the analysis of margarodite from St. Etienne, in which Delesse found

$\ddot{\text{Si}}$	$\dddot{\overline{\text{Al}}}$	$\dddot{\overline{\text{Fe}}}$	$\dot{\text{Mg}}$	$\dot{\text{Na}}$	$\dot{\text{K}}$	$\dot{\text{H}}$	Fl
46.23	33.08	3.48	2.10	1.45	8.87	4.12	trace=99.33

In a former paper we have mentioned the difficulty of obtaining a correct formula from the analyses of margarodite, owing to slight differences in the protoxides. The relation of the oxygen of the silica to that of peroxides in most of the analyses is as 3 : 2.

The specimens examined were received from Professor Silliman, jr.

18. The Chesterlite Talc—a Mica.

Associated with the chesterlite a micaceous mineral is found, which has been called talc. It occurs in implanted crystals, in minute tuft-like aggregations on dolomite; there is frequently an iron stain upon the surface, due to the decomposition of some of the minerals with which it is associated; the crystals are seldom over a line in diameter. Its chemical composition is that of a mica; but owing to the small amount examined it is impossible to say positively whether it be muscovite or margarodite, although from its association we are inclined to consider it muscovite.

Silica	45.50
Alumina	34.55
Peroxide of iron	trace
Lime	2.31
Magnesia	1.08
Potash	8.10
Soda	2.35
Water and carbonic acid	5.40
	99.29

A large portion of the lime and magnesia is doubtless due to the dolomite with which it is associated. The specimen was received from Mr. Thos. F. Seal.

19. Rhodophyllite, identical with Rhodochrome.

The violet-colored mineral which occurs at Texas, Pa., and was circulated among mineralogists as "violet talc," has been analyzed by Dr. Genth,* of Philadelphia, who found for it a

* Proc. Acad. Nat. Sci., Phil., vi, 122.

distinct composition, and gave it the above name. Its physical characters correspond with rhodochrome and kammererite, but as there had been no analysis published of the first, and as Dr. Genth's results did not agree with those obtained for kammererite, he doubtless felt himself justified in considering it a new species. A short time after his results appeared an analysis of rhodochrome was published by Hermann; its identity with those of rhodophyllite induced us to re-examine the latter. The results on two analyses are:

	1	2
Silica	33.26	33.30
Alumina	10.69	10.50
Sesquioxide of chromium	4.78	4.67
Peroxide of iron	1.96	1.60
Magnesia	35.93	36.08
Soda and potash	.35	.35
Water	12.64	13.25
	99.61	99.75

These will be seen to correspond with the analyses (1, 2) of rhodophyllite by Dr. Genth, and the analysis of rhodochrome (3) and chrome-chlorite (4) by Hermann.

	$\ddot{Si}$	$\dddot{Al}$	$\dddot{Fe}$	$\dddot{Cr}$	$\dot{Ni}$	$\dot{Mg}$	$\dot{Ca}$	$\dot{Li}$ $\dot{Na}$	$\dot{K}$	$\dot{H}$
1. Texas, Pa.	33.41		18.15		trace	35.86	trace	0.28	0.10	12.79
2. Texas, Pa.	32.98	11.11	1.43	6.85	trace	35.22	trace	0.28	0.10	13.12
3. L. Itkul...	34.64	10.50	2.00	5.50	......	35.47	......	......	......	12.03
4. Texas, Pa.	31.82	15.10	4.06	0.90	0.25	35.24	......	......	......	12.75

Our analyses give the formula $4\dot{R}^3\ddot{Si}+\dddot{R}^2\ddot{Si}+10H$.

Dr. Genth gives the same formula minus one atom of water. The amount of oxide of chromium varies in different specimens, and to this is due the various shades of color. Dr. Genth informs us that he observed a like variation in the specimens he examined. The chrome-chlorite examined by Hermann was undoubtedly one of the light-colored varieties.

Nickel as well as lime is found in some specimens, but both are impurities. The nickel is due to small particles of sulphuret of nickel which occur at the same locality, and in many instances are disseminated through this mineral; in some specimens these impurities are not readily detected by the eye. In all probability the carbonate and silicates of nickel found on the Texas chrome iron proceed from the decomposition of this sulphuret.

Mr. T. H. Garrett* has recently given an analysis of this mineral. His results differ materially from those obtained above.

20. Cummingtonite — a Hornblende.

This mineral was described by Dewey,† and analyzed by Muir.‡ The latter obtained for its composition

$\dddot{\mathrm{Si}}$	$\dot{\mathrm{Fe}}$	$\dot{\mathrm{Mn}}$	$\dot{\mathrm{Na}}$	$\dot{\mathrm{H}}$
56.54	21.67	7.80	8.44	3.18=97.63

Authentic specimens for examination were procured from the Lederer collection in Yale College. Its structure is fibrous, resembling anthophyllite; luster silky; color ash-gray. It occurs in mica slate at Cummington, Mass. Two analyses gave

	1	2
Silica	51.09	50.74
Alumina	.95	.89
Protoxide of iron	32.07	33.14
Magnesia	10.29	10.31
Manganese	1.50	1.77
Lime	trace	trace
Soda	.75	.54
Potash	trace	......
Water	3.04	3.04
	99.69	100.43

These give the formula $(\dot{\mathrm{Fe}}\,\dot{\mathrm{Mg}})^4\,\dddot{\mathrm{Si}}^3 = \mathrm{R}^3\,\dddot{\mathrm{Si}}^2 + \dot{\mathrm{R}}\,\dddot{\mathrm{Si}}$.

	Atoms.	At. weight.	Per cent.	Oxygen ratio.
Silica	3	1731.93	53.59	3
Protoxide iron	$2\frac{1}{2}$	1125.00	34.80	$1\frac{1}{3}$
Magnesia	$1\frac{1}{2}$	375.00	11.61	

This is the chemical constitution of hornblende, and from its physical characters it was long since referred to this species.

21. Hydrous Anthophyllite—an Asbestus.

Thomson gave this name to an asbestiform mineral, which is found associated with chlorite on New York Island.§ His analysis gives

$\dddot{\mathrm{Si}}$	$\dot{\mathrm{Mg}}$	$\dot{\mathrm{Fe}}$	$\dot{\mathrm{Mn}}$	$\dot{\mathrm{K}}$	$\dddot{\mathrm{Al}}$	$\dot{\mathrm{H}}$
54.98	13.38	9.83	1.20	6.80	1.56	11.45=99.20

We have received undoubted specimens of this mineral from Messrs. Vaux, of Philadelphia, and Silas R. Horton, of New

* American Jour. of Science and Arts, May, 1853.

† Amer. Jour. of Science and Arts, 1st series, viii, 59.

‡ Thomson's Mineralogy, i, 493.

§ Thomson's Mineralogy, i, 209.

York. The asbestiform mineral, carefully freed from the chlorite and other impurities, gave on two analyses

	1	2
Silica	58.20	58.47
Magnesia	28.96	29.71
Protoxide of iron	8.46	9.06
Soda	.88	.88
Potash	trace	trace
Ignition	2.26	2.26
Alumina	trace	trace
	98.76	100.38

These correspond to the formula $\dot{R}^4 \ddot{Si}^3$ or $\dot{R}^3 \ddot{Si}^2 + \dot{R} \ddot{Si}$.

	Atoms.	At. weight.	Per cent.	Oxygen ratio.
Silica	3	1732.00	61.16	3
Magnesia	$3\frac{1}{2}$	875.00	30.90	$1\frac{1}{3}$
Protoxide of iron	$\frac{1}{2}$	225.00	7.94	

This is the formula given for the last mineral, and the composition is that of an asbestus or magnesian hornblende.

22. Monrolite, identical with Kyanite.

This mineral was described by Professor Silliman, jr.,* as a hydrous silicate of alumina resembling wœrthite. Professor Silliman, however, observes that the water varied in several specimens examined from 3.09 to 1.84 per cent.; subsequent examinations† made by one of us showed that the water in the pure mineral was not over 1 per cent.

In the analysis recently made we find that the silica and alumina are the same as in kyanite, and that the high silica obtained by the analyst quoted was undoubtedly owing to the impurity of the mineral, as a careful examination with the magnifier shows plates of quartz interlaced with almost every specimen. The results of analysis on specimens carefully selected to avoid the quartz gave

	1	2
Silica	37.20	37.03
Alumina	59.02	61.90
Peroxide of iron	2.08	
Ignition	1.03	.85
	99.33	99.78

These correspond to the formula $\dddot{Al}^3 \ddot{Si}^2$.

*Amer. Jour. Science and Arts, 2d series, viii, 385.

† Dana's Mineralogy, third edition, p. 317.

23. Ozarkite, an amorphous Thomsonite.

This mineral was described by Prof. Shepard as a new species.* It occurs in irregular veins and masses in elæolite at Magnet-cove, Arkansas.

We are indebted to Mr. Markoe, of Washington, for a large quantity of the elæolite, from which we were able to obtain the mineral in a pure state. Its color is white; structure granular to compact; hardness 5; specific gravity 2.24 (Shepard); gelatinizes with hydrochloric acid. Two analyses gave

	1	2
Silica	36.85	37.08
Alumina	29.42	31.13 (Alumina and Peroxide of iron together)
Peroxide of iron	1.55	
Lime	13.95	13.97
Soda	3.91	3.72
Water	13.80	13.80
	99.48	99.70

This is the composition of thomsonite, and the mineral is a massive variety of that species.

The analyses give the formula $\dot{R}\,\ddot{S}i+3\dddot{\bar{A}}l\,\dddot{S}i+7\dot{H}$=silica 37.8, alumina 31.5, lime 13.00, soda 4.80, water 12.90.

Special examination was made for phosphoric acid, but in the pure mineral none could be found, although in some impure specimens that one of us had previously examined† it existed in the form of apatite in considerable abundance, and the specimens examined were pronounced a mixture of apatite and a zeolite.

This mineral was first referred to the zeolites by Mr. J. D. Whitney,‡ who on a qualitative analysis found it to be a hydrous silicate of alumina and lime, with a little soda.

24. Dysyntribite, a Rock of indefinite Composition.

The substance to which the above name was given by Prof. Shepard§ occurs in large masses in the northern part of the state of New York. It is of a green color, sometimes mottled with red. It resembles serpentine, but has a strongly argillaceous odor when moistened.

* Amer. Jour. Science and Arts, 2d series, ii.

† Amer. Jour. Science and Arts, 2d series, ix, 430.

‡ Jour. Bost. Soc. Nat. Hist., 1849, p. 42.

§ Rep. Amer. Assoc. Advan. Sci., vol. iv, 311.

Having reason to suspect that the substance was not perfectly homogeneous—from our first analysis not agreeing with Prof. Shepard's—various specimens were examined. The correctness of the supposition will be seen by comparing the following results:

	1		2	3		4	
Silica.........	44.80	44.77	44.94	46.70	46.60	44.74	44.10
Alumina.....	34.90	35.88	25.05	31.01	} 35.15	20.98	20.64
Iron	3.01	2.52	3.33	3.69		4.27	4.03
Manganese..	.30	.30	trace	trace	trace	trace	......
Lime	.66	.52	8.44	trace	trace	12.90	12.34
Magnesia....	.42	.53	6.86	.50	.50	8.48	8.57
Potassa	6.87	......	5.80	11.68	11.68	3.73	5.92
Soda..........	3.60	......	trace	trace	......	trace	......
Water	5.38	4.72	6.11	5.30	5.30	4.86	6.30
	99.94		100.53	98.88	99.13	99.96	99.90

There is a remarkable agreement in the percentages of silica. The mineral was found to lose about two per cent. of water by desiccation. Some specimens showed the presence of a small amount of phosphoric acid. Nos. 1 and 4 were received from Mr. S. W. Johnson; No. 2 from Mr. Silas R. Horton, and the exact locality of it is Diana, N. Y.; No. 3 was received from Prof. Hume, of Charleston, who obtained it from Prof. Shepard. The original analysis by Prof. Shepard gave

$\ddot{Si}$	$\bar{\ddot{Al}}$	$\dot{Fe}$	$\dot{H}$	$\dot{Ca}$ $\dot{Mg}$
47.68	41.50	5.48	4.83	traces

This substance bears a close relation to agalmatolite, which from the variable proportion of its constituents can not be considered a mineral, but is a rock. Some of the specimens of dysyntribite give the composition of pinite, but it is reasonable to suppose that a mineral varying so much in its alumina, magnesia, lime, and alkalies, may in different masses furnish a resemblance to a vast number of minerals.

25. Gibbsite.

Gibbsite was first described by Dr. Torrey * as hydrate of alumina. This composition, confirmed by Dewey and Thomson, was considered correct until Hermann † announced the discovery of over 37 per cent. of phosphoric acid in it, and that the mineral was a hydrous phosphate of alumina. To

* N. Y. Med. and Phys. Journal, i, 68.

† J. f. pr. Chem., xl, 32.

satisfy all doubts in this matter, Prof. Silliman, jr.,* examined it with direct reference to the occurrence of phosphoric acid, and in none of the specimens examined could he find more than a trace. Subsequently Mr. Crossley,† of Boston, analyzed it, and his results confirmed the previous analyses of American chemists; meanwhile Hermann‡ re-examined it, and found the phosphoric acid to vary in different specimens from 37.62 to 11.90 per cent., showing that he had not obtained a homogeneous mineral.

Considerable pains have been taken to obtain a number of authentic specimens for examination, some of which are direct from the locality, and others from our own collection. The results show that the mineral is a hydrate of alumina, ($\dddot{Al}\,\dot{H}^3$), with but a *trace* of phosphoric acid, and in some specimens not even a trace existed. Two analyses gave

	1	2
Alumina	64.24	63.48
Peroxide of iron	trace	trace
Water	33.76	34.68
Silica	1.33	1.09
Phosphoric acid	.57	trace
Magnesia	.10	.05
	100.00	99.30

The phosphoric acid was determined by molybdate of ammonia; the small amount of silica is due to the intimate mixture of the mineral with allophane. From the results of these examinations we are confident that Mr. Hermann has not at any time analyzed pure Gibbsite.

26. Emerald Nickel.

We notice in the last edition of Phillip's Mineralogy that Professors Miller and Brooke place this species among the doubtful ones, without, however, giving any reasons for so doing. To ascertain if any good reason existed for this doubt we have re-analyzed it, and find the same composition as given by Professor Silliman, jr., which was

Oxide nickel.	Carbonic acid.	Water.
58.81	11.69	29.49

*Amer. Jour. Science and Arts, 2d series, vii, 411.

†Amer. Jour. Science and Arts, 2d series, ix, 408.

‡J. f. pr. Chem., xlvii, 1.

We obtained

		Oxygen.
Oxide of nickel	56.82	12.10
Magnesia	1.68	.67
Carbonic acid	11.63	8.46
Water	29.87	26.56

which gives the formula $\dot{N}i^3\ \ddot{C}+6\dot{H}$.

	Atoms.	At. weight.	Per cent.	Oxygen ratio.
Oxide of nickel	3	1408	59.72	3
Carbonic acid	1	275	11.66	2
Water	6	675	28.62	6

We would prefer expressing its formula by an atom of carbonate of nickel plus two atoms of the hydrated oxide of nickel. This is rendered probable from the fact that whenever it is attempted to form a carbonate of the protoxide of nickel by precipitating a protosalt with an alkaline carbonate, a carbonate is obtained containing a certain amount of the hydrated oxide. For these reasons we would express emerald nickel by $\dot{N}i\ \ddot{C}+2(\dot{N}i\ H_3)$. It is without question a distinct species and a most beautiful and interesting mineral, both from the richness of its color and its association with chromic iron.

Since completing our examinations of this mineral we have observed an analysis of a mineral called emerald nickel by Mr. T. H. Garrett. The mineral examined was of a very impure description, and was supposed by the analyst to be a mixture of emerald nickel, meerschaum, and augite; of course it is impossible to furnish any correct idea of the composition of *pure* emerald nickel from results on such impure specimens. This mineral is often associated more or less intimately with a nickel serpentine or gymnite; but with an abundance of the mineral to select from, a hydrous carbonate of nickel can be obtained of a uniform composition. A quarter of an ounce of selected fragments sent by Mr. L. White Williams furnished us with about one gramme of the pure mineral.

27. Danburite.

This mineral has as yet been found in only one locality—Danbury, Conn. It was first described by Prof. C. U. Shepard,* and considered by him a hydrated silicate of lime and potash. Still later it was examined by Dr. H. Erni,† and boracic acid

* Amer. Jour. Science and Arts, 1st series, xxxv, 129.

†Amer. Jour. Science and Arts, 2d series, ix, 286.

was found to be an important ingredient in its constitution; in his analysis, however, a large amount of alkalies is indicated. As the results of our analyses show but a trace of these substances and no other base but lime, it is possible that Dr. Erni's alkali determination was made through mistake on some of the feldspar accompanying danburite, as it not unfrequently happens that the granular portions of the feldspar resemble the lighter varieties of danburite. This supposition appears reasonable, as the silica and lime of his determinations agree with those about to be stated. The composition as given by the authorities mentioned are

$\dddot{Si}$ $\dot{Ca}$ $\ddot{\bar{Al}}$

56.00 28.33 1.70 $\dot{Y}$? 0.85 $\dot{K}$ (with $\dot{Na}$?) and loss 5.12 $\dot{H}$8.0 =100. Shepard.

49.74 22.80 $\dot{Mg}$ 1.98 $\ddot{\bar{Fe}}$ and $\ddot{\bar{Al}}$ 2.11 $\dot{K}$ 4 31 $\dot{Na}$ 9.82 $\dddot{B}$ 9.24=100. Erni.

The results of the analyses just made are as follows:

	1	2	3	4
Silica	48.10	48.20	......	......
Alumina, Peroxide of iron	.30	1.02	......	......
Manganese	.56			
Lime	22.41	22.33	22.22	22.11
Magnesia	.40	undeterm.	......	......
Boracic acid	27.73	27.15	......	......
Ignition	.50	.50	......	......
	100.00	99.20		

This corresponds to the following constitution:

	Per cent.	Oxygen ratio.
4 atoms silica	49.42	4
3 " boracic acid	28.02	3
3 " lime	22.56	1

It is not easy to decide upon the formula by which danburite is to be expressed. The same difficulty occurs here as in the case of datholite, Berzelius considering the lime only as acting the part of base, while Rammelsberg regards the boracic acid also as performing the part of a base. The formula for danburite under these views would be expressed by

$$2\dot{Ca}\,\dddot{Si}^2+\dot{Ca}\,\dddot{B}^3 \text{ or } \dot{Ca}^3\,\dddot{Si}^2+\dddot{B}^3\,\dddot{Si}^2.$$

We are in favor of regarding the latter as representing the true formula.

If danburite be examined in its relation to datholite, it will be found to differ from the latter in having just one half the number of atoms of lime minus three atoms of water.

$$\dot{C}a^6 \ddot{S}i^2 + \ddot{B}^3 \ddot{S}i^2 + 3\dot{H} = \text{Datholite.}$$
$$\dot{C}a^3 \ddot{S}i^2 + \ddot{B}^3 \ddot{S}i^2 = \text{Danburite.}$$

This mineral forms then the second natural borosilicate of lime. Its physical characters have been fully described elsewhere.

The results of the analysis somewhat surprised us, as we confidently expected to find alkalies; of course much care was directed to the settling of this question, which from the simplicity of the composition of danburite was easily done in the following manner: to one gramme of the mineral strong hydrofluoric acid was twice added, the mass carefully heated and evaporated nearly to dryness on each addition; finally sulphuric acid was added in excess, the whole evaporated to dryness and ignited; the residue weighed 563 milligrammes, of which 544 were sulphate of lime, the remaining 19 being composed of alumina, iron, manganese, and magnesia. The 544 milligrammes of sulphate of lime, correspond to 22.44 per cent. of lime in the mineral used, proving clearly that the remainder was all volatilized under the action of hydrofluoric acid; which, according to previous experiments, was shown to consist of silicic and boracic acids. Three repetitions of the above experiment gave perfectly concordant results: in fact no more beautiful proof could have been had of the absence of all other bases but lime in any very sensible quantity. Moreover, the amount of lime thus indicated agrees perfectly with that obtained by the direct method of analysis.

The boracic acid was estimated by first ascertaining the exact amount of silica by a soda fusion, and then deducting this quantity from the entire loss of a given quantity of the mineral under the action of hydrofluoric acid.

The specimens examined were among the finest that have ever been found, and were procured by Mr. Brush at the locality.

28. Carrollite, a Copper Linnæite.

This mineral occurs at Finksburg, Carroll County, Maryland, and was described as a new species by Mr. W. L. Faber.* He gave as its chemical composition

S	Co	Ni	Cu	Fe	As	$\ddot{S}i$
27.04	28.50	1.50	32.99	5.31	1.82	2.14

Formula $2Co\,S + Cu^2\,S$.

* Amer. Jour. Science and Arts, 2d series, vol. xiii, 418.

Our attention was first called to this mineral from the unusual relation of the sulphur to the metals in its composition, it being, we believe, the first example of a natural subsulphuret that had been observed, the relation being $R^4 S^3$.

Having been furnished, through the kindness of Prof. Dana, with an abundance of the mineral to select from, which he had procured from original specimens in the hands of Prof. Booth, it was carefully separated from the copper pyrites with which it is associated. A re-examination of it gives the following for its physical and chemical characters: Hardness 5.5; sp. grav. 4.85;* luster metallic; color steel-gray; fracture uneven, without sufficient indication to make out clearly the nature of its cleavage. The results of three analyses are

	1	2	3
Sulphur	41.93	40.94	40.99
Cobalt	37.25	38.21	37.65
Copper	17.48	17.79	19.18
Nickel †	1.54	1.54	1.54
Iron	1.26	1.55	1.40
Arsenic	trace	trace	trace
	99.46	100.03	100.76

These correspond to the general formula $RS+R^2 S^3$, as will be seen by comparing the amount of sulphur required for the metals indicated in the three analyses with the quantities found.

	1	Sulph.	2	Sulph.	3	Sulph.
Cobalt	37.25	27.00	38.21	27.68	37.65	27.21
Copper	17.48	11.65	17.79	11.88	19.18	12.78
Nickel	1.54	1.12	1.54	1.12	1.54	1.12
Iron	1.26	.93	1.55	1.15	1.40	1.03
		40.70		41.83		42.14

Substituting cobalt for the copper, iron, and nickel, the entire amount of the cobalt would be represented in the three analyses by 56.37, 57.92, and 58.50 per ct. The formula requires

	Atoms.	Pr. ct.
Sulphur	4	42.06
Cobalt	3	57.94

This is the formula and constitution of linnæite, and the mineral in question is a copper linnæite similar in composition to the one from Riddarhythan, Sweden.

---

* It may be well to remark that there is a typographical error in the statement of the sp. grav. of linnæite in Dana's Mineralogy; it is never above 5.00.

† Only one estimate of nickel made.

The composition of this mineral is interesting, as it furnishes the only example in the mineral kingdom of the isomorphism of copper and cobalt, where the latter may be replaced to a greater or less extent by the former. Among artificial products exhibiting this replacement, we have the cupro-sulphate of cobalt with 10 and with 7 atoms of water, the latter crystallizing in oblique prisms, like the sulphate of cobalt with the same number of atoms of water, which is also the form of green vitriol.

29. Thallite, identical with Saponite.

This mineral was originally described by Dr. D. D. Owen,* by whom it was found on the north shore of Lake Superior, diffused in the amygdaloidal traps of that locality. At the time it was first noticed it was supposed to contain a new element, which was called thallium; the mineral itself was named thallite. Through the kindness of Dr. Owen we were furnished with some of the mineral, which was subjected to most careful analysis, the result showing nothing in its composition by which it differs from saponite; and all attempts to isolate a new earth from it were vain. A second portion of the thallite with some of the supposed thallia was sent us by Dr. Genth, of Philadelphia, which was labeled "not quite pure;" its composition, however, differed from the first principally in containing less water, as it was allowed to dry for a greater length of time—it being a common thing for saponites to lose more or less of their water by desiccation in the air.

The result of the examination of the thallite was given in a note in the last number of the American Journal of Science and Arts. Many of the reactions contained in the original description of thallite and thallia we have been unable to recognize, among them the evolution of chlorine by the action of hydrochloric acid, and the precipitation by a neutral solution of succinate of ammonia. The pea-green color of the concentrated hydrochloric acid solution of the thallia, prepared in the way mentioned by Dr. Owen, is easily explained by the presence of an exceedingly minute quantity of the chloride of chromium, as the smallest trace of this last metal will, under the circumstances, produce that color. The results of our analyses are as follows:

---

* Jour. Acad. Nat. Sci. Phila., ii, part 2d, 1852.

	1	2
Silica	45.60	48.89
Alumina	4.87	7.23
Oxide of iron	2.09	2.46
Manganese	trace	trace
Lime	1.07	......
Magnesia	24.10	24.17
Soda, Potash,	.45	.81
Water	20.66	15.66
	98.84	99.22

The specimen No. 1, as already stated, came directly from Dr. Owen; No. 2 from Dr. Genth. The original analysis of Dr. Owen, with the exception of the new earth, does not differ very materially from the above, when we consider that the saponites vary more or less in their composition. It is as follows:

$\dddot{Si}$	$\overset{...}{Al}$	$\overset{...}{Fe}$	new earth	$\dot{Mg}$	$\dot{K}$	Mn	$\dot{H}$
42	4.6	1.5	10–12	20.5	0.8	trace	18

If then the existence of a new earth in this mineral can not be established, it is clear that it must be a saponite, with which mineral it is identical in physical properties.

30. Hudsonite, a Pyroxene.

This species was described by Prof. L. C. Beck.* It has since been shown to be a pyroxene in which a portion of the silica is replaced by alumina. Beck and Brewer obtained for its composition

$\dddot{Si}$	$\overset{...}{Al}$	$\dot{Fe}$	$\dot{Mn}$	$\dot{Mg}$	$\dot{Ca}$	
36.94	11.22	36.03	2.24	......	12.71 = 99.14.	Brewer.†
37.90	12.70	36.80	......	1.92	11.40 = 100.72.	Beck.

A recent examination of some specimens has shown the presence of a considerable amount of alkalies. The mineral was received from Mr. Silas R. Horton, and is the same as was sent by him to Prof. Beck. The results of two analyses are

	1	2
Silica	39.30	38.58
Alumina	9.78	11.05
Protoxide of iron	30.40	30.57
Protoxide of manganese	.67	.52
Lime	10.39	10.32
Magnesia	2.98	3.02
Potash	2.48	4.16
Soda	1.66	
Ignition	1.95	1.95
	99.61	100.17

* Mineralogy of New York, p. 310.

† From Dana's Mineralogy, 3d edition, p. 267.

Considering the alumina as replacing silica, these give the oxygen ratio of pyroxene and the formula $\dot{R}^3$ ($\ddot{S}i$, $\bar{\bar{A}}l$)^2.

31. Jenkinsite.

The green mineral that occurs in velvety coatings on the magnetite of O'Neil's Mine, in Orange County, New York, has been described by Prof. Shepard,* as a new species. Its intimate association with magnetite renders it somewhat difficult to obtain it perfectly pure; but by placing the fine particles of jenkinsite in a vessel of water, and stirring the mass with a clean rod of soft iron that passes through an electro-magnetic coil in connection with a battery, every particle of magnetic iron is removed. Two different portions thus purified, procured at different times from Mr. Silas R. Horton and Mr. John Jenkins, gave

	1	2
Silica	38.97	37.42
Protoxide of iron	19.30	20.60
Protoxide of manganese	4.36	4.05
Magnesia	22.87	22.75
Alumina	.53	.98
Water	13.36	13.48
	99.39	99.28

From these we obtain the mean oxygen ratio for silica and protoxides, and water, 19.84 : 14.50 : 11.92=4 : 3 : $2\frac{1}{3}$, and the formula $\dot{R}^9\ \ddot{S}i^4+7\dot{H}$.

	Atoms.	At. weight.	Per cent.
Silica	4	2309	38.83
Magnesia	6	1500	25.23
Protoxide of iron	3	1350	22.70
Water	7	788	13.24
		5947	

The mineral is similar in composition to serpentine with one atom more of water, and the magnesia replaced in part by protoxide of iron and manganese. It also has a strong resemblance to *hydrophite*, both in chemical and physical properties.

32. Lazulite.

This species occurs in considerable abundance in Sinclair County, North Carolina, at present the only American locality. It is of interest to compare its composition with the European

* Amer. Jour. Science and Arts, 2d series, xiii, 392.

varieties, and for that reason the examination was made. The specific gravity was 3.122. Two analyses gave

		Oxygen.			Oxygen.	
Phosphoric acid	43.38=24.31			44.15=24.74		
Alumina	31.22	14.59		32.17	15.03	
Protoxide of iron	8.29	1.84	5.86	8.05	1.79	5.80
Magnesia	10.06	4.02		10.02	4.01	
Water	5.68	5.05		5.50	4.89	
$\dddot{S}i$	1.07			1.07		
	99.70			100.96		

No. 1 has the oxygen ratio=24.90 : 14.94 : 6 : 5.16, or very nearly 25 : 15 : 6 : 5. No. 2 has=25.56 : 15.54 : 6 : 5.04. From these we deduce the formula

$$2\,(\dot{M}g,\ \dot{F}e)^3\ \ddot{\bar{P}}+\ddot{A}l^5\ \ddot{\bar{P}}^3+5\dot{H}.$$

	Atoms.	At. weight.	Per cent.
Phosphoric acid	5	4460	44.02
Alumina	5	3209	31.67
Protoxide of iron	2	900	8.88
Magnesia	4	1000	9.87
Water	5	563	5.56

The formula differs from that of Rammelsberg by one atom less of alumina and of water; calculated by his formula, it would give the alumina much too high for our analyses. The phosphoric acid was separated from the alumina by fusing the mineral with carbonate of soda and silica, this being the most perfect method, in fact the only one to be safely relied on. It appears to be identical in specific gravity and composition with the variety from Gratz examined by Rammelsberg.

33. Kyanite.

Associated with the lazulite just described is a very beautiful white kyanite. Its composition is

Silica	37.60
Alumina	60.40
Peroxide of iron	1.60
	99.60

This corresponds to the formula $\ddot{A}l^3\ \dddot{S}i^2$=silica 37.47, alumina 62.53.

34. Elæolite.

The elæolite of Magnet-cove in Arkansas passed under the name of "flesh-red feldspar" until recognized by Prof. Shepard.* It has the following physical and chemical properties: Hard-

* Amer. Jour. Science and Arts, 2d series, ii, 252.

ness 6; specific gravity 2.65; color flesh-red; luster greasy; structure massive. Chemical composition:

Silica	44.46
Alumina	30.97
Peroxide of iron	2.09
Lime	.66
Soda	15.61
Potash	5.91
Ignition	.95
	100.65

From this we have the oxygen ratio for the silica, peroxides, and protoxides, 9 : 6 : 2, and the formula $R^2\ddot{S}i+2\ddot{\bar{A}}l\ddot{S}i$. The mineral examined was furnished by Mr. Markoe, of Washington; it is the one alluded to in our last paper as containing the compact thomsonite, under the name of ozarkite. Since publishing the analysis of the latter we have procured a specimen of the elæolite containing the thomsonite in handsome radiated crystallizations.

35. Spodumene.

Several analyses of this species from Norwich and Sterling, Massachusetts, were given by Mr. Bush in vol. x of American Journal of Science and Arts.

In these analyses the alkalies were determined as sulphates, and from the amount of sulphuric acid the relative amount of alkalies was calculated. Since these examinations it has been found that this process is liable to great inaccuracy; and we have consequently made a re-examination with direct reference to this point; it has been made altogether on the variety from Norwich, which occurs in beautiful crystals. The results of the re-examination show that the lithia and soda determinations in the analyses referred to are erroneous. The method used in this instance is that recommended by Rammelsberg—the separation of the chloride of lithia from the chlorides of potash and soda by a mixture of alcohol and ether. These analyses gave

	1	2	3
Silica	64.04	63.65	63.90
Alumina	27.84 }	28.97	28.70
Peroxide of iron	.64 }		
Lime	.34	.31	.26
Magnesia	trace	trace	trace
Lithia	5.20	5.05	4.99
Soda	.66 }	.82	.80
Potassa	.16 }		
Ignition	.50	.50	.60
	99.38	99.30	99.25

From these is deduced the oxygen ratio for the protoxides, peroxides, and silica, 1 : 4 : 10, and the formula $\dot{R}^3 \ddot{Si}^2 + 4\dddot{Al} \ddot{Si}^2$, corresponding with the results recently obtained by Rammelsberg* for the composition of the spodumene from Uton and the Tyrol. The silica, however, is somewhat lower than that of the European specimens, which is probably owing to the greater purity of the American mineral, it being found in crystals.

Since completing the above examination of the spodumene from Norwich, we have noticed a recent examination of the variety from Sterling, Massachusetts, by Rammelsberg, in which he states his having found 4.5 per cent. of potash, and a ratio nearer 1 : 5 : 12 than 1 : 4 : 10. This difference is accounted for by that author on the ground that the mineral is somewhat decomposed, and that the original ratio was 1 : 4 : 10. The fact is Rammelsberg did not examine the pure spodumene from Sterling, as will be seen by his own account of the specimens analyzed. They were white, yellowish, or bluish gray in color, possessing little luster, penetrated with small fissures, and in these fissures, as well as in many places on the surface, were scales of mica with yellow flakes of oxide of iron; the specific gravity was 3.073. Now the spodumene, as we have procured it perfectly pure from Sterling, Massachusetts, is of a greenish-gray color, fine luster, and very compact, cleaving with perfection, a specific gravity of 3.182, and with a composition according perfectly with the requirements of the ratio 1 : 4 : 10, containing even a little more lithia than the variety from Norwich. In an early examination by Brush, published several years ago, the silica was given too low, he not having separated the small amount of silica always associated with the alumina. A recent analysis gives as the composition of this spodumene:

		Oxygen ratio.
Silica	64.50	33.41 or 10.66
Alumina	25.30	12.54 or 4
Peroxide of iron	2.55	
Lime	.43	3.43 or 1.09
Magnesia	.06	
Lithia	5.65	
Potash and soda	1.10	
Loss by ignition	.30	
	99.89	

* Pogg. Ann., lxxxv, 544.

It will be seen that this forms no exception, and when examined pure has the same composition as spodumene from other localities. In fact, so far as the American spodumene is concerned, we believe that the specimens from Norwich are the most beautiful that have been found any where, and we shall take pleasure on some early occasion in furnishing Rammelsberg with good specimens from both Norwich and Sterling.

36. Petalite.

In connection with the analyses of spodumene it was thought interesting to examine the American petalite. For this purpose a specimen from Bolton, Massachusetts, was selected. The results of analyses are:

	1	Oxygen.		2	Oxygen.	
Silica	77.95	40.50		77.90	40.47	
Alumina	16.63	7.77	7.95	15.85	7.41	7.56
Peroxide of iron	.62	.18		.51	.15	
Lime	trace	......		trace	......	
Magnesia	.21	.08		.26	.10	
Lithia	3.74	2.07	2.27	3.52	1.94	2.18
Soda	.48	.12		.53	.14	
Potasa	trace	......		trace	......	
Ignition	.60	......		.70	......	
	100.23			99.37		

No. 1 gives oxygen ratio 20 : 3.92 : 1.12; No. 2 gives oxygen ratio 20 : 3.74 : 1.08, or quite nearly 20 : 4 : 1, from which we have the formula $\dot{R}^3 \ddot{Si}^4 + 4\dddot{Al} \ddot{Si}^4$ = silica 78.37, alumina 17.44, lithia 3.40, soda 0.79.

37. Boltonite, identical with Chrysolite.

Boltonite was first described as a new species by Professor C. U. Shepard. He made the specific gravity from 2.8 to 2.9. It was subsequently examined by Professor Silliman, jr., who found 3.008 as its specific gravity, with a hardness of from 5 to 6. His analysis gave for its constituents

Silica	46.062
Alumina	5.667
Magnesia	38.149
Protoxide of iron	8.632
Lime	1.516

With this knowledge of the mineral I undertook its examination on specimens in the gangue furnished me by Professor Shepard. Examination of different portions separated mechanically from the gangue made it very evident that the mineral was more or less mixed with other substances which had escaped observation, for no two analyses agreed; and it was soon discovered that it was impossible (from the specimens in my possession at least) to separate boltonite in a state of purity without the aid of other means than had been adopted.

Boltonite, as is well known, occurs at Bolton, Mass., disseminated in irregular masses and grains in a white limestone. If a piece of the mineral in its gangue be placed in cold dilute hydrochloric acid, the limestone is readily dissolved and a mass left which is seen to consist of asbestus, dolomite, a little mica, small crystals of magnetic iron, and a greenish or yellowish-green mineral; if the acid be now heated, the dolomite will be entirely dissolved with a little of the last-mentioned mineral.

In order to obtain the boltonite as pure as possible for analysis the following method was adopted. Pieces were separated by the hammer as thoroughly as possible from all other substances; these were subsequently placed in dilute hydrochloric acid and boiled for some time; the acid being washed away and the substance dried, it was crushed in a mortar to fragments from the twentieth to the tenth of an inch in diameter; these were again introduced into dilute acid and heated for a short while; the acid was thoroughly washed away and the mineral dried. The small fragments (now like coarse gravel) were placed on a piece of glazed paper, the hand laid flat upon it, and the mineral rubbed so as to grind the particles against each other for the purpose of ridding their surfaces of a little cohering silica arising from its partial decomposition; with a small gauze sieve the finer particles are separated, and from that remaining in the sieve we are enabled with the aid of a glass without any difficulty to pick out the pure boltonite. This method requires a little patience, but no extraordinary care; and, however unpromising the original specimens may have been, there is no difficulty in obtaining a material the results of whose analysis are constant. From a larger selection of specimens than that used there doubtless could be obtained pieces, perfectly pure, of some size. After being satisfied with this method of obtaining the pure mineral, three different portions were prepared and examined, the first two being of the greenish variety and the third of the yellow variety, which color is doubtless due to a peroxidation of a minute quantity of the protoxide of iron entering into the constitution of the mineral. Mr. L. Saemann, in a communication made to the American Association some time since, attributed this change to magnetic iron undergoing decomposition; but this, however, does not appear to me to be the case, for the reasons that crystallized magnetic iron is a mineral difficult of decomposition, and the color is not in fissures, as would be the case if the peroxide arose from a substance foreign to the composition of the mineral, but enters into its most intimate structure.

The hardness of boltonite is found to be, as already stated, between 5 and 6. The specific gravity was taken on three specimens; Nos. 1 and 2 on a gramme each of fine particles; No. 3 on a piece of 0.150 gramme, all possible precautions being used to arrive at correct results.

No. 1......... 3.270 No. 2......... 3.208 No. 3......... 3.328

No. 3 is to be regarded as by far the most reliable, as in taking the specific gravity of fine grains it is almost impossible to detach the last particles of air, and consequently the specific gravity they indicate is below the true number. The analyses of three portions gave

	1	2	3
Silica......	42.56	41.95	42.41
Magnesia.........................	51.77	51.64	50.06
Protoxide of iron	2.35	3.20	3.59
Alumina..........................	.10	.25	
Loss by heat........	2.22	1.58	not estimated.
	99.00	98.62	

Nos. 1 and 2 were the greenish variety, No. 3 the yellowish. The oxygen ratios of the silica and protoxides are

	1	2	3
Silica	22.11	21.77	22.03
Magnesia	20.35	20.30	19.67
Protoxide of iron	.52	.71	.75

This being as 1 to 1 within a small fraction, the formula therefore is $(\dot{Mg}\dot{Fe})^3\ddot{Si}$, or of the general form $\dot{R}^3\ddot{Si}$, which of course proves it to be *chrysolite*, a fact sustained in every respect by its physical characters.

38. Iodide of Silver.

In this re-examination of American minerals it was not originally designed to include those of South America, but my recent examination of the minerals obtained by Lieut. Gilliss, of the U. S. Chili Expedition, has afforded an opportunity of analyzing certain minerals that it was well to investigate, and among these were one or two fine specimens of iodide of silver. A re-examination of this mineral is especially interesting, from the fact that its composition is still in doubt, owing to the discrepancy between the original analysis of Vauquelin on the mineral from Zacatecas in Mexico and that of Domeyko on the mineral from Chañarcillo in Chili.

	Vauquelin.		Domeyko.	
Iodine	22.6	$Ag.^2$ I.	46.89	Ag. I.
Silver	77.4		54.25	

The constitution of the native chlorides and bromides of silver would lead to the supposition that Domeyko's analysis was the correct one, and this is strengthened by its resemblance to the artificial iodide of silver.

The specific gravity was found to be 5.366, being a little lower than that given by M. Domeyko. The analysis of an exceedingly pure specimen gave me

	1	2
Iodine	52.934	53.109
Silver	46.521	46.380
Chlorine	trace	trace
Copper	trace	trace
	99.455	99.489

clearly showing its constitution to be Ag. I = Iodine 53.85 Silver 46.15 = 100, leaving no doubt of its perfect analogy to the natural chloride and bromide of silver. The other properties of this mineral are not mentioned, as they are all fully stated in all works on mineralogy.

38. Copiapite.

This mineral was also furnished me by Lieut. Gilliss, it having been brought from Chili. It consists of most beautiful silky fibers or fibrous masses of a pearly luster. Its color is white, with a very fine tinge of yellow. From the specimens in my possession there was no difficulty in picking out a portion in a state of great purity. Its specific gravity is 1.84. Examined under the microscope its form appears to be a hexagonal prism. Cold water has but little action on it, merely causing the crystals to separate and the mass to swell out to a very much increased bulk. If the water be boiled, decomposition ensues, with a deposition of the oxide of iron and the formation of a soluble sulphate. On analysis it afforded

	1	2
Sulphuric acid	30.25	30.42
Peroxide of iron	31.75	30.98
Water	38.20	not estimated.
Undissolved	.54	
	100.74	

The analyses correspond to the formula $\overline{\mathrm{F}}\mathrm{e}\,\ddot{\mathrm{S}}^2+11\dot{\mathrm{H}}$. This is the same formula as that obtained by Rose, with an additional half atom of water, his formula being $2\overline{\mathrm{F}}\mathrm{e}\,\ddot{\mathrm{S}}^2+21\dot{\mathrm{H}}$. Protoxide of iron was looked for, but none found.

40. Owenite,* identical with Thuringite — with an announcement of a New Locality.

Owenite was first described by Dr. F. A. Genth as a distinct species, who gave a minute and accurate analysis in the American Journal of Science and Arts, vol. xvi, 2d series, page 167. It was found on both sides of the Potomac River, near Harper's Ferry. The physical characters being already fully and accurately given, it is needless to repeat them here, merely remarking that its specific gravity as taken by me is 3.191. It is readily soluble in hydrochloric acid; notwithstanding, analysis No. 2 was made by fusion with carbonate of soda. Results of analyses as follows:

	1	2	Genth.
Silica	23.58	23.52	23.21
Peroxide of iron	14.33	......	13.89
Alumina	16.85	16.08	15.59
Protoxide of iron	33.20	32.18	34.58
Protoxide of manganese	.09	......	trace
Magnesia	1.52	1.68	1.26
Lime	......	......	.36
Soda	.46	......	.41
Potash	trace	......	.08
Water	10.45	10.48	10.59
	100.50		99.97

After this examination it was rendered strongly probable that owenite and thuringite were similar if not identical minerals, yet in the analysis of thuringite by Rammelsberg alumina is not mentioned as one of its constituents. This

* The identity of these two minerals had already been announced by me in a letter to one of the editors of the Amer. Jour. Science and Arts, xvii, 131, but no details were then given.

view was sustained by the apparently perfect accordance in the physical characters of the two minerals, coupled with the fact that the amount of silica and water in the two, as already examined, was the same, and also the sum of the oxides of iron and alumina in the owenite were equal to the sum of the oxides of iron in the thuringite. To settle the question, it became necessary to re-examine thuringite, of which I obtained a specimen from Mr. Markoe, coming from the original locality; it was slightly altered by the action of the air, but this could interfere only with the correct estimate of the protoxide of iron. Its specific gravity was 3.186, and its composition

Silica	22.05
Peroxide of iron	17.66
Alumina	16.40
Protoxide of iron	30.78
Magnesia	.89
Soda, Potash	.14
Water	11.44
	99.36

The peroxide of iron is a little higher and the protoxide a little lower than in the analysis of owenite, but this arose from the partial decomposition of the specimen. The correct analysis of thuringite is that first given, and the formula deduced by Mr. Genth from it is to be looked upon as the correct one—namely, $2\dot{R}^3\,\ddot{Si}+\bar{R}^3\,\ddot{Si}+6\dot{H}$, corresponding to the oxygen ratio for $\dot{R}$, $\bar{R}$, $\ddot{Si}$, $\dot{H}$, 1 : 1.5 : 1.5 : 1.

In looking over some minerals placed in my hands by Mr. Markoe, I have found a specimen of thuringite coming from the Hot Springs of Arkansas. Its identity is made out without the slightest difficulty, as all its physical characters correspond most perfectly with the thuringite, its specific gravity being 3.184 and composition

Silica	23.70
Peroxide of iron	12.13
Alumina	16.54
Protoxide of iron	33.14
Magnesia	1.85
Manganese	1.16
Soda, Potash	.32
Water	10.90
	99.74

An interesting fact connected with this mineral, as shown by this investigation, is, that although not crystalline, or at least very obscurely so, yet coming as it does from three localities so widely separated as Thuringia, the Potomac, and Arkansas, it is nevertheless found quite unmixed with any other mineral, as the analyses indicate.

41. Xenotime of Georgia.

In examining, a few years ago, some of the residue of gold washings from Clarksville, Georgia, in the possession of Prof. Gibbes, of Charleston, I observed some small octahedral crystals associated with zircon, titaniferous iron, and kyanite. Two or three of the most perfect were selected; and, having no goniometer at hand, they were sent to Mr. Teschemacher, who referred them, after a partial examination, to zircon. Prof. Gibbes subsequently examined their form, and pronounced them xenotime (Amer. Jour. Science and Arts, 2d series, xiii, 143). Since then, from material that had been placed in my hands by that gentleman, nearly a gramme of the substance has been procured, and upon that the following examination has been made.

Some of the crystals are exceedingly short prisms, surmounted by four-sided pyramids, but most of them are without the prism, the summits coming together forming a flattened octahedron. The measurements made were: over the pyramidal edge 123° 10′; over basal 81° 30′; face of pyramid on prism 131° 40′. The above measurements can be made with perfect accuracy; not so the faces of the prisms on each other; and, as far as I could make it out, I am inclined to think that they are not square prisms, but rhombic prisms of 93°. Its hardness is 4.5, specific gravity 4.54, and the physical characters those given for xenotime.

It was decomposed by fusion with carbonate of soda and silica, and analyzed with the following results:

Phosphoric acid	32.45
Yttria	54.13
Oxide of cerium (with a little La and D)	11.03
Oxide of iron	2.06
Silica	.89
	100.56

This analysis will be seen to differ from that of the xenotime of Hitteroe, Sweden, by Berzelius, in that a portion of the yttria is replaced by the oxide of cerium; the formula represented by the analysis is, however, the same, namely: $(\dot{Y}, \dot{C}a)^3 \ddot{\ddot{P}}$.

Great care was taken in the separation of the oxide of cerium, which after being peroxidized by heat yielded but little to dilute nitric acid, indicative of the presence of but a small quantity of the oxides of lanthanum and didymium.

42. Lanthanite.

This mineral was first observed in America by Mr. W. P. Blake, and described in the American Journal of Science and Arts, September, 1853; it was obtained by Mr. Blake from Bethlehem, Lehigh County, Pennsylvania, where only one specimen had been found. It was handed to me for examination, and ascertained to be carbonate of lanthanum; the analysis made was given in the original description of the mineral. Since then I have made another analysis on a portion remaining in my possession; and, although not differing from the former one, it is thought proper to insert it in this paper.

	1	2
Water	24.21	24.09
Carbonic acid	21.95	22.58
Protoxide of lanthanum (with some oxide of didymium)	55.03	54.90
	101.19	101.57

No. 2 is the analysis already published in the paper before mentioned. In both instances there was an excess, owing to the peroxidation of a portion of the lanthanum—a circumstance that can not be avoided, nor do we know how to allow for it in our calculation. This mineral has the same formula as the artificial carbonate, namely, $\dot{L}a\ddot{C}+3\dot{H}$=carbonic acid 21.11, oxide of lanthanum 52.94, water 25.95.

The only other known locality of this mineral is Bastnäs in Sweden; it is there found only as a coating to cerite, and doubtless was not obtained in a perfectly pure state by Hisinger, who gave as its formula $\dot{L}a^3\ddot{C}+3\dot{H}$. I have no doubt as to the minerals being identical, and that whenever the Bastnäs variety is obtained crystallized it will prove to have the same composition as the Bethlehem variety.

43. Mangano-magnesian Alum from Utah.

This alum was observed a few years ago by Dr. Gale among specimens brought from the Salt Lake, in Utah, by Mr. Stansbury. It occurs at a place called Alum Point, and was considered altogether a manganese alum, of which Dr. Gale gave what he then stated he considered an imperfect analysis (American Journal of Science and Arts, vol. xv, 2d series, 434):

$\dddot{S}$ 18.0 $\dot{Mn}$ 8.9 $\dddot{\bar{Al}}$ 4.0 $\dot{H}$ 73.0

Being desirous of having it more carefully analyzed, Dr. Gale placed in my hands the specimen which is the subject of the present investigation. It was not received as it occurs at the locality, but had been recrystallized, and consisted of delicate needle-shaped crystals, adhering in small masses. It dissolves very readily in water; in fact so soluble is it that it is difficult to decide the amount of water requisite for its complete solution. It crystallizes from solution in the form of delicate crystals, with a plumose aggregation. On analysis it furnished:

Alumina	10.40	10.65
Magnesia	5.94	5.65
Manganese	2.12	2.41
Sulphuric acid	35.85	35.92
Oxide of iron	.15	.09
Potash	.20	.20
Water	46.00	46.75
	100.66	101.67

This analysis shows an amount of protoxides a little too high for the requisites of the formula of alum; but this, however, is of frequent occurrence in the natural alums, owing to admixture of impurities. This variety of alum has been before observed by Stromeyer, and was brought from a cave in Southern Africa. Its formula is $(\dot{Mg}, \dot{Mn})\dddot{S}+\dddot{\bar{Al}}\dddot{S}^3+24\dot{H}$.

44. Apophyllite.

The specimen of this mineral examined came from Lake Superior. It is eminently lamellar in its structure, and was placed in my hands as being possibly diaspore; its luster is, however, much more pearly than this latter mineral. Its specific gravity is 2.37 and its constitution

Silica	52.08
Lime	25.30
Potash	4.93
Fluor	.96
Water	15.92
	99.19

45. Schreibersite (of Patera).

This meteroic mineral occurs in the American meteorites in more abundance than has usually been supposed, as was fully shown in a communication made to the American Association for the Advancement of Science, in April, 1854; and as that memoir will be published in full in the American Journal of Science and Arts, nothing farther than the mere statement of the analysis of this mineral is here given. G.=7.017.

	1	2	3
Iron	57.22	56.04	56.53
Nickel	25.82	26.43	28.02
Cobalt	.32	.41	.28
Copper	trace not est.	......	......
Phosphorus	13.92	......	14.86
Silica	1.62	not estimated.	
Alumina	1.63		
Lime	trace not est.		
Chlorine	.13		
	100.66		99.69

Nos. 1 and 2 were separated mechanically from the meteoric iron; No. 3 chemically. The silica, alumina, and lime were almost entirely absent from No. 3; and in the other specimen it is due to a siliceous mineral that I have found attached in small particles to the schreibersite, and of which I have preserved one or two small specimens. The formula of schreibersite I consider to be $Ni^2 Fe^4 P$.

	Per cent.
Phosphorus, 1 atom	15.47
Nickel, 2 "	29.17
Iron, 4 "	55.36

Further particulars of this mineral will be found in the paper already referred to.

46. Protosulphuret of Iron.

This sulphuret is the one found in the meteoric irons of this country. The specimen examined came from Tennessee; its specific gravity is 4.75. Its composition is different from that of magnetic pyrites, although some authors consider the magnetic pyrites a protosulphuret, an inference not sustained by analysis. The mineral in question afforded me

Iron	62.38	Copper	trace
Sulphur	35.67	Silica	.56
Nickel	.32	Lime	.08

The formula Fe S requires sulphur 36.36, iron 63.64=100. Further remarks on this mineral will be found in the paper on meteorites.

47. Cuban.

This variety of copper pyrites was first noticed by Breithaupt as occurring among the copper-ores of Cuba. Desiring to re-examine it, specimens were obtained from Prof. Booth; they were massive and not perfectly pure, furnishing an insoluble residue consisting of silica and oxide of iron, which are very probably combined. Its specific gravity was 4.180, and its composition

	1	2	3
Iron	37.10	......	......
Copper	18.23	19.10	19.00
Sulphur	39.57	39.20	39.30
Residue (silica and oxide of iron)	4.23	......	......
	99.13		

This seems to substantiate the formula already received (agreeing with the analysis of Professor Booth), $Cu\,S+Fe^2\,S^3$, pyrites being $Cu^2\,S+Fe^2\,S^3$=sulphur 42.28, copper 20.82, iron 36.90.

MINERALS OF THE WHEATLEY MINE, PENNSYLVANIA.

Before describing the minerals of this mine it is well to say a word with reference to its location, and also to quote some remarks on the geology of the surrounding country by Prof. H. D. Rogers. Although this is departing from the plan usually adopted in this series of papers, still the occurrence of all the minerals here described at one locality can not but render the geology of the place interesting to mineralogists.

This mine is situated in Chester County, near Phœnixville, Pennsylvania, and is one of several interesting developments of a thorough and very able exploration of this region by Mr. Charles M. Wheatley. At the request of Mr. Wheatley, Prof. Rogers made a geological examination of the metalliferous veins of this district, and the following remarks are taken from his report:

"These veins belong to a group of lead and copper-bearing lodes of a very interesting character, which form a metalliferous zone that ranges in a general east and west direction across the Schuylkill River, near the lower stretches of the Perkiomen and Pickering creeks in Montgomery and Chester counties, and bids fair to constitute at no distant day a quite productive mineral region.

"The individual veins of this rather numerous group are remarkable for their general mutual parallelism, their average course being about N. 31°—35° E. by compass, and not at all coincident with that of the belt of country which embraces them. They are true lodes or mineral injections, filling so many dislocations or fissures transverse to the general direction of the strata which they intersect. The metalliferous belt ranges not far from the boundary which divides the gneissic or metamorphic rocks of Chester County from the middle secondary red shale and sandstone strata.

"This vein varies in thickness from a few inches to about two and a half feet, and we may state its average width at not less than eighteen inches. It is bounded by regular and well-defined, nearly parallel walls, the prevailing material of which is a coarse, soft granite, composed chiefly of white feldspar and quartz.

"It would seem to be a pretty general fact that such of these veins as are confined entirely or chiefly to the gneiss bear *lead* as their principal metal; whereas those which are included solely within the red shale are characterized by containing the ores of copper. But the zinc ores—viz., zinc-blende and calamine—prevail in greater or less proportions in both sets of veins, existing perhaps in a rather larger relative amount in the copper-bearing lodes of the red shale.

"The *gneissic strata* of the tract embracing this group of lead-bearing veins seem to differ in no essential features from the rest of the formation ranging eastward and westward through this belt of country. Here, as elsewhere, they consist chiefly of soft, thinly-bedded, micaceous gneiss; a more dense and ferruginous hornblendic gneiss; and thirdly, a thicker bedded granitic gneiss, composed not unfrequently of little else than the two minerals, quartz and feldspar.

"Penetrating this quite diversified formation are innumerable injections of various kinds of granite, greenstone-trap, and other genuine igneous rocks. The granites, as throughout this region generally, consist for the most part of a coarse binary mixture of quartz and opaque white feldspar, tending easily to decomposition. This rock abounds in the form of dykes and veins, sometimes cutting the strata of gneiss nearly vertically, but often partially conforming with its planes of bedding for a limited space, and then branching through or expiring in it in transverse or tortuous branches. A not uncommon variety of granitic dyke is a simple syenite composed of quartz, greenish semi-translucent feldspar, and a smaller proportion of dark-green hornblende. A soft, white, and partially decomposed granite is a very frequent associate of the stronger lead-bearing veins, particularly in their more productive portions; but this material belongs, in all probability, not to the ancient granitic injections of the gneiss, but to those much later metalliferous intrusions which filled long parallel rents in that formation with the lead-ores and their associated minerals.

"The gneissic strata and their granitic injections throughout this district display a softened, partially decomposed condition, extending in many places to a depth of twenty fathoms.

"Of the dozen or more lead and copper lodes of greater or less size brought to light in this quite limited region of five or six miles length and two or three miles breadth, the greater number are remarkably similar in their course, ranging N. 32°—35° E. and S. 32°—35° W.; and, what is equally worthy of note, they dip with scarcely an exception toward the same quarter, or south-eastwardly, though in some instances so steeply as to approach the perpendicular.

"There is no marked difference in the general character of the vein-stones of the several mineral lodes, nor any features to distinguish as a class those of the red shale from those of the gneiss."

The minerals found in these veins are quite numerous, and among them there are specimens of species hardly equaled in beauty by those coming from any other locality. Professor Silliman, in his report on the minerals of this mine exhibited at the Crystal Palace, says that the specimens of sulphate and molybdate of lead are the most magnificent metallic salts ever obtained in lead-mining, and unequaled by any thing to be seen in the cabinets of Europe.

48. Anglesite.

This mineral is found abundantly and in beautiful crystals at this locality. The magnificence of many of the specimens can only be realized by seeing those in Mr. Wheatley's cabinet. The crystals are remarkable for their size and transparency—in some instances they weigh nearly half a pound, being as transparent as rock crystal in nearly every part. Crystals with terminations at both ends have been obtained five and a half inches in length by one and a half in thickness; perfectly limpid crystals an inch in length are quite common.

The following are some of the forms:

1.—$0, \infty, 1\text{-}\breve{\infty}$

2.—$0, \frac{1}{2}\text{-}\bar{\infty}, \infty\text{-}\bar{\infty}, 1, \infty, 1\text{-}\breve{2}, 1\text{-}\breve{\infty}, \infty\text{-}\breve{\infty}$

3.—$0, \frac{1}{4}\text{-}\bar{\infty}, \frac{1}{2}\text{-}\bar{\infty}, \infty, 1\text{-}\breve{2}, 2\text{-}\breve{4}$

4.—$0, \frac{1}{2}\text{-}\bar{\infty}, \infty\text{-}\bar{\infty}, \frac{3}{4}\text{-}\bar{\frac{3}{2}}, 1, \infty, 1\text{-}\breve{2}, 2\text{-}\breve{4}, 1\text{-}\breve{\infty}$

Sometimes the crystals of this mineral are full of cavities, and of a milk-white color; but these do not differ in composition from the colorless and transparent forms. It also occurs in circular crystals.

It is sometimes colored. There is a black variety produced by the more or less perfect admixture of the salphurets of lead and copper (containing traces of silver) in the mass of the crystals, whose form is not altered. There are crystals of a delicate green color arising from carbonate of copper, and others of a yellow color due to oxide of iron.

The transparent and colorless variety is remarkably pure. Its sp. grav. is 6.35. On analysis it afforded

	1	2
Sulphuric acid	26.78	26.61
Oxide of lead	73.31	73.22
Silica	.20	....
	100.29	99.83

according very precisely with the formula $\dot{P}b\,\dddot{S}$.

I would call attention to the method of analyzing this sulphate, as described in another paper, for it was analyzed in the moist way by dissolving it first in citrate of ammonia.

The anglesite of this mine is found variously associated. It is common to find it in geodic cavities in galena, the cavities being lined with hematite varying in thickness from $\frac{1}{20}$ to $\frac{1}{2}$ an inch or more, and often this hematite contains anglesite intimately mixed in the mass. It may occur in crystals occupying a portion of the geode, or it may fill its entire capacity, assuming the form of the cavity. It is also found compacted in the galena without the appearance of any cavity or the presence of any other mineral; acicular crystals occur diffused through the galena. Observed also on copper pyrites, with a thin layer of hematite intervening between the crystal and the pyrites—on crystals of zinc-blende in quartz—on quartz associated with pyromorphite—on galena with crystals of sulphur—on calc-spar without any associate. One very interesting specimen consists of a flattened crystal an inch square, having a delicate crystal of calc-spar over an inch and a half in length perforating the center and around which the sulphate appears to have formed. It is also found on fluor-spar without associate.

Some of the most beautiful specimens are where large crystals of anglesite are covered with crystals of carbonate of lead, these latter frequently penetrating the anglesite.

49. Cerusite.

The crystals of this mineral, though not as large as those of anglesite, are yet exceedingly beautiful, both in size as well as transparency. The twin crystals are often two inches broad, transparent, and presenting the appearance of the spread wing of a butterfly. Some of the single crystals are an inch in length and half an inch thick.

A transparent crystal weighing five grammes gave a specific gravity of 6.60, and on analysis furnished

Carbonic acid..	16.38	$\}=\dot{Pb}\ddot{C}$
Oxide of lead..	83.76	
	100.14	

It occurs in hematite, coating galena in a manner similar to the anglesite, and associated with it; also in connection with pyromorphite, which often colors the entire body of the crystals of cerusite. It is found on galena without the association of any other mineral; on green and blue carbonate of copper; on pyromorphite, which often covers the entire surface of the cerusite crystals, imparting to them an opaque yellowish-green color; on oxide of manganese, in snow-white crystals, without any other associate; on hematite in a similar manner. Mammillary masses of the hematite sometimes pass through the crystals. Some few specimens have been found consisting of crystals of galena, with a number of very fine hemitrope crystals of cerusite on the surface. The cerusite is occasionally covered with an exceedingly thin coat of oxide of iron, giving the crystals a dark-red appearance; and some of them again with a very thin layer of pyromorphite, as delicate as if it had been put on with a brush.

The cerusite is sometimes colored—black, green, and yellow—in a manner similar to that mentioned under anglesite.

50. Wulfenite.

This mineral is found in small crystals of every shade of color from a light-yellow to a bright-red. It has been found in some abundance, forming, from the manner of its occurrence, very beautiful specimens. The crystals present a variety of modified forms, tabular and octahedral, one of which is here figured.

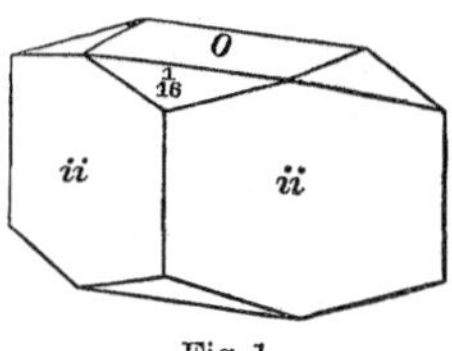

Fig. 1.

Other forms are 0, 1.

0, $\frac{1}{3}$, 1.

0, $\frac{1}{16}$, ∞-∞. (Fig. 1.) Specific gravity of a dark-yellow variety, 6.95.

The composition of both the yellow and red varieties was mined. The difference of color is due to the presence of *vanadic acid* in the red varieties, and the intensity of color is proportional to the amount of vanadic acid, which in no instance is much more than one per cent.

The analyses afforded

	Yellow variety.	Red variety.
Molybdic acid	38.68	37.47
Vanadic acid	....	1.28
Oxide of lead	60.48	60.30
	99.16	99.05

The second corresponds very nearly to 97 per cent. of molybdate and 3 per cent. of vanadate of lead. As the last substance varies in quantity, it can not be regarded as giving a distinct specific character to the mineral. This mineral has been described as a chromo-molybdate of lead, but by the most careful examination only a trace of chromium can be detected. In fact, the quantity is so minute as to require further examination in larger quantities to place the matter beyond a doubt.

Wulfenite occurs alone on crystallized and cellular quartz, or associated with pyromorphite, whose beautiful green color is often very much enhanced by the contrast of the yellow and red crystals on its surface.

Sometimes the wulfenite forms the mass, and crystals of pyromorphite are sparsely disseminated over the surface. It is also found in decomposed granite—on carbonate of lead and oxide of manganese—also associated with vanadate of lead.

51. Vanadate of Lead (Descloizite?).

This species has never before been remarked among American minerals, although the chloro-vanadate (vanadinite) was first discovered in Mexico. This adds another to the list of curious minerals from the Wheatley Mine. It was noticed about a year ago in the form of a dark-colored crystalline crust, covering the surface of some specimens of quartz and ferruginous clay associated with other minerals. Observed with a magnifying glass, it is seen to consist principally of minute lenticular crystals, grouped together in small botryoidal masses. The crystalline structure is perfect. Thus seen, the color of the mass is of a dark-purple, almost black. When seen by transmitted light, the color is dark hyacinth-red and translucent. The streak is dark-yellow. From the difficulty of obtaining any quantity of sufficient purity, nothing accurate can be stated with reference to its specific gravity and hardness; and for the purpose of analysis I was obliged to use material which, although containing pure crystals of the vanadate, was yet mixed with crystals of molybdate of lead and other impurities.

The chemical analysis is an imperfect one, yet the best that can be made from the mineral as it has been found. It is as follows:

Vanadic acid	11.70
Molybdic acid	20.14
Oxide of lead	55.01
Oxides of iron and manganese.. } Alumina }	5.90
Oxide of copper	1.13
Sand	2.21
Water	2.94
	99.03

If we subtract the amount of oxide of lead requisite to form wulfenite with the molybdic acid present, we have left 22.82 per cent., which is combined with 11.7 of vanadic acid, making a compound corresponding to: vanadic acid, 66.1; oxide of lead, 33.9=100.

This result is not considered precise. It corresponds, however, more nearly with the composition of descloizite, as given by Damour ($\dot{P}b^2\, \dddot{V} = \dddot{V}$ 29.3, $\dot{P}b$ . . . 70.7), than with dechenite by Bergmann ($Pb\, \dddot{V} = \dddot{V}$ 45.34 $\dot{P}b$ 54.66).*

The composition of descloizite can not be considered as having been fairly made out; for Damour's results are deduced, as mine have been, from a very impure material, and may on future examination prove to be $\dot{P}b^3\, \dddot{V}^2$; corresponding in composition to the chromate of lead called melanochroite. This mineral has as yet been found only in small quantity at this mine, associated with oxide of manganese and wulfenite, the crystals of this latter substance being more or less covered with minute crystals of the vanadate.

52. Pyromorphite.

There are several shades of color belonging to this mineral; a green so dark as to be almost black, olive-green, pea-green, leek-green, greenish yellow, and all intermediate shades. It

*Descloizaux has since verified this view by its crystalline form.

is a very abundant ore at the Wheatley Mine, and large quantities of it are smelted. Specimens of great beauty are found occurring in botryoidal masses, with columnar structure, in perfect hexagonal prisms, with the summits more or less modified. Crystals are found one half inch in diameter. Some of the crystals are hollow, with only a hexagonal shell. Sometimes the crystals are agglomerated in a plumose form.

A dark-green variety gave a specific gravity of 6.94. No analysis was made of this mineral, as it will be embraced in an examination of the American pyromorphites.

It is found in decomposed granite, on quartz crystals, occasionally covering their entire surface; in cellular quartz with molybdate of lead; in large masses of grouped crystals with small crystals of yellow and red molybdate inserted on crystals of sulphate and carbonate of lead, and forming a coating to large surfaces of galena.

53. Mimetene.

The specimens of this mineral that have been found, although few in number, are remarkable for their beauty of crystallization. Some of the crystals are nearly colorless and perfectly transparent; others of a lemon-yellow, either pure or tinged with green. The form is that of a perfect hexagonal prism, the edges of the summit most commonly truncated, often to such an extent as to terminate the crystal with a hexagonal pyramid. The crystals are sometimes as small as a hair, and a quarter of an inch or more in length, and again they are so broad and short as to form hexagonal plates half an inch across.

A specimen of the lemon-yellow variety was examined; it gave a specific gravity of 7.32, and was found to contain

Arsenic acid	23.17
Chlorine	2.39
Oxide of lead	67.05
Lead	6.99
Phosphoric acid	.14
	99.74

corresponding to arsenate of lead 80.21, chloride of lead 9.38= $\dot{P}b^3 \ddot{\ddot{A}}s + \frac{1}{3}Pb\,Cl$.

This specimen of mimetene is seen to be almost free from phosphoric acid, containing only about one tenth of a per cent., in this respect resembling that from Zacatecas as analyzed by Bergmann.

This mineral is found in granite or quartz. It is also associated with pyromorphite, and sometimes the two run together so as to present no distinct line of demarkation between them; some of the specimens consist of the two minerals, the pyromorphite occupying one entire surface, and mimetene the opposite surface, and between various shades of the mixture. It has been found with galena and carbonate of lead.

54. Galena.

The compact, fibrous, and crystallized varieties of galena occur at this mine. Fine crystals are found, either a perfect cube, cube with modified edges and angles, octahedron and rhombic dodecahedron often very much flattened out and occasionally rounded to an almost globular form; these rounded crystals are usually covered with pyromorphite. The galena is sometimes cellular, arising from partial decomposition, the exterior portion presenting a black drusy appearance, the interior of a bright steel color; this variety is particularly rich in silver, and also contains crystals of sulphur.

The galena is argentiferous, giving an average yield of thirty ounces to the ton. It is found associated with quartz, calcite, and fluor-spar, frequently inserted in the crystals of these substances; it is also a common associate of all the minerals of this locality. Some of the cubical crystals have their surfaces partly decomposed and covered with a layer of crystals of carbonate. Specimens are found of very large cubical and octahedral crystals, forming slabs several square feet in surface, completely covered with a layer of leek-green phosphate. The cavities of the galena frequently contain sulphur.

55. Copper.

Native copper is found only in delicate films on hematite or quartz crystal, and forms an interposing layer between the hematite and copper pyrites.

56. Copper Pyrites.

Copper pyrites is found in some cases in sufficient quantity to be worked as an ore; some of the masses are of considerable size, weighing three or four hundred pounds. Fine crystals are obtained, both tetrahedral and octahedral. It affords on analysis

Sulphur	36.10
Copper	32.85
Iron	29.93
Lead	.35
	99.23

It occurs alone and associated with the other sulphurets. It is found in various parts of the vein, there being no special point of deposit.

57. Malachite.

Malachite occurs in small reniform masses, consisting of fibrous crystals, and of a bright-green color; also in silky tufts of a very light-green color, which are associated with azurite and carbonate of lead. Its specific gravity is 4.06. An analysis gave

Carbonic acid	19.09
Oxide of copper	71.46
Water	9.02
Oxide of iron	.12
	99.69

affording the formula $\dot{Cu}\ddot{C}+\dot{Cu}\dot{H}$. It is associated with the various ores of copper and lead of the Wheatley Mine, and sometimes so thoroughly diffused through the sulphate and carbonate of lead as to give them a uniform green tint. It is not found in any quantity.

58. Azurite.

This mineral, although rare, is found in beautiful crystals, some measuring from one fourth to one half inch across, of a deep-blue color, and highly polished faces. Its specific gravity is 3.88. An analysis gave

Carbonic acid	24.98
Oxide of copper	69.41
Water	5.84
	100.23

giving the formula $2\dot{Cu}\ddot{C}+\dot{Cu}\dot{H}$. This species occurs in similar associations with the malachite; it is, however, rarer.

59. Zinc-blende.

Blende is found in considerable quantity, both massive and crystallized. Some of the crystals are exceedingly beautiful and of large size, being three or four inches in diameter, and with very brilliant surfaces. The colors are dark hair-brown and black, the brown being transparent. The specimens from this locality are hardly surpassed by those from any other mine. A specimen that was analyzed gave the following results:

Sulphur	33.82
Zinc	64.39
Cadmium	.98
Copper	.32
Lead	.78
	100.29

It is proposed to examine yet other specimens, to see if there may not be larger amounts of cadmium contained in some of them.

This mineral occurs in fluor-spar, calc-spar, and quartz, more or less mixed with the other sulphurets. In some instances it is very peculiarly interlaced in the rocks; thus we have specimens consisting as it were of four layers; namely, granite, then compact crystallized quartz three-fourths of an inch thick, then the blende an inch thick, on that a layer of crystals of calc-spar, and on this last fluor-spar.

60. Calamine.

Calamine is found in delicate crystals of a silky luster, forming in some instances snow-white tufts on fluor-spar, blende, and carbonate of lime. It is also found on cellular quartz. Some of the specimens are quite handsome, having a blue and yellow color from the presence of carbonate of copper and oxide of iron. No analysis was made of any of the specimens.

61. Brown Hematite.

This ore occurs in concretionary masses of a dark liver-color and compact structure, associated with nearly all the minerals of this mine; it very commonly forms a lining to cavities in galena, in which are found crystals of anglesite and cerusite;

sometimes it lines cavities in the rock that are completely filled with cubical galena. Acicular concretions of the hematite are found traversing crystals of anglesite and cerusite. A specimen of the purest hematite gave for its composition

Peroxide of iron	80.32
Oxide of copper	.94
Oxide of lead	1.51
Water	14.02
Silica	3.42
	100.21

62. Fluor-spar.

The remarkable feature of the fluor-spar of this mine is the absence of color, all the specimens yet found being colorless and transparent. The crystals are very perfect and beautiful, yet small; it is sometimes in globular concretions of crystalline structure radiating from the center. The cube, which is the more common crystalline form, is sometimes very much modified by the truncation of the edges and angles. A specimen that was examined gave a specific gravity of 3.15 and the following composition:

Fluorine	48.29
Calcium	50.81
Phosphate of lime	trace
	99.10

It is associated with calc-spar, and in some instances in a remarkable manner, mentioned under the head of calc-spar. Galena and blende are interspersed through it. Its occurrence in the mine was first noticed at the depth of three hundred feet, and since then it has been found abundantly.

63. Calc-spar.

There are a variety of interesting forms and associations of this mineral. The two most common are the dog-tooth spar and the hexagonal prism with a three-sided summit, and occasionally the hexagonal prism with flattened summits like aragonite. Sometimes slabs of this mineral are found, with a surface of eight or ten square feet completely covered with prismatic crystals an inch or two in length, and from half an inch to an inch in thickness; they are mostly vertical, but occasionally horizontal with double terminations. These crystals are sometimes of a remarkable character, being eight or ten inches in length and only a quarter of an inch in diameter, preserving a tolerably perfect hexagonal shape throughout the entire length; again, these slender forms are built up of small hexagonal prisms, their faces projecting from the side. It sometimes happens that these slender crystals are crossed by one of the same diameter and less length, firmly attached in the manner of a cross.

But of all remarkable crystallizations is one where the small prisms are so arranged as to form a perfect double spiral arranged around an axis (fig. 1). The specimen is three inches in length and three eighths of an inch in diameter, with the space of one fourth of an inch between each turn of the spiral. The spiral arises from one small prism crossing another at middle at a small angle of divergence (40°–50°), and so on in succession. These slender crystals are sometimes curved in a very remarkable manner.

Fig. 2.

Another thing to be remarked in connection with the calcite of this mine is its singular associations; thus, we find groups of hexagonal prisms where a small cubical crystal of fluor, about the one twentieth of an inch, is inserted in a small pit in the summit of almost every crystal (figure 3) without the occurrence of fluor-spar on any other parts of the crystal. These crystals appear to have been formed by successive crystallizations. Dog-tooth spar seems to have been first formed with these small crystals of fluor-spar on their extremities, and then by a subsequent process the calcite has closed around the dog-tooth spar in the form of a hexagonal prism with a three-sided summit. The summit never closes entirely at the center, the fluor-spar remaining visible on one side; and where there is no crystal of fluor-spar the extremity of the dog-tooth spar is frequently seen.

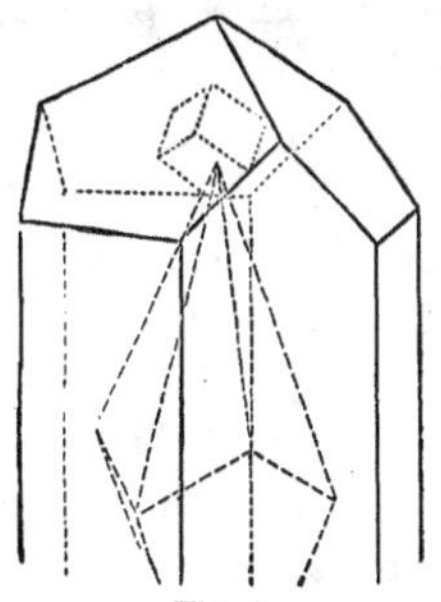

Fig. 3.

Other groups of calcite crystals have minute crystals of iron pyrites in the three faces of the summit, arranged near and perfectly parallel to the alternate edges, as seen in figure 4. Every crystal in the group is thus furnished with a set of crystals of pyrites.

In another group of crystals the pyrites, in equally small crystals, are found in three lines on the summit of every crystal

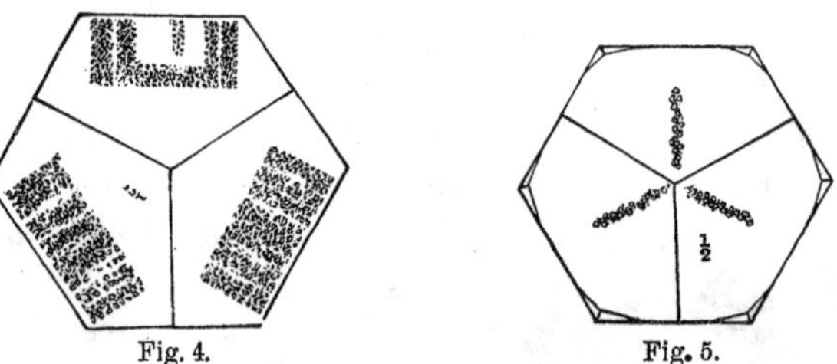

Fig. 4. Fig. 5.

running from the apex toward the edges, exactly bisecting each face, as seen in figure 5.

In this instance, as well as in the former, the pyrites are inserted entirely beneath the surface of the crystal, which is perfectly smooth.

The calcite is found in large crystals in dolomite, and is associated with most of the ores of the mine. It sometimes gives rise to pseudomorphs of molybdate of lead and carbonate of lead. These pseudomorphs are mere shells, however, retaining the form of the calcite.

64. Sulphur.

Sulphur occurs in the form of small pale greenish-yellow crystals. They are transparent and disseminated through cellular galena, which appears to have undergone partial decomposition. The galena in which it occurs is frequently associated with copper and iron pyrites, and in some rare instances with carbonate and phosphate of lead.

The other minerals occurring in the Wheatley Mine are finely crystallized *quartz*, *oxide of manganese*, *iron pyrites*, *sulphate of baryta*, *indigo copper*, *black oxide of copper*, and *dolomite*.

Of the other mineral veins in this region none have yielded the beautiful mineral species furnished by the Wheatley vein. The Perkiomen vein, five miles from the Wheatley vein, has furnished fine *capillary copper*, *indigo copper*, fine acicular crystals of *sulphate of baryta*, *crystallized copper*, and some crystals of *sulphate*, *carbonate*, and yellow *molybdate of lead;* but these last were small, and bear no comparison to those described,

It was hoped that something might be learned concerning the formation of the minerals of this vein; but the difficulties and uncertainty attendant upon the study of questions of this kind make it prudent to postpone any views that might be suggested. It may, however, be well to remark that in opening the vein, and descending from the surface for the first thirty feet, the phosphate of lead was very abundant, with some galena and carbonate. A little lower down the phosphate was less and the carbonate more abundant. Wulfenite and anglesite began to appear at one hundred and twenty feet. The phosphate and carbonate still continued with the galena, with fine large crystals of anglesite and considerable wulfenite. At one hundred and eighty feet phosphate very much diminished; carbonate and sulphate in fine crystals. Arsenate was found here. At two hundred and forty feet blende, calamine, and fluor-spar appear with considerable dolomite and but little phosphate of lead, galena forming almost the whole lead-ore. Anglesite is found, but in smaller crystals. These observations may hereafter lead to some conclusions as to the manner of the formation of these minerals, but at present I prefer dismissing the subject without further remarks.

TWO NEW MINERALS.

MEDJIDITE (SULPHATE OF URANIUM AND LIME)—LIEBIGITE (CARBONATE OF URANIUM AND LIME).

The minerals here alluded to were found associated with a specimen of pitchblende from the neighborhood of Adrianople, Turkey; it was quite impure, and a portion of it contained crystals of copper pyrites. On the surface of the pitchblende, besides the two minerals in question, there existed crystals of sulphate of lime and a little oxide of iron.

MEDJIDITE (SULPHATE OF URANIUM AND LIME).

This mineral is of a dark-amber color, transparent, of imperfect crystalline structure, fracture vitreous, although the surfaces exposed are sometimes of a dull-yellow color, arising from the loss of water. It is found on the surface of the pitchblende associated in some places with crystals of sulphate of lime. Hardness about 2.5; specific gravity not known.

Chemical Characters.—Heated gently it loses its water, becoming of a lemon-yellow color. Heated to redness it blackens (being converted into oxide of uranium and sulphate of lime). It is insoluble in water, but dissolves readily in the smallest quantity of dilute hydrochloric acid; (in this way, had it been necessary, I might have separated it from any adhering sulphate of lime.) The acid solution gives a tolerably abundant precipitate with hydrochlorate of baryta, and a red-brown precipitate with ferrocyanuret of potassium; bicarbonate of ammonia forms a precipitate that is redissolved by an excess of the ammoniacal salt; oxalate of ammonia added to this solution demonstrates the presence of lime. Farther examination detected no other substance.

So far as the small quantity at my disposal enabled me to make out its composition, it would appear to be a salt similar to the following one (liebigite) with less water, and sulphuric instead of carbonic acid, the acid being derived from the decomposition of the pyrites associated with the pitchblende. Its composition is represented by $\ddot{U}\dddot{S}+\dot{C}a\ddot{C}+15\dot{H}$.

This mineral is called *medjidite* in honor of the reigning Sultan of Turkey, Abdul-Medjid, who exhibits a most decided patronage of both the arts and sciences, certainly much more than any of his predecessors.

Liebigite (Carbonate of Uranium and Lime).

This mineral is not found crystallized, but occurs in the form of a mammillary concretion, having an apparent cleavage in one direction. It is of a beautiful apple-green color, transparent, with a vitreous fracture. Hardness between 2 and 2.5; specific gravity not ascertained.

Chemical Characters.—The mineral admits of ready separation from the pitchblende, and owing to its color and transparency is easily freed from the smallest portion of foreign matter. Heated gently, it loses water, acquiring a yellowish-gray color; heated to redness, it blackens without fusing, and on cooling returns to an orange-red color; heated strongly before the blowpipe, with or without charcoal, it blackens, and on cooling remains so; with borax it gives a yellow glass in the oxidizing and a green glass in the inner flame. Its property of blackening when heated to redness, and assuming an orange-red color on cooling, made me suppose that it might contain vanadic acid; but, as will be seen a little farther on, its reactions proved this not to be the case. As yet I believe that this property is not known to belong to any other natural combination of uranium.

The mineral dissolves readily in dilute acids and with violent effervescence, affording a yellow solution that gives a yellow precipitate with ammonia and its carbonate; but the latter in excess redissolves most of the precipitate, and what remains is found to be carbonate of lime. The carbonate of ammonia solution when boiled redeposits a yellow precipitate. In a neutral solution sulphureted hydrogen produces no precipitate, but the hydrosulphate of ammonia furnishes one of a brown color, and the ferro-cyanuret of potassium one that is reddish-brown.

The above properties show the presence of water, carbonic acid, lime, and uranium; farther examination gave no evidence of the presence of any other substance.

The amount of this mineral in a state of purity was too small to allow me to make as minute a quantitative analysis as I should desire; but, owing to the simplicity of its composition, the true nature has been very nearly if not exactly made out. The water was estimated by heating it to 400° Fah.; the carbonic acid by what was lost in dissolving it in hydrochloric acid in a small apparatus properly arranged; to the acid solution bicarbonate of ammonia was added, which redissolved all the precipitate first found; oxalate of ammonia when added to this precipitated the lime (which was afterward estimated as a sulphate); the solution filtered from this precipitate was boiled and the uranium deposited itself as a double salt that was heated to redness, and the oxide estimated in the form of olive-colored oxide. (Peligot's atomic weight for uranium was the one employed, 750; oxygen 100.)

The mean of two analyses, one of 85 and the other of 65 milligrammes, is

		Atoms.	Calculated.
Water	45.2	20	45.5
Carbonic acid	10.2	2	11.1
Lime	8.0	1	7.1
Peroxide of uranium	38.0	1	36.3
	101.4		100.0

The composition of the mineral is represented by $\dddot{U}\ddot{C}+\dot{Ca}\ddot{C}+20\dot{H}$. The pitchblende upon which the liebigite is found was analyzed, and at some future time I may have occasion to allude to this analysis with some remarks upon the salts of uranium; for the present suffice it to state that the pitchblende contains lime associated with the oxide of uranium, a circumstance that, along with the tendency of oxide of uranium to form double salts, accounts for the formation of both the liebigite and medjidite.

I have thought proper to give this double carbonate the name of the distinguished chemist of Giessen, as a demonstration of the high value I set upon his memoirs and important contributions to the science of chemistry in general.

Liebigite I have also found in one of the European localities of pitchblende — namely, Johanngorgenstadt. The first specimen I found was in the cabinet of the School of Mines at Paris; there is another in the Yale College cabinet, and a third in Prof. Shepard's collection. It exists as a thin yellowish, apple-green coating, semi-transparent, easily detached from the surface of the pitchblende, and is instantly recognized by heating it on a piece of platinum foil, when it loses water, blackens at a red heat, and on cooling becomes orange-red. It also effervesces violently with the strong acids, furnishing a test of uranium and lime. A European locality being now known, there will not be much difficulty in obtaining specimens. I also think that I have found the medjidite from specimens of the same locality; but, not having such a marked test for this mineral I could not decide on the little I obtained from a specimen which is in the cabinet of the Garden of Plants.

OXIDE OF COBALT AND MAGNESIAN OPAL.

Oxide of Cobalt from Silver Bluff, S. C.

The existence of this oxide was first brought to my notice by Prof. Ellet. It is accompanied by the oxide of manganese, and with it stains a coarse gravel found in the primitive region of this state. These stains are in the form of streaks, varying in length and breadth. The sand has the appearance of coarse gunpowder, and arises from the disintegration of mica granite. Hydrochloric acid readily dissolves this black matter with an evolution of chlorine. The solution is of a pinkish color, and affords a green salt when evaporated to dryness, from which it is evident that it must contain both the oxides of manganese and cobalt. It is perfectly free from iron, arsenic, and nickel; but at the same time it is impossible to obtain a solution of it free from iron, owing to the presence of ferruginous matter in the sand from which it is derived. This black matter has evidently originated from the disintegration of cobaltiferous oxide of manganese, minute particles of which are still to be traced, mixed with the sand. The relative proportions of the oxides are not always the same. One analysis gave

Oxide of cobalt	24.00
Oxide of manganese	76.00
	100.00

The method by which these oxides are separated is that recommended by Prof. Liebig, and one that deserves the notice of all analytical chemists engaged in the separation of the oxide of cobalt from other oxides. I allude to the method with the cyanide of potassium. The locality where this is found is one of the two or three in this country where cobalt is found under any form whatsoever, and the only one where it exists in the form of an oxide free from arsenic.

Magnesian Opal from near Harmanjick, Asia Minor.

I obtained for the composition of this opal, which occurs with serpentine and carbonate of magnesia, silica, 92.00; water, 4.15; magnesia, 3.00. It occurs also on the island of Mytilene.

MARL FROM ASHLEY RIVER, S. C.

The composition of this marl is remarkable for the large amount of phosphoric acid contained in it. The epoch to which this formation of marl is referable is not yet fully decided, but it probably consists of beds of different ages, the newest being as old as the eocene of Virginia and Maryland. It has been explored to the depth of three hundred and nine feet in boring a well. Specimens from depths varying from one hundred and ten to three hundred and nine feet have been sent to Prof. Bailey, of West Point, who has already subjected them to microscopic examination, and a short account of his results will be found in the accompanying note.* A fuller

**Extract of a Letter from Prof. Bailey to J. L. Smith.*—"Charleston is built upon a bed of animalcules several hundred feet in thickness, every *cubic inch* of which is filled with myriads of perfectly preserved microscopic shells. These shells, however, *do not*, like those beneath Richmond and Petersburg, etc., belong to the siliceous infusoria, but are all derived from those minute calcareous-shelled creatures called by Ehrenberg polythalamia, and by D'Orbigny the foraminifera. You are aware that Ehrenberg proved chalk to be chiefly made up of such shells; and you will doubtless be pleased to learn that the tertiary beds beneath your city are filled with more numerous and more perfect specimens of these beautiful forms than I have ever seen in chalk or marl from any other locality. The following are some of the results I have obtained:

"1. The marls from the depth of one hundred and ten to one hundred and ninety-three feet are certainly *tertiary* deposits, for I have found them to contain polythalamia of the family plicatilia of Ehrenberg (agathestegens of D'Orbigny), which family, as far as is yet known, occurs in *no formation older than the tertiary.*

"2. The beds from the depth of one hundred and ninety-three to three hundred and nine feet contain so many species in common with the beds above them, that although I have not yet detected the plicatilia, I yet believe they must also belong to the tertiary formation.

"3. The forms found in these beds agree much better with those detected by me in the eocene marls from Panumkey River, Va., than they do with miocene polythalamia from Petersburg, Va., and I am consequently inclined to believe that they belong to the *eocene* epoch.

detail may be expected from him at some future time; and when it does come it will no doubt be a rich feast for the naturalist of this country, prepared, as it will be, by a skillful hand.

Organic remains are quite abundant in this marl, although in an imperfect state. Its composition is somewhat peculiar, and a knowledge of it may be of some general importance. It varies in the proportion of its ingredients, but seems to be constant as regards their character. From among several analyses the following is selected as being an average one:

Carbonate of lime	65.8
Carbonate of magnesia	2.4
Silica	15.6
Alumina	10.0
Phosphate of lime with a small quantity of phosphate of magnesia	5.0
Phosphate of iron, Fluoride of calcium, Crenate of iron, Crenate of lime, Ammonia, Organic matter	1.2

"4. All the marls to the depth of two hundred and thirty-six feet present the polythalamia in *vast* abundance, and in a state of surprising preservation. The most delicate markings of the shells are perfectly preserved, and some of the forms are so large that they may be easily seen with a common pocket-lens.

"5. The lithological characters of the marls from two hundred and thirty-six to three hundred and nine feet differ from those above; and although the polythalamia are still abundant, and many of the species appear to be the same as in the strata above, yet they are less easy to observe on account of the greater compactness of the marls, and the adherence of crystalline calcareous particles to the shells.

"6. The marls which you sent from the Cooper River, thirty-five to thirty-eight miles above Charleston, also abound in polythalamia; and so many of the species are identical with those found beneath Charleston that they most probably belong to the same formation. This place on the Cooper River may be the outcrop of the very slightly-inclined beds which exist under Charleston. [In this conclusion Prof. Bailey is correct.—J. L. S.]

"7. The polythalamia, to whose labors South Carolina owes so large a portion of her territory, are still at work in countless thousands upon her coasts, filling up harbors, forming shoals, and depositing their shells to record the present state of the sea-shore, as their predecessors, now entombed beneath Charleston, have done with regard to ancient oceans. The mud from Charleston harbor is filled not only with beautiful polythalamian shells, but is also very rich in siliceous infusoria."

The proportion of phosphate of lime is large, and may be owing to the impalpable remains of animals that must have frequented these early seas in myriads, or it may be peculiar to the little shelly remains of polythalamia that form such a large portion of this bed. The fluoride of calcium, which I believe is here pointed out for the first time as existing in marls, does not owe its origin to any spiculæ of bony matter present in the specimen examined, at least none that the microscope could detect; so we must attribute it either to osseous matter triturated to an impalpable powder, or, what is more probable, suppose that it forms a part of the calcareous covering of those animalculæ just alluded to, the remains of which form the foundation of the city of Charleston. The ammonia is held mechanically in the pores, associated perhaps with carbonic acid, and is easily rendered apparent by dropping caustic potash on the marl. This has been found present in all the marls I have examined, and the fluoride of calcium in several.

SOURCE OF FLUORINE IN FOSSIL BONES.

The analyses of fossil bones communicated by MM. Girardin and Preisser to the Academie des Sciences, October, 1842, afford the only well-detailed numerical results of the composition of this class of bones that we possess. The authors appear to have bestowed considerable care upon their research, and their estimate of the proportion of fluoride of calcium present was as follows:

	Per cent.
A metacarpal bone of a bear from the cavern of Mialet...	1.09
Tusk of an elephant..	2.64
Vertebra of a plesiosaurus..	2.11
The great bone of the pœkilopleuron bucklandii...........	1.50
Rib of an ichthyosaurus..	1.02
Head of the same ichthyosaurus..................................	1.65
Bone of the lamentin from the tertiary formation in the environs of Valognes..	9.12

This fluoride would appear to form a distinguishing mark between fossil and recent bones, although Berzelius has found it to exist in these latter, and Marchand, in some recent experiments, mentions the same fact; still many other chemists have not succeeded in detecting it. MM. Girardin and Priesser suppose that it was owing to some accidental circumstances that Berzelius was enabled to discover it in the cases that he examined, they having in no instance found it, although carefully sought for. My experience tends to confirm Berzelius in his statement, having in several cases obtained most decided evidence of its presence in recent bones, but in very minute quantity. In many instances I failed to detect it, and attribute the failure more to the minuteness of the quantity than to the total absence of it.

I would here remark that in examining for fluorine in the ordinary way, by testing the effects of the hydrofluoric acid (liberated by the action of sulphuric acid) upon waxed glass with characters traced out, the process requires some precaution when the quantity present is supposed to be very small;

for I have been able in several instances to obtain a permanent delineation of the characters traced without the presence of fluorine. In these cases it is caused by the action of the vapors of either sulphuric or hydrochloric acid upon certain kinds of glass that contain a large quantity of metallic oxides, or upon glass the surface of which has been altered by the action of the air. There is, however, no apparent corrosion in these cases.

The existence of fluorine in fossil bones, and its doubtful—or, as some say, absolute—non-existence in those of recent animals, have induced MM. Girardin and Preisser to conclude that it did not belong originally to the bones of fossil animals, but has found its way there by infiltration after their death; and they appear to have come to this conclusion without having examined the chemical character of the formations from which the various bones were taken.

I have had an opportunity of throwing some light upon this subject from the examination of two bones taken from the same calcareous deposit and within two feet of each other, the one cellular and the other compact. The cells of one of these bones were filled with small concretions of calcareous matter, evidently arising from the infiltration of some of the material forming the bed in which they lie. These concretions, it would seem, ought certainly to contain a portion of whatsoever matter had been infiltrated, as all infiltrations must have passed in together. These concretions, carefully detached from the bone, were examined especially for fluorine, but not the slightest trace was found; while on the contrary a very small quantity of the compact part of the same bone gave decided indications of the presence of this substance. This fact must certainly lead to the conclusion that the fluoride of calcium in the body of the bone was not infiltrated; for had it been otherwise it would have been associated with matter known to be infiltrated, as the calcareous nodules.

The same cellular bone was examined as a whole—that is to say, without detaching the calcareous matter—in comparison with the compact bone from the same locality; and in the former there was found less fluoride of calcium than in the latter, contrary to what would have been the case had this fluoride been infiltrated.

	Per cent.
Compact bone, fluoride of calcium........................	2.45
Cellular bone, " "	2.00

The deposit from which these bones were taken was analyzed, and fluoride of calcium detected, pertaining to mollusca or vertebrated animals; and were it necessary to suppose that it could have existed in only one of these, I should unhesitatingly attribute its origin to the vertebrated animals, particularly on account of their abundant provision of phosphates. Bones were also examined that contained fluorine when the deposit from which they were taken showed no traces of this element.

Dr. Daubeny has lately examined the question of the existence of fluorine in recent bones, and decided it in the affirmative.

It is not surprising that we should find the phosphates and fluorides associated in the animal kingdom; for in the mineral kingdom fluorine is a very common attendant upon the phosphates, as for instance in the apatites, wagnerite, wavellite, uranite, etc.; and I think if we search the mineral kingdom we shall not find so constant an association of any two elements as fluorine and phosphorus. All the phosphates of the alkalies and earths contain fluorine.

If then this element is associated with the phosphates, they must exist together in the soils arising from the disintegration of the rocks containing these minerals, and the plants growing upon these soils would upon taking up the phosphates naturally appropriate the accompanying fluorides, which two classes of salts would subsequently pass to the same portion of the animal feeding upon these plants—namely, to the bones.

The reason why the existence of fluorine in recent bones is doubtful may be owing to the fact that the great mass of the phosphate of lime originally in the soil has from various causes disappeared, and with it the fluoride of calcium; and that the portion of this latter still remaining is so small that notwithstanding the double condensation that it undergoes through the agency of plants and animals, it is not in sufficient quantity to come readily within the reach of our tests.

CHROME AND MEERSCHAUM OF ASIA MINOR.

In my journey to the south of Broosa (Anatoly, Asia Minor) I crossed a formation of serpentine and other magnesian rocks of considerable extent. Fifty miles from this city I discovered *chromate of iron* disseminated in these rocks; and ten or fifteen miles farther south (near the city of Harmanjick) there is an abundant deposit of this mineral. A circumstance worthy of remark is that this chromate of iron (the first that has been discovered in Asia Minor) is found in serpentine as elsewhere. This important fact can explain to a certain extent the formation of this chromate. It is well known that serpentine contains all the elements of chromate of iron, which during the consolidation of this rock might separate themselves by the force of segregation, so well known to operate in many geological phenomena. Two facts which seem to confirm this supposition are; first, the existence of the chromate of iron in masses and not in veins; and secondly, the pale color of the serpentine associated with the chromate. One small specimen that I have consists of a white rock composed principally of carbonate of magensia, in which small specks of chromate of iron are visible. It is possible that this carbonate is the result of the decomposition of the serpentine at the surface by the action of water containing carbonic acid. It is only at this locality that I found crystals of the chromate octahedral, but very small.

This discovery is of great importance to the arts and to the Turkish Government, which proposes exploring the mine.

In quitting the locality of chrome and going north-east, I traversed in several places the serpentine containing veins of carbonate of magnesia, quite pure; and this occurs until we arrive at the plains of *Eski-Shehr*. It is from different parts of this plain that comes the meerschaum most esteemed in the arts. Its geological position is very different from what I had expected. The plain in which it is found is a deposit of drift; a valley filled up with the *debris* of the neighboring mountains, consolidated by lime in which I found no fossils.

The meerschaum is found in this drift in masses more or less rounded; the other pebbles are fragments of magnesian and hornblende rocks.

I have examined with care the neighboring mountains which surround the plain, and have found that the rocks are of the same nature as the pebbles in the plain, except those of the meerschaum; but on the other hand I found carbonate of magnesia in the mountains which is not to be found in the plains. And this makes me suppose that the meerschaum owes its origin to the carbonate of magnesia of the mountains decomposed after its separation by water containing silica.

If this supposition be true we should naturally find meerschaum which not being completely altered contains the carbonate of magnesia. A chemical examination of several specimens has served to establish this fact. Some of the specimens, taken at the depth of ten feet, when placed in hydrochloric acid, give rise to an effervescence that will continue for some time; the piece will not change its form, it only absorbs the acid; the solution will be found to contain chloride of magnesium nearly pure. Another proof that the meerschaum probably owes its origin to the carbonate of magnesia is that I have found attached to the meerschaum serpentine similar to that found in contact with the carbonate of magnesia of the mountains.

The meerschaum of *Eski-Shehr* differs completely from several other specimens that I have seen, coming from other localities, and which exist in the fissures of rocks; it is certain that the quality of the first is most esteemed.

LESLEYITE OF CHESTER COUNTY, PENN.

AND ITS RELATIONS TO THE EPHESITE OF THE EMERY FORMATION NEAR EPHESUS, ASIA MINOR.

Several years ago a small amount of mineral from Chester County, Penn., was handed to me for examination by Dr. Isaac Lea, of Philadelphia. The specimen was too impure to warrant any conclusion upon analysis. Its character and associates, however, led me to suppose that it was the same mineral described by me as associated with the emery of Asia Minor, and to which I gave the name *Ephesite.* In the mean time Dr. Lea described his mineral as a new species, calling it *Lesleyite;* and in a recent number of the American Journal of Science and Arts, S. P. Sharples has given an analysis of it that at once brought to my recollection my original opinion that it was close to Ephesite, and on recurring to my examination of this mineral, making due allowance for the impurities contained in it, the opinion was confirmed.

I then obtained from Dr. Lea another specimen of his mineral, and proceeded to analyze both it and the Ephesite for mutual comparison. Much labor was bestowed in selecting the pure mineral from each, the greater part of a day having been consumed in procuring the necessary quantity for analysis. They are similar in their associations and identical in color and luster and general physical properties. They are both very difficult to decompose by carbonate of soda, even when aided with caustic potash; so that in both analyses the silica obtained was fused a second time, and much alumina separated from it.

My original description of the mineral will be found under Emery, in the American Journal of Science and Arts, 2d series, vol. x, 1850, as follows:

"It is of a pearly-white color, and lamellar in structure; cleavage difficult. It scratches glass easily, and has a specific gravity of from 3.15 to 3.20.

Heated before the blowpipe, it becomes milk-white but does not fuse. At first sight it might be taken for white disthene. It is decomposed with great difficulty by carbonate of soda, even with the addition of a little caustic soda."

The lesleyite has identically the same properties. On analysis the two minerals were found to be composed as follows:

	Ephesite.	Lesleyite.
Silica	30.70	31.18
Alumina	55.67	55.00
Lime	2.55	.45
Soda	5.52	1.20
Potash	1.10	7.28
Water	4.91	4.80
	100.45	99.91

The alkalies in the two varieties are reversed, the Ephesite containing principally soda and the Lesleyite potash.

This close relation of the two minerals is an interesting fact as regards the associate minerals of corundum found in different parts of the world.

In regard to the reddish variety of Lesleyite examined by Roepper, the analysis can not be considered as giving very satisfactory results, for the mineral may have been impure, and the difficulty in decomposing by the soda fusion may give very erroneous results in a silica determination.

TETRAHEDRITE, TENNANTITE, AND NACRITE,

OF THE KELLOGG MINES OF ARKANSAS.

A short time since Prof. E. T. Cox, of Indiana, sent to me an antimonial copper-ore containing silver, one fragment being the termination of a crystal having a number of small but beautiful faces; another was a minute crystal of a different form. In the hands of Prof. Cox a blowpipe analysis had given about five per cent. of silver in some of the mineral.

The crystalline fragments were first examined, and they enabled me clearly to trace out tetrahedrite in one and tennantite in the other. The faces on the tetrahedrite were small, but beautiful and very numerous; from the number on the fragment examined there would not have been less than from sixty to seventy had the crystal been perfect. It corresponds very nearly to the crystal figured in Dufrenoy's Mineralogy, plate 124, fig. 441, which he speaks of as coming from Moschellandsberg, a locality that I am not able to discover. Good measurements were made on a few of the faces.

P on P 70°: P on b^3 159° 30′; P on a^2 144° 30′.

Specific gravity of different specimens varied from 4.78 to 5.08; the latter was the specific gravity of the above crystal. The analysis of two specimens, No. 2 being a part of the crystal, gave

	1	2
Antimony	26.50	27.01
Sulphur	26.71	25.32
Copper	36.40	33.20
Iron	1.89	.82
Zinc	4.20	6.10
Silver	2.30	4.97
Arsenic	1.02	.61
	99.02	98.03

The quantity of No. 2 analyzed did not exceed three hundred milligrammes.

There are two minerals, consisting of minute micaceous scales, on the quartz containing this gray copper. One of them

I could not obtain in sufficient quantity for examination; from an imperfect examination I conclude that it is muscovite. The other mineral—a soft, unctuous, talc-like mineral—is nacrite, composed as follows:

Silica	65.02
Alumina	26.11
Oxide of iron	2.20
Manganese	trace
Potash and soda	1.18
Water	4.98
	99.49

These minerals came from an exceedingly interesting mine in Arkansas that is as yet almost unexplored. I have obtained a description of it from Professor Cox, and I think it would be well to give it here; for, besides being likely to prove of considerable commercial value when properly explored, there will doubtless be found many interesting mineral species there.

The Kellogg mines are situated ten miles north of the city of Little Rock, in Pulaski County, Ark. The country in the vicinity is rolling; the highest hills are about two hundred and seventy feet above the water-level of the neighboring streams. The surface rocks are thick, and thin beds of sandstone alternating with shales occupying the base of the coal measures. The rocks are but little disturbed, and are for the most part horizontal. There are no metamorphic rocks showing themselves at the surface nearer than Little Rock, on the south side of the Arkansas River. Innumerable veins of milky quartz are seen traversing the sandstones and shales.

About seventeen years ago lead-ore was discovered at these mines by Mr. Kellogg; companies were organized and mining operations carried on extensively for about one year, when the flattering accounts of the gold discoveries in California caused the miners to leave, and the work, which had been badly conducted, was abandoned. Many tons of the ore, which is an argentiferous galena (containing sixty to two hundred ounces of silver to the ton), were extracted from the mine, and finally the greater part was shipped to England and sold at a good price. A smelting furnace has been erected on the grounds, but for lack of skill the proprietor never succeeded in working the ore profitably; consequently the impression was produced that the ore could not be smelted, but there is no good reason for such an opinion.

Since the mines have been abandoned, the old shafts, ranging in depth from fifteen to seventy feet, are all filled in, and the country has become covered with a dense undergrowth of brush and briars. About one year ago Prof. Cox revisited these mines for a company who had in view the lease or purchase of them; it was during this visit that the gray copper above referred to was discovered. This ore had previously escaped the observation of others who had explored these mines. It is impossible at present to see the ore in place, and those who previously worked the mine give conflicting statements as to the manner in which the ore is found.

The vein-rock and associated minerals with the galena are white quartz, spathic iron, zinc-blende, copper pyrites, gray copper, tennantite, and nacrite.

The mines are now in the hands of a new company, and the latest information from their operations is that matters look well; the vein now being worked is nearly three feet wide, principally lead-ore, the balance being zinc-blende; twenty hands are at work, and the shaft is down forty-five feet. My opinion is that in time this mine will become of considerable importance, and lead to further developments of argentiferous galena in that region.

NOTES ON THE CORUNDUM OF NORTH CAROLINA, GEORGIA, AND MONTANA,

WITH A DESCRIPTION OF THE GEM VARIETY OF THE CORUNDUM FROM THESE LOCALITIES.

The corundum formations in North Carolina and Georgia are the second in importance in the United States that have been brought to my notice; and the one in North Carolina is by far the most interesting in this country, and perhaps of any yet known, in the extent of the formation, the distribution of the corundum, and the purity of the mineral.

This mineral was first discovered in North Carolina in 1846—about the time I was engaged in developing the geology of emery in Asia Minor and the Grecian Archipelago; and upon communicating to American geologists my discoveries in relation to the associate minerals of the emery in Asia Minor, and directing them to search for the same in connection with the corundum found in different parts of America, the same associates were discovered in connection with the North Carolina corundum as well as that from other localities.

At this time there had been discovered but one detached block, but no other specimen could be discovered in that locality. There the matter rested until 1865, when C. D. Smith (to whom I am indebted for valuable information contained in this paper), assistant of Prof. Emmons, geologist of North Carolina, had brought to him by one of the inhabitants of the country west of the Blue Ridge Mountains a specimen of rock which was recognized as being corundum, and on visiting the spot this geologist discovered the corundum *in situ*, and a number of specimens were collected. Since that time public interest has increased in relation to this substance, and it has been discovered in such quantities as to make it an object of interest to the arts as a substitute for emery, and very rapidly other localities were brought to light along a distance of forty miles.

The colors of the corundum as found along this zone of outcrops are blue, gray, pink, ruby, and white. Sometimes it has broad cleavage faces, and then again it occurs in hexagonal prisms. One hexagonal prism weighed over three hundred pounds. There is a difference in the cleavage and the associate minerals at different localities.

In the development in North Carolina the corundum occurs in chrysolite or serpentine rocks, and outside of serpentine it has not been found. These chrysolite rocks belong to a regular system of dikes, which have been traversed for the distance of about one hundred and ninety miles. This system of dikes lies on the north-west side of the Blue Ridge, and has a strike parallel to the main mass of the ridge, and has an average distance from the summit of the ridge of about ten miles. It continues this strike to the head of the Little Tennessee River, say from Mitchell to Macon County, one hundred and thirty miles. Here the ridge curves around the head of the Tennessee and falls back about ten miles to the north-west. In conformity with this elbow in the ridge the disturbing force shifts to the north-west and re-appears at Buck's Creek, having relative position to the Blue Ridge.

The serpentine appears at intervals along this whole line of one hundred and ninety miles. There is a corresponding system of dikes traversing the southern slope of the Blue Ridge, but not so regular and compact as the system on this north-west side, nor are the outcrops so frequent. The main mass of the ridge bears no evidence of having been disturbed at all, at least none have been found. From Mitchell County to Macon the serpentine is usually inclosed in a hard crystalline gneiss-rock, which bears rose-colored garnets, kyanite, and pyrite. After its shifting to the right it occurs in hornblendic beds and gneiss. At Buck Creek and thence south-westward the hornblend beds assume very large proportions, and instead of common feldspar have in them albite, making an albitic syenite. At Buck Creek (which is named Cullakenih) the chrysolite covers an area of about three hundred and fifty acres. One or two observers have fallen into the error of confounding the two dike systems, whereas they have no connection whatever.

According to them the northern system cut through the Blue Ridge at right-angles, and then turn back on the opposite side of the ridge.

Now there is no such phenomena connected with these outcrops. They evidently belong to separate systems. The outcrops along the northern system occur at intervals ranging from one to fifteen miles. The belt or zone along which these outcrops occur never exceeds four miles in width on the northern side of the ridge. On the opposite side the system is not so well defined, and the outcrops are rarer.

Upon these serpentine beds there exists chalcedony, chromite on some of them, chlorite, talc, steatite, anthophyllite, tourmaline, emerylite, epidote on some of them, zoisite, and albite, with occasionally asbestus and picrolite, as also actinolite and tremolite. The corundum at some places seems to occur mostly in ripidolite in fissures of the serpentine. At Cullakenih the corundum with its immediate associates is in chlorite, except the red variety, which is in zoisite, containing a minute quantity of chrome.

Throughout all the range of rocks for the great extent referred to corundum forms a geognostic mark of this chrysolite-rock just as it does of the calcareous rock bearing corundum described by me in Asia Minor. They belong to the same geological epoch, and overlie the gneiss, etc.

The closest investigation shows that the chrysolite in North Carolina takes the place of calc-rock in Asia Minor; that these are invariably the gangue rock in the two different quarters of the globe; but, as remarked above, the contiguous rock shows them both to be of the same geological period, overlying directly the primary rocks; and both of them are also identical geologically with the Chester emery formation of Massachusetts.

While all the localities of corundum and emery I have examined exhibit certain marked and prominent characteristics common to them all, and evince unmistakable evidence of geological identity, yet each locality has its peculiar characteristics. In all cases, however, the masses of corundum give evidence of having been formed by a process of segregation, as described in my memoir on the Asia Minor emery.

In *Asia Minor* the Gumuch-dagh emery has but little black

tourmaline associated with it, and instead chloritoid in crystals or lamellæ; also its diaspore is rare, but when found is prismatic, affording the finest perfect crystals yet seen, from which M. Dufrenoy made his last study of the crystallography of this mineral; and the emery is associated with calcareous rock overlying gneiss. The Kulah emery from the same part of the world is equally in calcareous rock, and has very little chloritoid or chloritic mineral associated with it.

The Naxos and Nicaria emery of the *Grecian Archipelago* is also in connection with calcareous rock, but has no chloritoid associated with it, but in its place black tourmaline is abundant.

While in the above localities the rock bearing the corundum is calcareous, that in *Chester*, *Mass.*, is in talcose slate, and saponite with hornblendic gneiss immediately on one side of the vein, and is accompanied with a large amount of magnetic oxide of iron. Tourmaline also abounds in this corundum, and like the Asiatic variety contains rutile, ilminite, etc.

In the localities forming the subject of this memoir the following minerals are deserving special notice.

CORUNDUM.

This mineral occurs in finer and more beautiful variety than in any yet known locality. The masses in many instances are very large, weighing six to eight hundred pounds, having fine large cleavages, and are remarkably free from foreign ingredients. The crystals are also fine, and in some instances of great size and beauty. Two of them discovered by M. Jenks, and now in the possession of Prof. Shepard, have been described by him. They are respectively three hundred and twelve and eleven and three fourths pounds in weight. The largest is red at the surface, but within of a bluish-gray. The general figure is pyramidal, showing, however, more than a single six-sided pyramid, whose summit is terminated by rather an uneven and somewhat undefined hexagonal plane. The smaller crystal is a regular hexagonal prism, well terminated at one of its extremities, the other being drusy and incomplete. The general color of this crystal is a grayish-blue, though there are spots, particularly near the angles, where it is of a pale sapphire tint. Its greatest breadth is six inches, and its length over five. Some of the lateral planes are coated in patches with a white pearly margarite.

The smaller crystals are often transparent at their extremities. It is, however, in color that the corundum of this locality excels. It is gray, green, rose-color, ruby-red, emerald-green, sapphire-blue, and all intermediate shades to colorless. Many pieces of the blue and red have been cut and polished, presenting very good characters as gems, without being of the finest quality.

DIASPORE.

While this mineral is found so abundantly with the corundum of Chester, Mass., I have not been able to find it associated with these localities. Several specimens of supposed diaspore have been submitted to me, but on close examination it was found to be colorless kyanite.

CHLORITE.

This mineral abounds in this locality, and, as has been stated, is the gangue-rock of the corundum; it not only surrounds the corundum, but permeates it. There are several varieties, varying in color from a yellowish-green to a dark-green, and differing a little in composition. Two specimens from the same locality were composed as follows:

	Large plates.	Friable.
Silica	27.00	29.15
Alumina	21.60	10.50
Oxide of iron	16.63	23.50
Magnesia	22.00	25.44
Water	12.30	10.04

MARGARITE (EMERYLITE).

This curious mica—curious so far as that since my first pointing it out as a characteristic of the emery formation in Asia Minor and the Grecian Archipelago—has been found wherever corundum is, and in the case of Chester emery was the means of leading to its discovery. At the present localities it is abundant and mixed with the rocks and the associate minerals of this locality. Chemical analysis was made of the specimen with the following result:

Silica	32.41
Alumina	51.31
Lime	10.98
Soda	2.43
Water	2.13

ZOISITE.

This mineral occurs in two forms—a black variety and a light-green variety. These minerals have been called by some Arfverdsonite, but neither of them have the composition of that mineral. Their compositions were as follows: Green variety is of a very pale chrome-green, containing and compared with that from Lake Geneva

	Light green.	Lake Geneva.	Black variety.
Silica	45.70	43.59	45.90
Alumina	24.01	27.72	13.34
Peroxide of iron	4.56	2.61	11.46
Lime	13.44	21.00	12.20
Magnesia	8.03	2.40	12.53
Soda	2.91	3.08	3.39
Water	.60	......	.66
Oxide of chrome	.52	......	......

ANDESITE.

This mineral occurs mostly in a granular form. Its composition is

Silica	64.12
Alumina	24.20
Soda	9.28
Lime	2.80
Oxide of iron	.14

The other minerals associated with this emery formation are magnetic oxide of iron, chrome iron, rutile, asbestus, talc, actinolite, black tourmaline, chalcedony, anthophyllite, spinel, albite, and picrolite.

On the existence of the Ruby and Sapphire in North Carolina and Montana Territory.

The corundum locality that I have described in North Carolina furnishes masses of corundum from which small pieces can be detached, of good blue or of ruby color, perfectly transparent, and nearly free from flaws. When cut and polished they are gems of no mean value. I have not seen the most perfect of them that have been cut, but I have some polished specimens of fine color but with many flaws.

The question naturally arises, what may be the prospects of obtaining the gem from that locality in sufficient quantity to warrant exploration? Up to the present time its occurrence is so different from those in known localities in the East Indies that we are rather inclined to the opinion that it will only be occasionally that pieces of corundum will be found of sufficient purity and beauty to be of much value as gems; for it is well known that very small defects, if they do not destroy altogether the value of the gem, depreciate its value to a very great extent.

About a year ago a quantity of rolled pebbles were sent to me from the territory of Montana, which upon examination I found to consist principally of corundum; they were like the rolled pebbles from the ruby localities in the East Indies, each one being a little crystal in itself, more or less abraded on the angles, and being of a compact, uniform structure. They were flattened hexagonal prisms with worn edges. They were either colorless or green, varying in shade from a light to a dark-green; some were bluish-green, and there were not any red ones among them; there were some red pebbles, but on examination they proved to be spinel.

These pebbles are found on the Missouri River near its source, about one hundred and sixty miles above Benton; they are obtained from bars on the river, of which there are some four or five within a few miles of each other. In the mining region of this territory on these bars considerable gold is found, being brought down the river and lodged there, and are now being worked for the gold. The stones are found scattered through the gravel (which is about five feet deep), and upon the bed-rock in some of the claims they are abundant and in others scarce. Occasionally they are found in the gravel and upon the bed-rock in the gulches from forty to sixy feet below the surface, but they are very rare in these localities. The greatest quantity of them are found upon the Eldorado bar situated on the Missouri River about sixteen miles from Helena; one man could collect on this bar from one to two pounds per day.

I have had some of the stones cut, and among them one very perfect stone of three and a half carats, and of good green color, almost equal to the best oriental emerald.

My opinion is that this locality is far more reliable to look for the gem variety of corundum than any other in the United States I have yet examined.

REPORT ON DUPONT'S ARTESIAN WELL AT LOUISVILLE.

This work was commenced in April, 1857, from the bottom of a well that had a depth of twenty feet; the boring tools employed made a hole five inches in diameter to the depth of seventy-six feet from the surface; the boring was now reduced to three inches, and thus continued to the bottom of the well. The depth of well is two thousand and eighty-six feet, flow of water three hundred and thirty-thousand gallons in twenty-four hours, rise above the surface one hundred and seventy feet.

The rock struck, which geologically belongs to the Devonian series, is for thirty-eight feet shell limestone, then for forty feet coralline limestone, at which depth the Upper Silurian is reached. Without being able to make out with any degree of certainty the amount of Upper Silurian passed through, we suppose it to be over twelve hundred feet. At the depth of sixteen hundred feet a sandstone was reached, doubtless of the Lower Silurian, and ninety-seven feet deeper was encountered the first stream of water which reached the surface. This flowed out abundantly and with much force. The quantity not being sufficient, the boring was continued. After this it was unnecessary to use the bucket to take out the material detached by the borer, the force of the water bringing up the fragments very readily. The water increased in quantity in going deeper, the increase being more marked at eighteen hundred and seventy-nine feet, and still more at nineteen hundred feet, where pieces of rock weighing an ounce or two came up with the water. The water increased every ten or twenty feet to the depth of two thousand and thirty-six feet; here a very hard magnesian limestone was encountered six feet in thickness, after which the sandstone re-appeared, and for the next fifty feet there was no increase of water.

The following table exhibits the series of rock as far as it is possible to make it out by the fine fragments taken out at different depths, beginning at the top:

76 feet—sand and gravel.
100 feet—tolerably pure limestone, with fragments of fossils.
12 feet—soft limestone mixed with clay.
52 feet—tolerably pure limestone mixed with fossils.
5 feet—limestone with ferruginous clay.
81 feet—gray limestone.
157 feet—limestone mixed with clay.
149 feet—tolerably pure limestone with many portions quite white.
13 feet—clay shale with little calcareous matter.
207 feet—limestone with a little blue clay shale.
33 feet—same, little darker and more shale.
Next 94 feet—pure, very white limestone with fossil alternating with very dark limestone, color probably from organic matter, with some dark shale.
26 feet—shaly limestone.
40 feet—very light and hard pure limestone.
1 foot—white clay.
546 feet—gray limestone, alternating hard and soft.
41 feet—sand rock, white.
4 feet—same, very fine and hard, with little limestone.
60 feet—same, with more lime.
72 feet—same, less limestone.
308 feet—same, sandstone with but little lime.
6 feet—magnesian limestone, very hard.
50 feet—sandstone again.

At the urgent request of many citizens of Louisville the boring was now stopped to give a fair test of the medical virtues of the water that was pouring forth at the rate of two hundred and thirty gallons per minute, or about three hundred and thirty thousand gallons in twenty-four hours. The water by its own pressure rises in pipes one hundred and seventy feet above the surface. The boring was accomplished in sixteen months, and the depth reached is two thousand and eighty-six feet. In order to conduct the water to the surface and prevent its passing off into the gravel beds below, a tube five inches in diameter leads from the surface to the rock, a depth of seventy-six feet, into which it is driven with a collar of vulcanized gum-elastic around it. No tubing is found necessary for any other part of the boring.

When the size of the bore (three inches in diameter) and its depth are considered, the flow of water from the well is unequaled by any other artesian well yet constructed that flows above the surface; for although the Grenelle well at Paris delivers six hundred thousand gallons in twenty-four hours, it has at the bottom an area six times as great as the Dupont well, and a few hundred feet up seven times as great. A corresponding diameter to Dupont's well would, according to just and reasonable calculations, furnish about two million gallons in twenty-four hours; also the elevation of the water above the surface is greater than that of any other artesian well, and it is only exceeded in depth by the St. Louis well, and that to an extent of one hundred and thirteen feet.

The water comes out with considerable force from the five-inch opening, and a heavy body thrown into the mouth of the well is rejected almost as readily as a piece of pine wood. By an approximate calculation its mechanical force is equal to that of a steam-engine with cylinder ten by eighteen inches, under fifty pounds' pressure, with a speed of fifty-five revolutions per minute, a force rated at about ten-horse power. The top of the well is now closed, and the water conducted about thirty feet to a basin with a large jet d'eau on the center, from which there is a central jet of water forty feet in height, with a large water-pipe, from which the water passes in the form of a sheaf. When the whole force of water is allowed to expend itself on the central jet, it is projected to the height of from ninety to one hundred feet, settling down to a steady flow of a stream sixty feet high.

Temperature of the Water.—The water as it flows from the top of the well has a constant temperature of 76½° F., and is not affected either by the heat of summer or the cold of winter. The temperature at the bottom of the well is several degrees higher than this, as ascertained by sinking a Walferdin's registering thermometer to the bottom, which indicated 82½° F. Taking as correct data that the point of constant temperature below the surface of Louisville is the same as at Paris—namely, 53° F.—at ninety feet below the surface we have an increase of one degree of temperature for every sixty-seven feet below that point. The increase in Paris is one degree for every sixty-one and two tenths feet. The temperature of the water is sufficient for comfortable bathing during most of the year—a circumstance that will be of considerable importance if it ever be turned to the use of baths. The reason of the difference of six degrees between the water at the bottom of the well and at the top is that the iron pipe leading from the surface to the rock passes through a stratum of water sixty feet thick, having a temperature of fifty-seven degrees.

The Source of the Water.—The question naturally arises, if the vein of water supplying this well has a connection with some distant source higher than the surface of Louisville, where is that source? From all that we have been able to learn of the geology of this county, taking Louisville as a center, the first rocks encountered corresponding to the sandstone (in which the water of the artesian well was struck) are in Mercer, Jessamine, and Garrard counties, near Dix Creek, to the east of Harrodsburg. The rocks there are said to be cavernous and water-bearing. The elevation is about five hundred feet greater than Louisville, and about seventy-five miles in a straight line from this city.

This being the most probable source of the water, whence come its mineral constituents? These are obtained from the rocks through which it percolates in its way from its source to the point below Louisville where it has been tapped, and where it will doubtless flow in undiminished quantity for centuries to come, as wells having such deep sources as this are usually inexhaustible.

Nature of the Water.—The water is perfectly limpid, with a temperature, as already stated, of 76½°, which will be invariable all the year round. Its specific gravity is 1.0113. The solid contents left on evaporating one *wine gallon* to dryness are 915½ grains, furnishing on analysis

	Grains.
Chloride of sodium (common salt)	621.5204
Chloride of calcium	65.7287
Chloride of magnesium	14.7757
Chloride of potassium.	4.2216
Chloride of aluminum	1.2119
Chloride of lithium	.1012
Sulphate of soda	72.2957
Sulphate of lime	29.4342
Sulphate of magnesia	77.3382
Sulphate of alumina	1.8012
Sulphate of potash	3.2248
Bicarbonate of soda	2.7264
Bicarbonate of lime	5.9915
Bicarbonate of magnesia	2.7558
Bicarbonate of iron	.3518
Phosphate of soda	1.5415
Iodide of magnesium	.3547
Bromide of magnesium	.4659
Silica	.8857
Organic matter	.7082
Loss in analysis	8.1231
Chlorides of rubidium and cæsium	traces
	915 4582

GASES IN ONE GALLON.

Sulphureted hydrogen	2.0050
Carbonic acid	6.1720
Nitrogen	1.3580

The analysis was performed by the usual methods; but as chloride of lithium was sought for and found, it may be of interest to detail the method of research in this particular, as a guide to similar investigations of other mineral waters in this country. Ten gallons of water were evaporated to about two pints (there was an abundant deposition of salts); to this was added one gallon of ninety-five per cent. alcohol; it was then thrown on a filter, and the salts on the filter washed with alcohol of the same strength; the filtered liquor was evaporated nearly to dryness; in the present instance the residue consisted of a few ounces of a thick, syrupy liquid; to this was added one pint of absolute alcohol; additional salts were precipitated; the liquid was again filtered and evaporated nearly to dryness; to it were added eight ounces of distilled water and two ounces of milk of lime (pure lime made by igniting carbonate of lime prepared by carbonate of ammonia); the lime was added for the purpose of precipitating the magnesia and alumina; again filtered and washed; the filtered liquid was somewhat concentrated, and while warm carbonate of ammonia was added to precipitate the lime; it was then filtered and evaporated to about a fluid-ounce and treated with a little lime-water and carbonate of ammonia alternately, to insure the absence of the last traces of magnesia and lime.

Before going further it would be well to state that the treatment of alcohol separates the great mass of salts that are held in solution by the water, and which interfere with the detection of so minute a constituent as the lithium salt; by the alcohol we reduce the salts to small amounts of chlorides of magnesium, aluminum, calcium, sodium, pótassium, and lithium; by the lime the first two are got rid of, and by the carbonate of ammonia the lime is precipitated.

The solution, now containing the chlorides of sodium, potassium, lithium, and ammonium, is evaporated to dryness, and the residue heated to dull redness, by which the ammonia salt is expelled and a little organic matter destroyed; the residue is next dissolved in water, and a drop or two of the liquid tested for a sulphate; should this be present it must be got rid of by exact neutralization with chloride of barium (a slight excess of the chloride of barium will not interfere with the other steps in the analysis). In the examination of the water in question no trace of sulphate was found at this stage of the process; so it was again evaporated to dryness in a small capsule over a water-bath; there were now a few grains of residual matter. To this was added an ounce of a mixture of equal parts of pure ether and absolute alcohol, the capsule was covered with a small receiver and allowed to stand for eighteen hours, the liquid was then thrown on a small filter, and the filter washed with a little of the mixture of ether and alcohol. The alcoholic ether solution, evaporated to dryness, furnished the chloride of lithium, recognized by its well-known characteristics. Although this process requires considerable time and some careful manipulation, its results are both accurate and satisfactory. The evaporation of two hundred gallons of the water, and the examination of the concentrated mother-water, enabled me to detect rubidium and cæsium by the aid of the spectroscope.

The water of this artesian well has very valuable medical properties, and those readers who are curious to examine into these points will obtain all the required information by sending to Louisville for the medical report.

REMARKS ON THE ALKALIES

CONTAINED IN THE MINERAL LEUCITE AND ON THE COMPOSITION OF WARWICKITE.

In examining recently many of the silicates containing alkalies my attention has been called to leucite, and it is on that mineral especially that I would now remark, reserving for another time my observations on the other silicates.

The specimens of leucite examined came from four localities, Vesuvius, Andernach, Borghetta, and Frescati. They were about as good specimens as are obtained from those localities, although all of them were not equally pure. The alkalies found in each calculated as potash were

Vesuvius	21.85
Andernach	20.06
Borghetta	20.68
Frescati	20.38

The specimen from Andernach was analyzed for the silica, etc., and found to contain silica 54.75, alumina 23.08, and 1.55 of oxide of iron; this last seemed to be mechanically disseminated through the crystals.

I say above in relation to the alkalies "all calculated as potash," for the reason that there is a notable quantity of rubidium and cæsium present in all the specimens above mentioned. In fact, by the method adopted in testing for these alkalies, abundant indications are obtained of the presence of rubidium and cæsium (the last not so readily) even when operating on but half a gramme of the mineral. I am now engaged in working out a method of estimating quantitatively rubidium and cæsium in the presence of other alkalies; by this method, not yet perfected, the quantity of these alkalies in leucite is found to be about nine tenths of one per cent. of the entire mineral.

Of course it is not at all remarkable that the potash in the different specimens of leucite should be the same; but it is a matter of interest to know that from whatsoever locality it comes this minute quantity of rubidium and cæsium occurs with it. On some future occasion I hope to be able to bring together certain generalities in this connection of more or less interest to mineralogists.

I have also detected rubidium in half a gramme of margarodite and Warwick mica, and have failed to detect it in apophyllite, thomsonite, pectolite, elæolite, chesterlite, cancrinite, and other silicates.

WARWICKITE.

This interesting compound has been known for some time to American mineralogists, having been first described by Prof. C. U. Shepard under the name of *warwickite*, and considered as a hydrated silico-titanate of magnesia, iron, and alumina. It was afterward described by Mr. T. S. Hunt under the name of *enceladite*, and in his analyses (Amer. Jour. of Science and Arts, 2d series, xi, 352) considered a trititanate of magnesia.

In the re-examination of American minerals, in which Mr. Brush and myself were engaged, this mineral came up in turn for examination, and to our amazement it is found to contain a large amount of boracic acid, doubtless upward of twenty per cent. Approximative analyses are already made, but owing to the difficulty of obtaining it of sufficient quantity in a perfect state of purity, its final examination may be delayed for some time ; and it is for that reason thought advisable to publish the present note on the subject. It is essentially a borotitanate of magnesia and iron ; the metallic acid, however, has some anomalies about it not yet cleared up. This is the first borotitanate known, and as such highly interesting ; the smallest portion of it when acted on with sulphuric acid will give the strongest indication of the presence of boracic acid.

DETERMINATION OF ALKALIES IN MINERALS.*

1. In the examination for alkalies in the class of minerals alluded to in this article it is usual to devote a separate portion of the mineral to their special determination, without having reference to any of the other ingredients contained in the mineral. This method of proceeding naturally recommends itself, because a fusion with carbonate of soda is so greatly superior for the determination of all other ingredients that even the attempt to control the result of the soda fusion by making use of the one for the alkalies, to arrive at the other substances as well as the alkalies, will in many instances embarrass the analyst as to his results.

2. It is only in cases of absolute necessity that one portion of the mineral should be used to estimate all its constituents, and this condition of things will be alluded to in another part of this paper, as reference is now had to the quantitative determination of the alkalies, discarding whatever else the mineral may contain.

3. In the determination of the alkalies in silicates not soluble in acids three important points present themselves:

I. The means necessary to render the silicate soluble.

II. The separation of the other ingredients from the alkalies, more especially magnesia.

III. The removal of the sal ammoniac unavoidably accumulated in the process of analysis.

In all three of these the processes adopted will be found to differ essentially from those now in use, and they are made known only after much experience by the author, in which their advantages have been most fairly tested, comparatively

* This memoir embraces many important points connected with mineral analysis. The minute practical details for laboratory use are given in another article in this collection of papers, and one written after twenty years' experience with the method.

with methods already employed. In order that these processes may serve equally well in the hands of others, they will be given with some detail.

I. Method of rendering the Silicate soluble.

4. To render the silicate soluble various plans have been proposed, all of which have their objections. Among the agents used for the purpose are baryta and several of its compounds; viz., the nitrate, carbonate, and chloride.

5. The first of these is undoubtedly the best decomposing agent of the four, could we use a platinum crucible to heat the mixture of it and the mineral; as it is, a silver crucible is necessary, and this is not always capable of standing the requisite heat. According to Rose, "the silver crucible must be very strong, for if thin the action of a red heat might crack it, and a portion of the fused mass would ooze out through the crevices." It also may happen that a heat higher than the point of fusion of silver is necessary to a complete decomposition of the mineral.

6. All that is here said of caustic baryta is equally applicable to nitrate of baryta.

7. The chlorides of barium and calcium have been lately proposed by Prof. Henry Wurtz, but its decomposing properties are very feeble, as the chlorine in combination with the barium is not liberated at a white heat, and few silicates are able to produce the decomposition. It may succeed with some of the feldspars, but decomposes very imperfectly even the micas. So it is rather a risk to employ it with an unknown substance.

8. The carbonate of baryta is the compound of baryta most generally employed for silicate decompositions; still this is attended with much difficulty, owing to the infusibility of this salt and the impossibility of driving off the carbonic acid by heat alone; and even if this latter were possible, the objection pertaining to caustic baryta would then arise.

9. The following extract from Rose's Analytical Chemistry (translation by Normandy, in a note by the translator) presents fairly the difficulties attending this method of decomposing the silicates:

"The heat applied is so intense that some precautions must be taken. The platinum crucible containing the mixture should be exposed first to the heat of an argand lamp, and when the mass begins to agglutinate the crucible should be closed and its cover tied down with platinum wire, then placed in a Hessian crucible closed up also; the whole is placed upon an inverted crucible and submitted to the action of the blast of a wind-furnace, beginning first gradually with a red heat, piling on more coke, so as to fill up the furnace, and increasing the heat to the highest possible pitch, until the Hessian crucible begins to soften. It is absolutely necessary to the success of the operation that the Hessian crucible should be closed as well as possible, which is best done by luting the cover with fire-clay; the Hessian crucible and its cover, having fused together, can not be separated except by breaking, etc."

It will be seen in reading this extract that the heat required is not ordinarily at the command of most chemists; in fact, no other variety of furnace than a Sefstroem can be depended on for a complete decomposition.

10. Caustic lime and its salts have also been recommended and long used for the more imperfect decomposition of silicates, as for obtaining lithia from spodumene and lepidolite. Lime or its carbonate, well mixed with many silicates finely pulverized, will decompose them completely at a white heat, but no one salt of lime is capable of meeting the demand of the entire range of alkaline silicates.

11. In consideration of these difficulties Berzelius proposed the use of hydrofluoric acid, and this method, when applied with the numerous precautions required, will serve to decompose all silicates; still, according to Rose, there are siliceous compounds that can not be completely decomposed by hydrofluoric acid. Besides, this acid is a most disagreeable one to manipulate with, whether we employ Brunner's apparatus or Laurent's method, or, what is always the best, the *concentrated acid* previously prepared. I may also add that the necessity of using sulphuric acid after the decomposition is made is another objectionable feature in this process.

12. The above furnishes a hasty review of the methods we are now possessed of for decomposing the silicates in order to determine their alkalies; their merits can be contrasted with those of the method about to be described.

13. The decomposing agent which I present as a substitute for all others, and as capable of meeting the demands proposed in the commencement of this article, is *a mixture of carbonate of lime and fluor-spar.*

14. Carbonate of lime I have used for more than six years

for decomposing certain of the alkaline silicates, and more successfully than carbonate of baryta; still in numerous instances the decomposition was far from complete, and the method unsatisfactory. Notwithstanding these failures, I felt convinced that lime was the most powerful decomposing agent that could be conveniently employed for this purpose, as it could be used in its caustic state in a platinum crucible without injuring the latter, although exposed to the highest temperature. When its carbonate is used a red heat sufficed to drive off the carbonic acid and bring the mineral under the action of caustic lime—a circumstance that does not take place with carbonate of baryta; and it is well that it does not, for otherwise the platinum crucible would be seriously injured.

15. It was evident that the only obstacle in the way of lime decomposing the silicates as thoroughly as caustic potash was the impossibility of fusing the mixture, and thereby bringing the pulverized mineral and lime intimately in contact. This difficulty overcome, I felt confident of success. Without detailing the various methods resorted to, it will suffice to state that the object in view was to use some flux along with the mixture of the silicate and lime, which would render the mixture fluid at a bright red heat. The two substances which recommend themselves after many experiments are the *fluoride and chloride of calcium*, neither of which have any marked decomposing action on the silicates; in fact, their action is simply that of fluxes, which enable the lime and silicate to come in contact in a liquid state, effecting nothing beyond that. It is with the *fluoride of calcium* that we have to do in this part of the paper, leaving the details on the use of the chloride of calcium until further experiments are made to test fairly its value.

16. The manner in which I proceed is as follows: Pulverize the silicate to a sufficient degree of fineness; it is not required that the levigation be carried to any great extent; mix intimately in a glazed porcelain mortar a weighed portion of the mineral with one part of pure fluor-spar and four or five parts of precipitated carbonate of lime;* introduce it into a platinum

* The fluor-spar used is the transparent variety, free from all impurities. It is easily and abundantly procured in this as well as in all other countries. The carbonate of lime is made by dissolving calc-spar or pure marble in

crucible capable of holding three times the bulk of mixed powder. The platinum crucible should then be placed in one of earthenware, with a little magnesia on the bottom. (I always prefer the crucible made in France, called Beaufay's crucible, to inclose platinum crucibles when heated in a furnace, as their form and cleanliness make them superior to the Hessian crucible for this purpose.) The crucible may then be covered and introduced in any form of furnace where a bright red heat can be procured.

17. I have been using a common open portable furnace, heaping charcoal over the top of the crucible; and so easily does the effect take place that in no instance has there been a failure of complete decomposition with as simple a means of heating as the above; and I have ascertained that an alcoholic lamp with a large circular wick, such as Jackson's lamp, urged with a bellows, will answer for making a complete decomposition of zircon in twenty-five minutes. This circumstance is not stated to recommend the use of a lamp for every mineral decomposition when a simple portable furnace and charcoal are so accessible, and their effects so much more to be depended upon than a lamp. From thirty minutes to one hour's exposure to the heat is recommended.

18. It was an important point to test first how far this mixture could decompose the silicates without distinction as to their containing alkalies; for it was a very simple conclusion that if those silicates most difficult of decomposition, and containing no alkalies, were completely decomposed by this process, all others must naturally give way under its action. The silicates experimented on were *zircon*, *kyanite*, *beryl*, *topaz*, *spodumene*, *margarite*, *margarodite*, and *feldspars of different descriptions*. All were readily decomposed by the method just

---

hydrochloric acid (the common acid may be used), adding an excess of the carbonate; lime-water or milk of lime is then poured on the solution until it is alkaline. By this means any oxide of iron, alumina, or magnesia will be thrown down. To the hot filtered solution a solution of carbonate of ammonia is added, and the precipitate washed several times with distilled water. It is best to prepare one's own carbonate of lime, for as a general rule little reliance can be placed on the carbonates of lime, baryta, strontia, etc., sold as being precipitated by carbonate of ammonia, for in more than one instance I have found the carbonate of baryta, sold as a carbonate of ammonia precipitate, to contain soda.

described, and without any particular care in levigating them. One gramme of the zircon, for instance, after being crushed in the diamond mortar, was rubbed up for fifteen minutes in a large agate mortar, and used. Its complete decomposition was not only shown by its solution in hydrochloric acid, but by the amount of zirconia obtained, which was 64.8 per cent., with little iron. This concludes the first point to be considered in this article—namely, the means necessary to render the silicates soluble. The next point is the separation of the alkalies.

II. Separation of the other Ingredients from the Alkalies.

19. The platinum crucible, with its fused contents, is laid on its side in a capsule of platinum or porcelain. The latter can be used with perfect safety to the accuracy of the result. A quantity of dilute hydrochloric acid is poured into the capsule, one part of acid to two of water. The whole is heated over a lamp, when the contents of the crucible are rapidly dissolved out; the crucible is taken out and washed over the capsule; the contents of the capsule are then evaporated to dryness over a sand-bath; and, if thought necessary, it may be completed over the lamp without danger of the spitting which occurs in the soda fusion. This evaporation to dryness is not absolutely necessary; but the advantage of it is that any great excess of hydrochloric acid is got rid of, and the precipitate in the next operation is less bulky than it otherwise would be.

20. To the dry mass a little hydrochloric acid is added, and then three or four ounces of water, or more, as the occasion may require It is then boiled for a short time in the same capsule, allowed to cool down a little, and then a concentrated solution of carbonate of ammonia is slowly added until there is an excess of the same. The solution becomes at first quite thick with the precipitate, but in a short time (especially with a little warming over the lamp) the precipitate accumulates in a more or less granular state, and afterward occupies less space in the filter than the alumina it might contain (in a feldspar, for instance) were this latter precipitated separately by ammonia; and this circumstance is of much importance in diminishing the length of the operations and the amount of water accumulated by filtering it from several precipitates.

21. It will be seen that thus far the operations have been carried on in the capsule in which the fusion was dissolved. The contents of the capsule are now thrown on a filter; but before doing this it is well to pour on a little of the solution of the carbonate of ammonia, and see if the clear part of the liquid be rendered turbid; in other words, ascertain if sufficient carbonate of ammonia had been originally added.

22. The solution that passes through the filter contains much sal ammoniac, the alkalies of the mineral, and a little lime. If magnesia be one of the ingredients of the silicate examined, some of this is also present; and in still rarer instances some of the earths soluble in carbonate of ammonia. This latter complicates in no degree the remaining steps in the analysis. It is best to let the filtrate pass into a glass flask. The washings of the filter are collected in another vessel and concentrated to a small bulk, added to the first filtrate, and the whole boiled for some time to drive off the carbonate of ammonia.* When no great haste is required in the matter the whole filtrate (first portions as well as the washings) are collected in a beaker and concentrated over a sand-bath. What remains now to do is to separate from the alkalies the substances above alluded to. I commence by getting rid of the sal ammoniac, and this brings me to the third part of this paper.

III. The Removal of the Sal Ammoniac Unavoidably Accumulated in the Process of Analysis.

23. This is probably one of the greatest annoyances to the analyst in his examination of minerals: first, from the manner in which the salt creeps up the sides of the vessel in which the evaporation to dryness is carried on; and secondly, from the great difficulty of preventing loss of the chlorides of the fixed alkalies during the volatilization of the sal ammoniac. A better idea is formed of this by an experiment with a known quantity of the alkalies mixed with sal ammoniac. An array of the precautions requisite to be taken can be seen in Rose's last edition (German), pages 6 and 7. Owing to these difficulties, which my experience has often led me to contend with,

* What remains in the filter is silica, alumina, fluoride of calcium, oxide of iron, carbonate of lime, etc.

the method about to be mentioned was contrived. It recommends itself both on account of its simplicity and certainty of operation.

24. Having some time back noticed the decomposing effect produced by heating sal ammoniac with nitric acid, the nature of the decomposition was investigated to see how far it could be made use of to decompose entirely the sal ammoniac. The result of the investigation was that the sal ammoniac could be completely decomposed at a low temperature into gaseous products, and it was immediately adopted in my analytical process with the greatest satisfaction, both as to accuracy of results as well as economy of labor.*

**Formation of almost pure Protoxide of Nitrogen by the action of Nitric Acid on Sal Ammoniac.*—The experiments made with the nitric acid heated with sal ammoniac to test the character of the decomposition have resulted in the discovery of a new method for procuring protoxide of nitrogen with the aid of a very low temperature. Among the experiments the following were quantitative. Two grammes of sal ammoniac were placed in a glass flask, and half an ounce of nitric acid poured upon it; the flask was connected with a small wash-bottle containing a little water, and from this latter a tube passed into a pneumatic trough filled with hot water; heat was applied to the flask, and before the temperature reached 140° Fah. a gas began to be given off, and at 160° it came off rapidly, and continued to do so after the lamp was withdrawn. A small amount of red fumes appeared in the flask that were condensed in the wash-bottle. The gas that passed over was collected in a receiver, and measured one thousand and eight cubic centimetres. The gas smelt of chlorine. The flame of a candle burnt with an increased brilliancy when introduced in it. The candle was reignited when extinguished if a burning coal remained on the end of the wick. No red fumes were formed when it came in contact with the air, and the gas was absorbable by cold water. The properties were those of protoxide of nitrogen. In another experiment the gases were collected at different stages of the process, in vials over hot distilled water, and a solution of caustic potash introduced and shaken up for some time. This latter was subsequently analyzed for the chlorine it absorbed, and in three different portions, collected at the beginning, middle, and end of the process; the proportions of the chlorine to the whole bulk of the gas were one fifty-seventh, one twenty-ninth, and one sixteenth. The amount of protoxide of nitrogen due to the ammonia in two grammes of sal ammoniac and its equivalent of nitric acid is eight hundred and eighty-seven cubic centimetres. The gas freed from chlorine, on being shaken up with cold water for some time, was found to be almost entirely absorbed by the water. What remained was a mixture of nitrogen and a little air. Some nitrous or hyponitrous acid forms during

25. The manner of proceeding is as follows: to the filtrate and washings concentrated in the way mentioned (22), and still remaining in the flask, pure nitric acid is added—about three grammes of it to every gramme of sal ammoniac supposed to exist in the liquid. A little habit will suffice to guide one in adding the nitric acid, as even a large excess has no effect on the accuracy of the analysis.

26. The flask is now warmed very gently, and before it reaches the boiling-point of water a gaseous decomposition will take place with great rapidity. This is caused by the decomposition of the sal ammoniac in the manner described in the note. It is no advantage to push the decomposition with too great rapidity. A moderately warm place on the sand-bath is best adapted for this purpose. With proper precautions the heat can be continued and the contents of the flask evaporated to dryness in that vessel; but it is more judicious to pour the contents of the flask, after the liquid has been reduced to half an ounce, into a porcelain capsule (always

---

the whole process if concentrated nitric be used. If, however, it be diluted, little or none is formed, and the gas is readily given off at about 212° Fah.

In all my experiments the protoxide of nitrogen constituted from seven eighths to twenty-four twenty-fifths of the gaseous products, and when washed from its chlorine by a little lime-water or soda possessed all the properties of pure protoxide of nitrogen; and I would recommend it as a convenient way of forming this gas, especially when not required for respiration.

The character of the decomposition which takes place is somewhat curious and unexpected. At first I supposed that the decomposition resulted in the formation of equal volumes of NO, Cl, and N; but it appears that such is not the case, and that all but a very small portion of the ammonia, with its equivalent of nitric acid, is converted into NO, the liberated hydrochloric acid mixing with the excess of nitric acid. A little of the sal ammoniac and nitric acid does undergo the decomposition first supposed, and in this way only can the small amounts of chlorine and nitrogen be accounted for. At the time this method was first tried I also tried the decomposing effects of nitrate of ammonia on sal ammoniac, that has been shown by Mauméné (Comptes Rendus, October 15, 1851) to result in the formation of chlorine and nitrogen; but the difficulty of controlling the decomposition once commenced, the puffing-up of the mixture, and the necessity of having the salts dry to begin with, render this method (which was proposed by the author for forming chlorine) useless in processes for removing the sal ammoniac in analysis.

preferring the Berlin porcelain) of about three and a half to four inches diameter, inverting a clean funnel of smaller diameter over it, and evaporating to dryness on the sand-bath or over a lamp. I prefer the latter, as at the end of the operation the heat can be increased to four or five hundred degrees.

27. By this operation, which requires no superintendence, one hundred grammes of sal ammoniac might be separated as easily and safely as one gramme from five milligrammes of alkalies, and no loss of the latter be experienced. What remains in the capsule occupies a very small bulk. This is now dissolved in the capsule with a little water (the funnel must be washed with a little water), small quantities of a solution of carbonate of ammonia added, and the solution gently evaporated nearly to dryness. This is done to separate what little lime may have escaped the first action of the carbonate of ammonia, or may have passed through the filter (22) in solution in carbonic acid. If any of the earths soluble in carbonate of ammonia existed in the mineral, those now become separated along with the lime.

28. A little more water is now added to the contents of the capsule, and the whole thrown on a small filter; the filtrate as well as washings are received in a small porcelain capsule. The liquid contains only the alkalies (as chlorides and nitrates), mixed with a minute quantity of sal ammoniac. This is evaporated to dryness over a water-bath, and then heated cautiously over the lamp to drive off what sal ammoniac may have formed (27), which is exceedingly minute if the process as pointed out be closely adhered to. It is not absolutely necessary to heat the capsule over the lamp to get rid of the sal ammoniac, for the little sulphate of ammonia which may be formed in the next step is easily removed in the final heating in a platinum vessel.

29. On the contents of the capsule, as taken either from the water-bath or as after being heated over the lamp, pure dilute sulphuric acid is poured (1 part acid, 2 water), and the contents boiled for a little time, when all the nitric acid and chlorine in combination with the alkalies will be expelled; the acid solution of the alkalies is now poured into a platinum capsule or crucible, evaporated to dryness, and ignited. In order to insure complete reduction of the bisulphates into the neutral

sulphates the usual method must be adopted of throwing some pulverized carbonate of ammonia into the platinum capsule or crucible, and covering it up so as to have an ammoniacal atmosphere around the salt, which will insure the volatilization of the last traces of free sulphuric acid. The alkalies are now in the state of pure sulphates, and may be weighed as such. The manner of separating the alkalies from each other will be mentioned further on.

30. Thus far the mineral has been supposed to contain no magnesia. If this alkaline earth be present, we take the residue as found in the capsule (26), dissolve it in a little water, then add sufficient pure lime-water* to render the solution alkaline; boil and filter; the magnesia will in this simple way be separated from the alkalies. The solution which has passed through the filter is treated with carbonate of ammonia in the manner alluded to (27), and the process continued and completed as described (28, 29).

Summary.—Fuse one part of mineral with one of fluoride of calcium and four to five of carbonate of lime; dissolve out the contents of the crucible with hydrochloric acid; evaporate to dryness and redissolve; precipitate with carbonate of ammonia; filter, boil, and concentrate the filtrate; add nitric acid; heat and evaporate to dryness; dissolve the dry mass in a little water and treat with carbonate of ammonia; filter and concentrate; then add sulphuric acid; boil for a little while; pour in a platinum crucible, evaporate to dryness, and ignite. If magnesia be present, treat with lime-water prior to the last application of carbonate of ammonia.

Conversion of the Sulphates into Chlorides.

31. In continuation of the subject, the next point to be considered is the conversion of the sulphates of the alkalies into chlorides. The method ordinarily adopted to accomplish this change is to precipitate the sulphuric acid by means of chloride of barium, care being taken to avoid the slightest excess of the

* If lime-water be made, it is well to make it of lime of the best quality, and the first two or three portions of distilled water shaken up should be thrown away as containing the small amount of alkalies sometimes present in lime.

latter. The annoyance attendant upon this exact precipitation is familiar to all who may have had occasion to make the trial.

32. Instead of the chloride of barium the acetate of lead is used; a solution of this salt is poured in excess upon the solution of sulphates; warming the latter slightly, the sulphate of lead readily separates; the whole can be immediately thrown on a filter and washed. A drop or two of the acetate of lead should be added to the filtrate to insure there being an excess of the lead-salt.

33. The filtrate is then warmed and sulpureted hydrogen added; care must be taken to see that there is an excess of sulphureted hydrogen, a test most readily performed by means of a piece of lead-paper. The liquid is thrown on a filter to separate the sulphuret of lead; the filtrate containing the alkalies as acetates is evaporated, and when nearly dry an excess of hydrochloric acid is added, and the whole evaporated to dryness over a water-bath, and finally heated to above 500°. A hot solution of the chloride of lead can be used instead of the acetate, rendering the addition of hydrochloric acid unnecessary.

34. It needs but little experience to convince one of the superiority of this method over that by the chloride of barium for converting the sulphates into the chlorides, its principal recommendation being the indifference with which an excess of the lead-salt can be added to precipitate the sulphuric acid, and the subsequent facility with which that excess of lead can be got rid of. It may be well to state that experiments were made to prove the perfect precipitation of the sulphuric acid from the sulphates of the alkalies by the salts of lead, and it is only after numerous comparative results that it is now recommended.

To distinguish the Alkalies from each other when Mixed.

35. To distinguish potash, soda, and lithia when mixed is attended with more or less difficulty, according to the proportions in which they are mixed; of the three, potash is the most easily recognized, next in order is soda, and lastly lithia; the presence of which, mixed in small amount with proportionally large quantities of the other alkalies, it is almost impossible to decide on with any accuracy without direct separation.

36. In the analysis of many minerals, the characters of which lead to the supposition of the presence of the alkalies, it is useless to precede the quantitative determination by one of a qualitative character, especially as the steps to be followed to separate the alkalies are the same in both cases, and to proceed in this way is economy of time. From my own experience concerning the constitution of the silicates there are doubtless but a very few of them without an appreciable quantity of alkalies in their constitution; and an easy method to examine with certainty the composition of such small quantities of alkaline chlorides must add to our analytical knowledge. With a little experience the method about to be described will be found very available.

37. To ascertain the nature of the alkalies present in the chlorides before proceeding to separate them we abstract a quantity so small as not sensibly to affect the weight of the mass. The smallest piece of the dried mass, that need not exceed one fiftieth of an inch in diameter, is placed on a slip of glass, and to it is added a drop of a watery solution of *very pure* chloride of platinum,* not too concentrated, and the plate gently warmed; if potash be present, a yellow deposit soon takes place, which by the microscope will be seen to consist of octahedral crystals of the double chloride of potash and platinum; the evaporation should be continued very gently with a heat not exceeding 120° to 130° until the liquid begins to dry on the edge; if this be now examined under the microscope and soda be present, beautiful needle-shaped crystals will be seen, both formed and forming, with an oblique angle of termination, the crystals frequently having re-entering angles, as represented in the figure. The border of liquid on the glass is the place to observe these crystals, and that while the pro-

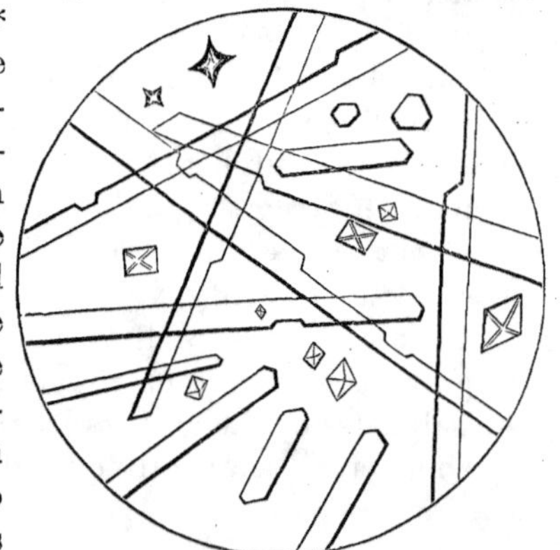

* The alcoholic solution does not give perfect results and should not be used.

cess of drying is going on. When the amount of soda is very small it is best to allow the solution on the glass to dry in the slowest possible manner. Should the quantity of soda be still smaller or the nature of the crystals doubtful, resort may be had to polarized light, when the prismatic crystals of chloride of platinum and sodium will be at once rendered visible by their beautiful colors, as they possess polarizing properties, whereas the crystals of chloride of platinum and potassium, besides differing in form, do not polarize light.

38. This method of detecting a small quantity of soda in the presence of potash I have employed since June, 1850, while engaged in the examination of the collection of urinary calculi belonging to the Dupeytren Museum at Paris; at that time it was employed daily in the laboratory of Messrs. Wurtz and Verdeil; the special reason for devising it was to examine the nature of the trace of alkali almost invariably found in the uric acid calculi after combustion.

39. The reason for making special reference to the date of the original employment of this method is to claim priority in its use, as Mr. Andrews announces it in a late number of the Chemical Gazette as a new method. Were not the method so well known and so constantly employed in the laboratory of Wurtz and Verdeil at the period above mentioned, I should not now set up any reclamation in the matter.

40. The amount of soda that can thus be detected is exceedingly small, as the liquid can be concentrated to the very smallest bulk. When the amount of potash is proportionally large compared with that of the soda it is better to put the chloride of platinum in a drop of the solution of the alkaline chlorides placed in a watch-glass, allow the potash-salt to settle, take a little of the clear liquid, place it on a slip of glass, evaporate slowly, and examine in the way already mentioned. We should avoid the use of alcohol as a solvent for the salts employed.

41. For the full appreciation of this method it requires some experience, and on first trials the extreme results will not be readily obtained; too much care can not be taken with reference to the evaporation; sometimes, if the evaporation be a little too speedy, no indication of the presence of soda will be evinced; so in all doubtful instances the glass should be laid aside

for an hour or two, when the excess of the chloride of platinum will attract moisture from the air and afford an opportunity for the chloride of platinum and soda to crystallize regularly. For the most perfect success in very minute quantities of soda too great an excess of chloride of platinum should be avoided. Those engaged in mineral analysis, who will employ this means of detecting the presence of the alkali, will find it of great assistance in facilitating their labors, especially when directed to very minute accuracy in their results, for I have reason to believe it rare to find in minerals any one of the alkalies perfectly free from one of the others.

42. When the chloride of lithium is present it interferes materially with this method of detecting small amounts of soda; for, owing to its very deliquescent nature, it abstracts moisture from the air, and dissolves the double chloride of sodium and platinum, or prevents altogether its formation into recognizable crystals. These investigations have not added any thing to what is already known concerning the detection of lithia mixed with soda and potash; the plan invariably adopted is to treat the mixed chlorides with a solution of alcohol and ether, and examine the part dissolved by the blowpipe. Details as to the manner of using the alcohol and ether solution are given under the next head.

Separation of the Alkalies from each other.

43. Under this head I have nothing to add to what is already known on the subject. It may be well, however, to mention the manner in which Rammelsberg's method of separating lithia has been employed, as it has not yet been fairly tested in this country. His method, it is well known, is based on the solubility of the chloride of lithium in a mixture of equal parts of absolute alcohol and ether, neither of the other chlorides being dissolved by this menstruum. A number of experiments were made on known quantities of the alkalies, and the results of some of them are as follows:

(*a*) Five hundred milligrammes of chloride of potassium treated with the mixture of ether and alcohol, ten grammes of the latter being used, yielded only three tenths of a milligramme to the liquid.

(*b*) Five hundred milligrammes of chloride of sodium treated in the same way yielded one half of a milligramme to the liquid.

(*c*) A mixture of chlorides of potassium, sodium, and lithium, in which the latter constituted two per cent. of the mass, was acted on by the ether and alcohol, filtered and evaporated to dryness; the residue was equal to 2.53 per cent. The quantity used was Cl K and Cl Na each 200 milligrammes, Cl Li 0.8 milligramme.

(*d*) A similar mixture containing 18.10 per cent. of chloride of lithium furnished a residue of 17.65 per cent.

(*e*) A similar mixture containing 67.20 per cent. of chloride of lithium gave a residue of 68.40.

44. By these results it may be seen that this method of separating lithia from the other alkalies may be perfectly relied on. It only remains to detail the precautions to be taken in order to insure accurate results.

45. The solution of alcohol and ether must be made of absolute alcohol mixed with its volume of pure ether. The chlorides must be dried thoroughly at 212° or a little above; if they have at any time been heated much higher, a drop or two of hydrochloric acid must be added to the chlorides, that are subsequently dried at the temperature just mentioned. The desiccation is best carried on in a small-sized capsule. To the dry mass a small quantity of the mixture of alcohol and ether is added and stirred with a small glass rod; the chlorides soon disintegrate; the capsule and its contents are placed on a glass plate and covered with a small bell-glass (a common tumbler answers the purpose very well, especially if the edge be ground); this is left to digest for twenty-four hours, and then thrown on a filter and washed with the *alcohol-ether* solution; the chlorides of sodium and potassium remain on the filter. These last can be dissolved off the filter by means of water, and separated in the ordinary way.

46. The alcohol-ether solution of chloride of lithium is evaporated to dryness, converted into sulphate, and weighed. The results thus obtained far exceed in accuracy those of any other method for separating lithia. The indirect method, by ascertaining the quantity of sulphuric acid contained in the mixed sulphates, is the next best, but like all indirect methods of analysis should never be employed except when it is absolutely necessary.

47. When the alkalies are presented in the form of chlorides before their quantity has been estimated in some other form it is best to proceed first to the separation of lithia, afterward weigh the chlorides not dissolved by the *alcohol-ether*, and lastly separate the potash-chloride from the soda-chloride, if both be present, by means of the bichloride of platinum. Experiments were made with a mixture of alcohol and chloroform, the results of which were not as satisfactory as those afforded by the alcohol-ether.

Substitution of Chloride of Ammonium for Fluoride of Calcium, to mix with Carbonate of Lime to decompose the Silicates.

48. It was mentioned in the previous paper on this subject how carbonate of lime could be rendered as powerful in its decomposing agency on the silicates as caustic potash, the effect being due to the use of some flux, fluoride and chloride of calcium being used for that purpose. I have since tested more carefully the merits of the chloride of calcium, and for various reasons prefer it to the exclusion of the fluoride. In the first place, it introduces chlorine instead of fluorine into the analysis; and secondly, the fusion is more easily detached from the crucible and dissolved by hydrochloric acid.

49. The manner of introducing the chloride of calcium into the mixture of mineral and carbonate of lime was a point of some little importance, as from the deliquescent nature of that compound it was inconvenient to weigh and mix it with the carbonate of lime and mineral. These difficulties are obviated by employing *chloride of ammonium* to form indirectly the chloride of calcium.

50. The process, which appears to leave hardly any thing to desire, is to take *one part of the finely-pulverized mineral, five to six of carbonate of lime, and one half to three fourths of chloride of ammonium*,* mix them intimately in a glazed mortar, intro-

*The chloride of ammonium is best obtained in a pulverulent condition by dissolving some of the salt in hot water and evaporating rapidly; the greater portion of the chloride of ammonium will deposit itself in a pulverulent condition; the water is poured off, and the salt thrown on bibulous paper allowed to dry; the final desiccation being carried on in a water-bath, or in any other way with a corresponding temperature.

duce the mixture in a platinum crucible, heat to bright redness in a furnace* from thirty to forty minutes.

51. There is no silicate which after having undergone this process is not easily dissolved by hydrochloric acid. For the action of the lime to have been complete it is not necessary that the mass should have settled down in a perfect fusion. The contents of the crucible are dissolved, and the analysis continued as pointed out in 19, 20, 21, etc.

52. This method insures the obtaining of every particle of the alkalies in the mineral examined, requiring no more precaution than any good analyst is expected to take in the simplest of his processes, and not the least of the advantages is the ready method of separating all the other ingredients and the small accumulation of water arising from the little washing necessary.

A more speedy Method of separating the Alkalies directly from the Lime Fusion for both Qualitative and Quantitative Determination.

53. As soon as the fusion with carbonate of lime and sal ammoniac gave evidence of the mineral being so thoroughly attacked, the question naturally arose as to the condition the alkalies were in after the fusion, and the possibility of dissolving them out by the agency of water alone, at least for the purpose of qualitative determination. Experiments directed to this object soon made it evident that the alkalies might be obtained from any silicate without resorting to the use of acid as a solvent for the fusion.

54. The mass as it comes from the crucible is placed in a capsule with water, and then heated in a sand-bath or over a lamp for two or three hours, renewing the water from time to time as it evaporates. The mass disintegrates very shortly after being placed in the water. The contents of the capsule are next thrown on a filter, and the water passes through containing the chloride of the alkalies, a little chloride of calcium and caustic lime. All else that the mineral may have contained

* An ordinary portable furnace with a conical sheet-iron cap from two to three feet high answers the purpose perfectly well, all the requisite heat being afforded by it.

remains on the filter, except baryta and strontia, if they be present in the mineral; but as these oxides are of rare occurrence in silicates no allusion will be made to them here.

55. To the filtrate add carbonate of ammonia, and boil for some time, when all the lime will be precipitated as carbonate; add a few drops of a solution of carbonate of ammonia to the hot solution, to be sure that all the lime is precipitated. Should this be the case, filter; the filtrate will contain the chlorides of the alkalies and chloride of ammonium. It is evaporated to dryness over a water-bath in a small platinum capsule; the capsule is carefully heated to expel the sal ammoniac, and finally warmed up to 700 or 800° F. It is then weighted with its contents, and the chlorides, if mixed, separated in the way mentioned (45 and 46). The amount of sal ammoniac to be expelled is quite small, not equaling the weight of mineral originally employed.

56. Nothing in analysis can be simpler or more speedy than this process. Its constant accuracy still lacked some little to render it perfect, as usually an amount of alkali remains behind, represented by from two tenths to one per cent. of the mineral used; certainly a small amount, but still too much to be omitted in an accurate analysis. This also must be arrived at, and it can be accomplished in the following manner.

57. After the fused mass has been treated with water filtered and washed as in 54, the filter and its contents are dried; the latter are detached from the filter and rubbed up in a glazed mortar with an amount of sal ammoniac equal to one half the weight of the mineral, and reheated in a platinum crucible exactly as in the first instance, treated with water, thrown on a filter and washed, the filtrate added to that from the first fusion, the whole treated with carbonate of ammonia, and completed as in 55.

58. This second fusion complicates the method but little, as the residue on the filter readily dries in a water-bath into a powder that is easily detached from the filter, and the small portion adhering to the latter may be disregarded, as the alkalies remaining rarely exceed more than one five-hundredth of the whole mass, and in most instances not more than one thousandth. In many analyses made one fusion sufficed for the entire extraction of the alkalies; but as a few tenths would

occasionally remain behind, we preferred the additional fusion to get at that small quantity, and to entitle it to rank as a method by which all but the merest trace of the alkalies could be extracted from the insoluble silicates.

59. The proportion of sal ammoniac added to the carbonate of lime as here recommended is arrived at after numerous experiments. By increasing the sal ammoniac, and thereby augmenting the amount of chloride of calcium formed, the mass fuses more thoroughly, but the water does not disintegrate it as completely as when the ammoniacal salt is less; also the accumulation of this latter at the end of the process is less, an object not to be disregarded. The advantage of thus estimating the alkalies in insoluble silicates is obvious. The long routine of separating silica, alumina, lime, etc., is done away with; the accumulation of chloride of ammonium is very trifling; and lastly, the alkalies are obtained directly in the form of chlorides. The method will vie in accuracy with any other, including the one already mentioned in the first part of this paper; and at the same time it is unequaled in simplicity, speed of execution, and constancy of results.

60. In examining for alkalies qualitatively, one fusion will of course be all that is necessary, and the action of the water need not be continued more than thirty minutes before filtering. This method answers even when boracic acid is present in the silicate. The manner of proceeding in such a case will be mentioned in a future paper on the determination of boracic acid in minerals.

61. There is nothing new in the attempt to dissolve out the alkalies by water from a silicate that had been heated with lime. M. Fuchs used the method for procuring lithia from lepidolite; but of course his efforts were entirely directed to procuring the lithia from the mineral, and not to estimating its quantity, as the method he followed could not have furnished such results. I have also lately learned that Mr. A. A. Hayes, of Boston, proposed and employed a mixture of chloride of calcium and caustic lime to decompose alkaline silicates, heating over a lamp, and subsequently treating the mass with water to extract the alkalies or chlorides. As little has been heard of this process, I presume the author found it defective. If we are correctly informed, the proportions used were three

parts of chloride of calcium and one of caustic lime. Experience proves that, however readily such a mixture may reduce the feldspars, it fails when tried on kyanite, zircon, micas, and other silicates difficult of decomposition. This arises from the fact that the chloride of calcium has but little decomposing effect on the silicates, its action being simply that of a menstruum in which the lime can act conveniently on the mineral.

62. The use of lime or its carbonate mixed with chloride of calcium or chloride of ammonium, for the purpose of effecting the decompositions alluded to, would be considered by me of questionable utility if the mixture were not so proportioned and employed as to decompose the most difficult silicates, if necessary; for unless this be done we can at no time be certain that the decomposition is complete. As a general rule for decomposing silicates by lime or soda, it is far better to use a charcoal fire than the flame of a lamp, as it is better to heat too high than not to heat sufficiently the mineral to be acted on.

Complete Analysis of an insoluble Silicate on one Portion of the Mineral.

63. The effort to accomplish an analysis of this description deserves no encouragement, from the almost invariable inaccuracy attending the results. If we have a given quantity of any one of the silicates alluded to requiring analysis, we had better subdivide it, however small the entire quantity may be, ascertain one set of ingredients by the soda and the other by the lime fusion; for the results thus obtained may be relied on as more accurate than those furnished by an analysis of the whole quantity through the agency of baryta or hydrofluoric acid.

64. Should it be desired to undertake the analysis on a single portion, I would recommend the silicate to be attacked with *carbonate of baryta mixed with the chloride*—three to four parts of carbonate and two of chloride. This mixture can be made to decompose all silicates at a much lower temperature than when the carbonate alone is used, but its action is not near so powerful as the carbonate of lime and sal ammoniac.

65. This terminates an account of my labors in the determination of the alkalies in insoluble silicates. The conclusions

have been arrived at from more than two years' experience and over a hundred alkali determinations by myself and others made on minerals of the most varied composition. In my laboratory an accurate alkali determination is one of the most simple and speedy analytical processes now conducted, and the presence of magnesia in no degree complicates the result. A little experience will no doubt bring others to the same conclusions. Many singular decompositions of salts have been noticed in the course of these researches; but as they do not bear directly on the object of this article they will be made known on another occasion.

Many analytical processes mentioned in this article can be applied when operating on soluble silicates.

CONVERSION OF THE SULPHATES OF THE ALKALIES

INTO THE CARBONATES, TARTRATES, &c., IN THE MOIST WAY.

Having had occasion more than once to convert small quantities of the sulphates of the alkalies into carbonates, I have for several years employed a process that has been found both certain and convenient; in some recent investigations it has been used, and as it has never been described it may not be unimportant to explain the nature of the process and its results. The agent used to produce the conversion is carbonate of baryta, made by precipitation; where precise results are required the carbonate should be prepared by carbonate of ammonia. The manner of producing the decomposition is as follows: Dissolve the sulphate of potash in water, using about twenty or thirty grammes of water to every gramme of the sulphate, and saturate the solution with carbonic acid by passing a current of carbonic acid into it; or, what is better, dissolve in the beginning the sulphate in water already saturated with carbonic acid; now add to this solution precipitated carbonate of baryta, in the proportion of about one and a half of the carbonate to one part of the sulphate. It is always best in adding the carbonate to rub it up in a mortar with a little water, so as to form a thick cream, for by so doing it mixes well in the solution.

This operation is performed in a bottle that can be well corked with a cork or gum stopper; now agitate the bottle frequently, or, what is still better, attach it to a piece of machinery that will agitate the bottle. Many laboratories have such, and it is a very useful one in many experiments. In a longer or shorter space of time the decomposition will be completed; pour the solution into a capsule and heat to the boiling-point; the solution will then contain only carbonate of potash.

The reaction is readily understood; the carbonic acid in the water dissolves a little carbonate of baryta, which is immediately precipitated in the form of sulphate, carrying down a portion of the sulphuric acid of the soluble sulphate, and replacing the same with carbonic acid; this is rapidly repeated through the agency of the free carbonic acid, until the decomposition of the sulphate is complete.

Among many experimental results I will give the following: Five grammes of the sulphate of potash dissolved in carbonic-acid water, to which was added seven grammes of precipitated carbonate of baryta, after four and a half hours' shaking (being attached to a suitable piece of machinery), on testing showed not a trace of sulphuric acid, care being taken to wipe the neck of the bottle near the end of the stopper before pouring out the liquid.

Other experiments, varying in proportion, gave similar results. I tried to substitute the natural for the precipitated carbonate of baryta, but with very unsatisfactory results.

DIRECTIONS FOR CONVERSION OF THE ALKALINE SULPHATES INTO TARTRATES, OXALATES, ETC.

As the tartrate and oxalates of baryta are but very slightly soluble in water, we can not form the alkaline salts of these acids by direct double decomposition of the sulphates of the alkalies and the tartrate, etc., of baryta, as in forming the alkaline chlorides from the sulphates; but it is easily done by the following indirect process.

Add to the alkaline sulphates in solution, in a porcelain capsule, carbonate of baryta rubbed up into a thick cream in the proportion of about five of the sulphate to seven of the carbonate of baryta; heat the mass and add little by little the requisite quantity of tartaric or oxalic acid; solution of the baryta and precipitation of the sulphuric acid take place rapidly, and the decomposition is soon completed.

I have used this process in forming the bitartrates in the process of separating potassium, rubidium, and cæsium, that were in the form of sulphates.

The carbonates of the alkalies can also be formed by first forming these organic salts from the sulphates, evaporating the solution to dryness and burning the residue; in fact, I frequently find it more convenient to convert the sulphates of the alkalies into their carbonates by this last instead of the first process. And finally I would remark that where magnesia is present with the sulphates this is also separated from the alkalies.

ACTION OF SOME OF THE ALKALINE SALTS UPON THE SULPHATE OF LEAD.

It has been for some time known that certain neutral salts possess the property of dissolving to some extent the sulphate of lead, which properly belongs neither to the acids nor bases constituting these salts. By referring to Berzelius' Chemistry it will be found that the acetate and nitrate of ammonia are among the number. "One part of the sulphate was dissolved in 47 parts of a solution of the acetate of specific gravity 1.036, and in 172 parts of a solution of the nitrate of specific gravity 1.144." In the *Annalen der Chem. und Phar.* (vol. xxxiv, 235) will be found the following statement under the head of *Reactionen:* "Sulphate of lead is easily dissolved, and in a large quantity, by a solution of neutral tartrate of ammonia; a concentrated solution forms after some time a stiff jelly like silica." This last is no doubt a double tartrate of lead and ammonia.

I had also observed some time previously that a solution of the citrate of ammonia, when poured upon the sulphate of lead and allowed to stand, altered the character of the sulphate, and this, with the other fact above stated, led to the examination of what was really the action of these as well as other ammoniacal salts in general upon the sulphate in question, and it was found that in every case it was decomposed.

Citrate of ammonia.—If a solution of citrate of ammonia be poured upon the sulphate of lead and shaken together, the clear solution will be found to contain the sulphate of lead, as shown by hydrosulphuric acid, and a salt of baryta (taking care in testing with the baryta to acidulate first with pure nitric acid, to prevent the formation of the citrate of baryta.) If they be allowed to remain several weeks in contact, the solution will be found to contain more lead, the sulphate having undergone decomposition, sulphate of ammonia and a double citrate being the result; as this latter salt is not very soluble, a large portion of it remains in the form of a precipitate. The rapidity of this change is in proportion to the concentration of the solution of the citrate. If instead of performing the experiment in the cold we boil a tolerably concentrated solution of the citrate with the sulphate of lead, a very large quantity of the latter will be dissolved, and the solution become perfectly transparent; if it be set aside and allowed to cool, in the course of a few hours an abundant white precipitate will be formed, and upon testing the clear solution sulphuric acid, ammonia, citric acid, and oxide of lead will be found present. The precipitate when washed affords citrate of lead and ammonia. I was at first inclined to think it simply a citrate of lead, attributing the ammonia present to some of the citrate not washed out; but from its possessing certain characters which do not belong to the simple citrate I consider it a double citrate of lead and ammonia. It contains not the slightest trace of sulphuric acid. It was not analyzed, from the difficulty of obtaining it perfectly pure, as the water used to wash it decomposes it, and as yet this difficulty has not been surmounted. So then the result of the action of the citrate of ammonia upon the sulphate of lead is first to dissolve it, and subsequently to decompose it, forming the sulphate of ammonia and citrate of lead and ammonia.

Tartrate of ammonia.—If a solution of this salt be added to the sulphate of lead and shaken with it in the cold, the clear solution will be found to contain both lead and sulphuric acid, and if set aside for a few weeks the precipitate will have changed its character, having assumed a crystalline nature; the solution will no longer contain lead, but the quantity of sulphuric acid present will be found to have increased. The precipitate now consists of tartrate instead of sulphate of lead, which is completely soluble in dilute nitric acid, affording no precipitate with a salt of baryta. If the mixture of the tartrate and sulphate be boiled, this change takes place more rapidly and in a manner somewhat different from the case of the citrate; the sulphate will not be dissolved in such large quantities, and moreover, by continuing to boil the solution after the sulphate has been completely dissolved, the tartrate forms during the ebullition and is precipitated in little shining crystals. If the ebullition be continued a sufficient length of time, the whole of the lead previously dissolved will combine with the tartaric acid. This is different from what takes place with the citrate, which when boiled upon the lead-salt dissolves it, and no length of ebullition will produce a precipitate. The action of the tartrate is first to dissolve the sulphate, decompose it in part. and form a double tartrate of lead and ammonia, which last salt is subsequently decomposed by continued contact with water, or still more rapidly by its solution being boiled.

Acetate of ammonia.—This salt also dissolves to some extent the sulphate of lead, but not so readily as either of the above salts. If the solution be boiled and evaporated to dryness, crystals of sulphate of ammonia are obtained, and an uncrystallizable salt of lead, probably an acetate of lead and ammonia; from the difficulty of separating the sulphate of ammonia from it, it is impossible to pronounce positively whether it is a double salt or simply an acetate of lead. We see in this reaction the existence of a soluble salt of lead and the sulphate of ammonia simultaneously in the same solution, without a precipitate being formed.

Oxalate of ammonia dissolves but slightly the sulphate of lead, owing no doubt to the impossibility of forming a double salt; but it will nevertheless decompose largely, the sulphate furnishing the oxalate of lead.

Muriate of ammonia, if boiled with the sulphate of lead, will decompose it instantly, furnishing the chloride of lead and sulphate of ammonia.

The *nitrate of ammonia* does the same, forming nitrate of lead and sulphate of ammonia.

The *carbonate* and *succinate of ammonia* produce similar effects.

The action of most of the corresponding salts of potash and soda was examined, and with very similar results. The fact is it would appear that those alkaline salts which dissolve the sulphate of lead decompose it without reference to the time occupied in the solution, as in the case of the carbonate of ammonia, which decomposes the sulphate at the very instant of its solution; and it is impossible to detect at any one time other than a trace of lead in solution, whereas the quantity of sulphuric acid is constantly increasing.

The explanation is clear: the sulphate of lead is a salt with a strong acid and feeble base; the alkaline salts used contain feebler acids and stronger bases; they dissolve the sulphate, thus affording an opportunity for the acids and bases to act upon one another under favorable circumstances, and to follow a natural law in chemistry: the stronger acid combined with the stronger bases, and *vice versa.*

From the foregoing facts some important hints might be afforded to analytical chemistry, for it will be at once seen that the presence of any of the alkaline salts in a solution from which it might be wished to precipitate lead in the form of a sulphate would affect the accuracy of the result. What is true of the sulphate of lead may be found also true for other insoluble salts. Moreover, this shows the importance—in the analysis of mineral waters, for instance—of weighing well the relative strength of the various acids and bases therein found, in order to ascertain what salts are present, and not to be contented with evaporating the water to dryness, and considering such salts as remain to be those existing in the water, for many of them may be formed during the evaporation. It is not at all improbable that before many years the examination of mineral waters will be based as much upon calculation as upon analysis, the former of course being guided by the latter and by certain laws not yet developed.

COMPOSITION AND PRODUCTS OF DISTILLATION OF SPERMACETI,

WITH SOME FEW REMARKS UPON ITS OXIDATION BY NITRIC ACID.

Of all the fatty bodies that have been examined there is perhaps not one whose composition has been so imperfectly arrived at as that of spermaceti, and it is a little remarkable that Chevreul, with the accuracy which distinguishes his researches upon the fats, should not have ascertained more nearly its true composition.

Chevreul made the examination of spermaceti in the same way as he did that of other fatty bodies, by digesting it with a solution of an alkali, and examining the products that combined with the alkali and those that did not. In the case of *spermaceti* he obtained a solid substance that did not combine with either *soda* or *potash*, resembling strongly in its external characters the fats, and upon analysis with bioxide of copper gave

Carbon	79.76
Hydrogen	13.95
Oxygen	6.29
	100.00

From this he calculated its formula ($C_{32} H_{33} O+HO$), and this substance he called, from some peculiarities in its composition when compared with that of alcohol and of ether, *æthal*, or rather hydrated æthal.

That part of the spermaceti which combined with the potash Chevreul considered to be composed of margaric and oleic acids, without apparently any strong grounds for so doing. He gives us no analyses of these acids, and the following is all that is said concerning them:

"The margaric acid of spermaceti is fusible at from 131° to 132° Fah., crystallizes in little radiating needles, is insipid and inodorous; at 140° Fah. it dissolves in all proportions in alcohol of 0.820, the solution reddening strongly litmus.

"I treated the margarate of potash with alcohol to see if I could obtain margaric acid fusible at 140° Fah. I submitted a portion of the same salt to five successive treatments, and obtained, first, a portion of the salt whose acid was fusible at 113° Fah.; and secondly, a portion whose acid was fusible at 132° Fah. This last portion, treated again six times with alcohol, gave, first, an acid fusible at 131° Fah.; and secondly, an acid fusible at 122° Fah. From this I conclude that it is margaric and not stearic acid which manifests itself in the saponification of spermaceti."

Chevreul gives the following analyses of the salts of this acid, to which I have affixed the atomic weight of the acid that each indicates, and which differ considerably one from the other.

Salt.			Atomic weight of acid.
Potash-salt	Acid	100.0	265.12
	Base	9.8	
Baryta-salt	Acid	100.0	275.36
	Base	27.8	
Strontia-salt	Acid	100.00	255.52
	Base	20.26	
Lead-salt bibasic	Acid	100.00	262.40
	Base	85.00	

What was considered to be oleic acid could not be entirely separated from the supposed margaric acid, and consequently it was impossible for Chevreul to study accurately its properties.

The elementary analysis of spermaceti, according to Chevreul, is

Carbon	81.66
Hydrogen	12.86
Oxygen	5.48
	100.00

Dumas and Peligot, considering the base of spermaceti to be athal, and its acids margaric and oleic, have proposed the following formula.

Spermaceti	2 ats. margarate of athal.	2 ats. margaric acid..$(C_{34}\ H_{32}\ O_3)2$	$C_{208}\ H_{205}\ O_{14}$*
		2 atoms athal..........$(C_{32}\ H_{33}\ O)2$	
	2 atoms oleate of athal.	1 atom oleic acid......$(C_{44}\ H_{39}\ O_4)$	
		1 atom athal...........$(C_{32}\ H_{33}\ O)$	
		1 atom water..........$(H\ O)$	

* The formula for margaric and oleic acids are those that have lately been given by Varrentrapp.

The athal here mentioned as the base is now considered as composed of a substance called *cetyl*, or *ceten*, and water. This cetyl, or ceten, is obtained from athal by the action of anhydrous phosphoric acid; it is a substance of an oily nature, and consists of equal equivalents of carbon and hydrogen.

1 atom ceten	$C_{32}\ H_{32}$
1 atom water	$H\ O$
1 atom anhydrous athal	$C_{32}\ H_{33}\ O$
1 atom water	$H\ O$
1 atom hydrated athal	$C_{32}\ H_{34}\ O_2$

Taking ceten as being the probable base of spermaceti, Dumas has also proposed the following formula:

Spermaceti.	2 ats. margarate of ceten.	2 atoms margaric acid..($C_{34}\ H_{33}\ O_3$)	$C_{208}\ H_{204}\ O_{13}$
		2 atoms ceten..........($C_{32}\ H_{32}$)	
		2 atoms water..........($H\ O$)	
	1 atom oleate of ceten.	1 atom oleic acid........($C_{44}\ H_{39}\ O_4$)	
		1 atom ceten..........($C_{32}\ H_{32}$)	
		1 atom water..........($H\ O$)	

But it will be seen that neither the percentage indicated by this nor by the last formula agrees with Chevreul's analysis of spermaceti.

			Analysis of Spermaceti by Chevreul.	
208 atoms carbon	1272*	80.06		
205 atoms hydrogen	205	12.90		
14 atoms oxygen	112	7.04		
	1589	100.00	Carbon	81.66
			Hydrogen	12.86
208 atoms carbon	1272	80.51	Oxygen	5.48
204 atoms hydrogen	204	12.95		
13 atoms oxygen	104	6.54		100.00
	1580	100.00		

What has been stated thus far is a short account of all that was known concerning the nature and composition of spermaceti previous to my attention being attracted to this subject, and what follows is a detail of my investigations.

Having undertaken some time since, at the suggestion of Prof. Liebig, to examine the products afforded by the distillation of spermaceti, I arrived at certain results which lead me to believe that the composition of this body was not properly made out, and therefore I undertook an examination of it after the most recent methods for the investigation of fatty bodies.

* The atomic weight here taken for carbon is that of Berzelius (6115), as Chevreul's calculation is made with the same.

The examination was directed to two points in particular; first, to the ascertaining whether *spermaceti contained oleic acid;* and secondly, whether *the solid acid obtained by Chevreul in his researches upon this body was margaric acid.*

The *saponification of the spermaceti* being the first step necessary in this examination, it was of some importance to make use of that method which would bring about the change the most easily. Chevreul digested the spermaceti with a strong solution of potash for a number of days to effect this change; but Dumas, in speaking of the easiest manner of obtaining athal from it, recommends that it should be saponified by fusing it with one half its weight of potash, and as by this latter means the process is completed in about one hour, it seemed to me the more preferable, and was consequently adopted.

Two ounces of spermaceti was fused with one half its weight of powdered potash, care being taken that the temperature did not rise above 230° Fah.; the mass soon became solid; it was then allowed to cool, and afterward treated with boiling water, which dissolved that portion of it which consisted of the acids arising from the saponification in combination with potash; the other portion, consisting of athal and undecomposed spermaceti, was held in suspension. To the fused mass, treated as just mentioned with boiling water, was added hydrochloric acid, which decomposed the soap in solution and liberated the acid which it contained, and this acid, being fusible at a temperature much below that of boiling water, melted and arose to the surface along with the athal and undecomposed spermaceti. This mixture upon cooling was again fused with pulverized potash, for the purpose of acting upon that part of the spermaceti which was not yet decomposed. After this second fusion it was again dissolved in hot water, which solution, holding athal in suspension, was treated with a solution of chloride of calcium, and by double decomposition a combination of the acids resulting from the saponification of spermaceti and lime was obtained, which though was mixed with athal.

The water was filtered away from the mixture of the lime, salt, and athal, and the mass, being dried, was treated with warm alcohol of 0.820, which dissolved the athal, and by repeatedly washing the lime-salt upon a filter with warm alcohol, and lastly with ether, until the liquid that passed through gave upon evaporation no residue, it was obtained perfectly free from athal. By this process a small portion of the lime-salt is dissolved, which can subsequently be obtained by treating the athal, from which the alcohol has been evaporated with ether, which leaves undissolved the lime-salt, and this is added to what remains upon the filter.

The lime-salt was dried, and decomposed by dilute hydrochloric acid, which furnished me with the acids arising from the saponification, and like most of the fat acids it floats about the water in flakes, which melt and collect at the surface if the water be heated.

Having now the acids free from undecomposed spermaceti and athal, the first part of the examination—that is to say, the examination for oleic acid—was carried on as follows.

Examination for oleic acid in spermaceti.—A portion of the acid was digested with water and the protoxide of lead, at a temperature of 212° Fah., and in the course of a short time a lead-salt was formed, which after being perfectly dried was treated with cold ether, that dissolved no portion of the salt—a circumstance that could not have occurred had the oleate of lead been present, as this salt is soluble in ether, and it is one of the means used to separate oleic acid from other fatty acids.

The above is the most direct way that we have of deciding upon the presence of oleic acid, and the indication which it affords in the present case was of too positive a character to admit for a moment the existence of this acid in the substance examined. But this single evidence, although sufficient of itself, has other indirect proofs to support it.

Redenbacker,* in his examination of the products of the distillation of oleic acid, observed the fact that if this acid or any substance containing it be distilled sebacic acid is invariably formed. To this test spermaceti has been subjected by both Redenbacker and myself, with similar results; that is is to say, that in the products afforded by the distillation of spermaceti no trace of sebacic acid is to be found.

The products furnished by the oxidation of spermaceti by nitric acid is another proof of the non-existence of oleic acid

---

* Redenbacker, properly speaking, was the first to generalize this fact, for it has been a long while since it was observed.

in this substance. Laurent and subsequently Bromeis have shown that when oleic acid is oxidized by nitric acid suberic acid is one of the most abundant products of this decomposition. Now if spermaceti be oxidized by nitric acid, no trace of suberic acid is furnished.

Having then the support of both direct and indirect evidence, I do not hesitate to affirm that *spermaceti contains no oleic acid.*

A question necessarily arising from this fact was, what was the acid that Chevreul had taken for oleic acid? To decide this the following steps were taken: That portion of the acid obtained from the lime-salt which had not been digested with the oxide of lead was treated with carbonate of soda, this forming a soda-salt, which, being dissolved in hot water, was decomposed by tartaric acid. The fat acid thus liberated from the soda was dissolved in warm alcohol, and upon allowing the solution to cool a considerable quantity of the acid crystallized out. The alcohol was poured off this crystalline deposit and concentrated by evaporation, from which another portion of the acid was allowed to crystallize. The alcohol was decanted a second time, concentrated, and allowed to cool, and by repeating this four or five times, and at last evaporating all the alcohol away, there was left a small quantity of a solid fatty mass, which evidently still contained a considerable portion of the same acid that had been crystallized from the alcoholic solution. This acid had a melting point of 68° F., and consisted of a mixture of a fluid and solid acid, but it was impossible to obtain the former in a state of purity, and as consequently no accurate examination of it could be made none was undertaken.

The fluid acid that composed a portion of this mass was in too inconsiderable a quantity to be considered an essential constituent of spermaceti, particularly too as its presence can be plausibly accounted for. Spermaceti as it exists in nature is mixed with an oil, from which it is separated by pressure for domestic use; now it is impossible that by simple pressure we should be able to deprive the spermaceti completely of this oil; but in Chevreul's analysis, as well as in mine, the spermaceti of commerce was treated with hot alcohol of 0.820; still there are many reasons for supposing that even by this means it is impossible to extract all the oil, either from the fact that the oil is not more soluble in alcohol than the spermaceti, or that the attraction that the oil and spermaceti have for one another is too strong to be overcome by this means. Of the truth of this latter supposition we have many similar examples, particularly among the fats—a circumstance which renders their examination to the present day incomplete and imperfect. At some future time my attention will be directed to the examination of spermaceti prepared in a different manner from that pursued in the present case, particularly with the object of ascertaining whether spermaceti can not be so purified as that its saponification will give rise to no fluid acid.

Thus then, as regards the existence of a fluid acid in spermaceti, all that can be said is that from the small quantity found, and from other reasons just stated, there are strong grounds for believing that it contains none, and that what has been found is due to an impurity which is not removed by alcohol of 0.820.

Solid acid resulting from the saponification of spermaceti.—I come now to the second part of the examination, and by far the most interesting—that of the solid acid obtained from the saponification of spermaceti; for it is this and athal that are the essential products resulting from the action of potash upon spermaceti.

The solid acid obtained in that part of the examination which was directed to ascertaining the presence of a fluid acid in spermaceti, and which was crystallized out of alcohol, was found to be nearly in a state of purity. This was dissolved in a mixture of equal parts of alcohol and ether, and allowed to crystallize out. This operation was repeated two or three times, and the crystalline deposit was then thrown upon a filter and washed with cold alcohol of 0.820. The acid thus obtained was pure, and possessed the following properties.

It melted at 130° F., and upon cooling crystallized in small needles, diverging from a number of centers, and when cool is white; it resembles somewhat in appearance wax, it being slightly translucent. It was dissolved in all proportions by alcohol of 0.820 at 140° F., and upon cooling crystallized out in small needles, which collected together in the form of moss and sometimes in that of cauliflower; from this the alcohol can be poured so as to leave it almost perfectly dry. Out of ether this acid crystallized with difficulty, owing to its excessive solubility in this menstruum. When heated to a high degree it volatilizes without leaving a residue. The alcoholic solution reddens litmus.

The physical properties of this acid will be seen to differ from those of margaric acid, which it has been supposed to be; but there is no striking difference between these two bodies in composition, as will be seen in the results afforded by the analysis of this body.

Exp. 1.—0.2815 gramme of the acid burnt with the bioxide of copper gave* 0.7725 gramme carbonic acid and 0.320 gramme water.

Exp. 2.—0.2325 gramme of the acid burnt with the bioxide of copper gave 0.637 gramme carbonic acid and 0.261 gramme water.

Exp. 3.—0.328 gramme of the acid burnt with the chromate of lead gave 0.890 gramme carbonic acid and 0.3685 gramme water.

These three analyses furnish the following proportions of carbon, hydrogen, and oxygen in 100 parts of the acid:

Carbon	75.44	75.31	74.64	75.13 †
Hydrogen	12.60	12.47	12.46	12.51
Oxygen	12.96	12.22	12.90	12.36
	100.00	100.00	100.00	100.00

Having thus found the relative proportions of the elements contained in this acid, it was necessary to examine one of its salts to ascertain its atomic composition, and for this purpose, as in most cases, its combination with the oxide of silver was chosen; but to form this salt it was necessary first to form its soda-salt.

Soda-salt.—A portion of the acid was digested with a solution of carbonate of soda until a complete combination had taken place, which is easily known by the acid no longer floating on its surface, it having all united with the soda, forming a salt soluble in water. The solution of this salt, which contained

* All my calculations are made with the atomic weight of carbon given by Liebig and Redenbacker (75.85 oxygen being considered 100, or 6.068 hydrogen being taken as unity).

† Composition of margaric acid after the analysis of Varrentrapp: Carbon, 75.35; Hydrogen, 12.33; Oxygen, 12.32.

an excess of carbonate of soda, was evaporated to dryness in a water-bath, and the dry mass pulverized was treated with absolute alcohol, which dissolved the soda-salt and not the carbonate of soda; from this the alcoholic solution was separated by filtration, and this last, evaporated to dryness, furnished the salt perfectly pure.

Silver-salt.—This salt was formed by a double decomposition of the salt just described and nitrate of silver. The soda was dissolved in water, and to this was added a solution of nitrate of silver, which produced a white flocculent precipitate, the salt in question. This precipitate was thrown on a filter and well washed with warm distilled water, and dried at 212° in the dark. This silver-salt when burnt in a porcelain crucible gave the following results:

Exp. 1.—0.332 gramme silver-salt gave 0.098 gramme silver.
Exp. 2.—0.389 gramme silver-salt gave 0.117 gramme silver.
Exp. 3.—0.5765 gramme silver-salt gave 0.1705 gramme silver.

Out of these the following percentage of silver and oxide of silver in the salt was calculated.

Exp. 1.—29.56 silver;	31.75 oxide of silver.
Exp. 2.—29.39 silver;	31.57 oxide of silver.
Exp. 3.—29.59 silver;	31.77 oxide of silver.

From the same analyses the atomic weight of the anhydrous acid was calculated to be from

Exp. 1.—250.00		
Exp. 2.—251.52	Mean	250.24
Exp. 3.—249.36		

The silver-salt was now analyzed with bioxide of copper to ascertain the quantity of carbon and hydrogen that it contained.

1.—0.4735 gramme silver-salt burnt with the bioxide of copper gave 0.910 gramme carbonic acid and 0.3595 gramme water.

2.—0.483 gramme silver-salt burnt with the bioxide of copper gave 0.934 gramme carbonic acid and 0.3705 gramme water.

From these analyses we find in 100 parts of the salt:

	1	2	Mean.
Carbon	52.84	53.15	53.00
Hydrogen	8.42	8.52	8.47
Oxygen	7.04	6.63	6.83
Oxide of silver	31.70	31.70	31.70
	100.00	100.00	100.00

Out of this the following formula is calculated:

Atoms.		Atomic weight.	In 100 parts. Calculated.	Found.
32	Carbon	194.18	53.16	53.00
31	Hydrogen	31.00	8.48	8.47
3	Oxygen	24.00	6.57	6.83
1	Oxide of silver	116.13	31.79	31.70
		365.31	100.00	100.00*

The anhydrous acid is constituted as follows:

Atoms.		Atomic weight.	In 100 parts.
32	Carbon	194.18	77.92
31	Hydrogen	31.00	12.44
3	Oxygen	24.00	9.64
		249.18	100.00

The atomic weight of this acid, found by burning the silver salt, was 250.24. The acid not in combination with a base, contains one atom of water, and has for its composition:

Atoms.		Atomic weight.	In 100 parts. Calculated.	Found.
32	Carbon	194.18	75.21	75.13
32	Hydrogen	32.00	12.39	12.51
4	Oxygen	32.00	12.40	12.36
		258.18	100.00	100.00

After the results afforded by these analyses it is impossible to confound this acid with margaric acid, and particularly too as its composition agrees with that of another acid, described by Dumas and Stass under the name of *athalic acid*, and which they obtained by acting upon *athal with potash* at a temperature of from 390° to 410° Fah. The acid then which has just been described, and which was obtained from the saponification of spermaceti, is *athalic acid.*

The athal that Chevreul mentions as being the base of spermaceti, and related to it as glycerine is to the other fats, was found to be of the same nature that he describes it to be.

	Atoms.			In 100 parts.
* Margarate of silver,	34	Carbon	206.31	54.57
	33	Hydrogen	33.00	8.65
	3	Oxygen	24.00	6.30
	1	Oxide of silver	116.13	30.48
			379.44	100.00

Conclusion as regards the Composition of Spermaceti considered as a fat.—Before coming to this conclusion a *resumé.* will be made of the results that have been arrived at in the different steps of this investigation. As regards *athal*, or what is considered the base of spermaceti, nothing has been brought to light to change in any way the statements made concerning its nature. *Oleic and margaric acids* have been proved not to exist in spermaceti. From the saponification of spermaceti, prepared as it was for these experiments, a small quantity of a fluid acid was obtained, but for reasons before stated considered as an impurity. The acid product arising from the saponification of spermaceti was found to consist almost entirely of *athalic acid.*

From these facts *spermacti*, considered as a fat (I make this qualification, as a little farther on it will be attempted to be shown that it is not, properly speaking, a fatty body), is composed of one acid and one base, the former being *athalic acid* and the latter *athal*, and it is therefore an *athalate of athal*, consisting of

		Atomic weight.
One atom anhydrous athalic acid..	$C_{32}\ H_{31}\ O_3$	249.18
One atom anhydrous athal	$C_{32}\ H_{33}\ O$	235.18
One atom spermaceti..................	$C_{64}\ H_{64}\ O_4$	484.36

That this is no doubt the true composition of spermaceti will be seen by the results afforded by the analysis of this substance prepared by crystallizing it out of absolute alcohol.

Exp. 1.—0.306 gramme spermaceti burnt with the bioxide of copper gave 0.8945 gramme carbonic acid and 0.370 gramme water.

Exp. 2.—0.2385 gramme spermaceti burnt with the bioxide of copper gave 0.691 gramme carbonic acid and 0.282 gramme water.

Exp. 3.—0.408 gramme spermaceti burnt with the chromate of lead gave 1.198 gramme carbonic acid and 0.486 gramme water.

Exp. 4.—0.314 gramme spermaceti burnt with the bioxide of copper and oxyen gave 0.913 gramme carbonic acid and 0.370 gramme water.

Exp. 5.—0.212 gramme spermaceti burnt with the bioxide of copper and chlorate of potash gave 0.625 gramme carbonic acid and 0.252 gramme water.

Comparing the percentage of carbon, hydrogen, and oxygen afforded by these experiments with that given by the supposed composition of spermaceti ($C_{64}H_{64}O_4$), we have in 100 parts:

	Atoms.	At. wght.	Calcul.	Found. 1.	2.	3.	4.	5.
Carbon........	64	388.36	80.18	80.36	79.66	80.70	79.91	81.08
Hydrogen....	64	64.00	13.22	13.53	13.12	13.23	13.40	13.21
Oxygen........	4	32.00	6.60	6.11	7.22	6.07	6.69	5.71
		484.36	100.00	100.00	100.00	100.00	100.00	100.00

Distillation of Spermaceti.—The products furnished by the distillation of spermaceti were examined some time since by Bussy and Lecanu; but they appear to have fallen into the same error with regard to them as was committed in the analysis of spermaceti, for they state that oleic and margaric acids were among the products.

To make a correct examination of the products of the distillation of spermaceti it was necessary that the substance should be in the greatest state of purity, as the presence of the smallest quantity of tallow, sometimes used as a means of adulteration, would serve to lead one into error. The manner of purification here employed was to dissolve the spermaceti in a mixture of two parts of alcohol of 0.820 and one part of ether, allowing it to crystallize out, and washing the crystals with boiling alcohol of 0.820.

If some of the spermaceti purified as just mentioned be placed in a small retort, and this last in mercury heated to its boiling-point, the spermaceti will be found to distill over slowly; and in fact this appears to be the lowest temperature at which the distillation takes place—a temperature of about 600° Fah. The matter distilled, possesses no longer the properties of spermaceti; its melting is at a temperature somewhat lower, and it has a strong acid reaction upon litmus-paper, as well as a peculiar smell, which though is not at all that of *acroleine.**

* If tallow be heated until it distills, it will be found to possess an odor which irritates both the nostrils and eyes, and the substance to which this odor belongs is called acroleine, and is a product of the decomposition of the glycerine in the tallow. It has been found that all fatty bodies that contain glycerine, when heated sufficiently high, give the same odor, and it has therefore become the test for the glycerine in them.

If the products afforded by the distillation be digested with water, and this water be examined, it will be found not to possess the slightest acid reaction; a fact of considerable importance, and one that has been mentioned in a former part of this article as an evidence of the non-existence of oleic acid in spermaceti; oleic acid or any of its compounds always furnishing by distillation sebacic acid, an acid soluble in water. The water, moreover, will be found to have taken up nothing, it having simply acquired a slight odor resembling that of the mass with which it was digested.

The steps taken to ascertain the nature of the products were the following. The mass obtained from the distillation was digested with a solution of potash for an hour or two, and to this, placed in a convenient vessel, was added ether, and the two agitated together, and then allowed to repose. The ether arose to the surface, containing in solution certain products. This was drawn off and a fresh portion added, and the agitation repeated. This operation was carried on until nothing remained that was soluble in this menstruum.

The ether was evaporated and a residue obtained consisting of an oily fluid holding spermaceti in solution. The separation of the oil from the spermaceti was attended with considerable difficulty; but by the aid of pressure at a very low temperature, and careful distillation, a small quantity of the oil was obtained tolerably pure.

0.222 gramme of the oil, burnt with the bioxide of copper, gave 0.688 gramme carbonic acid and 0.282 water, and this in one hundred parts gives

Carbon	85.04
Hydrogen	14.12
	99.16

These numbers show it to be a carbureted hydrogen, composed of equal equivalents of carbon and hydrogen; and this, together with such of its physical properties as I had been able to examine, led me into the belief that it was *ceten*, the carbureted hydrogen already spoken of as the supposed base of *athal.* Considering this oil to be *ceten*, its composition is represented by

Atoms.		Atomic weight.	In 100 parts. Calculated.	Found.
32	Carbon	194.18	85.85	85.04
32	Hydrogen	32.00	14.15	14.12
		226.18	100.00	99.16

The solution which had been treated with ether was now perfectly transparent, and contained potash in combination with the acid products resulting from the distillation. To this was added a solution of chloride of calcium, by which means an insoluble compound of the acids and lime was formed, and this, being decomposed by hydrochloric acid, furnished the acids for examination.

Although it was evident from what had been before done that no oleic acid could be present, yet to prevent any doubt I made a direct examination for this acid by digesting a portion of the acid mass with water and oxide of lead, and then treating the lead-salt thus formed with ether, which dissolved no portion of it. The portion of the acid product not digested with the oxide of lead was found to consist of a solid acid, mixed with a very small quantity of a fluid one, which I considered to be the same that has been before mentioned, as a probable impurity of spermaceti, and for the same reasons then stated it was impossible to make any examination of it.

The solid acid, which was obtained pure by repeated crystallization out of alcohol, exhibited the same physical properties, as well as chemical composition, as the acid obtained from the saponification of spermaceti, and which has been shown to be *athalic acid.*

0.2715 of the acid gave 0.739 gramme carbonic acid and 0.306 gramme water, making in one hundred parts

		Hydrated athalic acid.
Carbon	75.00	75.21
Hydrogen	12.52	12.39
Oxygen	12.48	12.40
	100.00	100.00

0.8625 gramme of the silver-salt, when burnt, gave 0.254 gramme silver, which indicates in one hundred parts 29.44 silver, 31.61 oxide of silver, and an atomic weight of 250.

0.481 gramme of the silver-salt burnt with the oxide of copper gave 0.932 gramme carbonic acid and 0.368 gramme water. The percentage afforded by this will serve to show the identity between it and *athalic acid.*

Athalate of silver.	Atoms.	Atomic weight.	In 100 parts. Calculated.	Found.
Carbon	32	194.18	53.16	53.28
Hydrogen	31	31.00	8.48	8.50
Oxygen	3	24.00	6.57	6.61
Oxide of silver..	1	116.13	31.79	31.61
		365.31	100.00	100.00

In the distillation of spermaceti there are other products found than those just mentioned; but they appear only toward the latter end of the process, and result from an elementary decomposition. They are water, carbonic acid, carbonic oxide, and gaseous carbureted hydrogen, carbon being left behind in the retort; and these products are very small in quantity, except when the vessel is very deep and the heat strong.

If proper care be taken, spermaceti can be distilled almost completely, there being left behind an exceedingly small black residue. A circumstance which facilitates this complete distillation is having kept the spermaceti for some time at the temperature of about 550° to 600° Fah.

The results of the investigations upon the distillation are, first, that it is impossible to distill spermaceti without more or less of it undergoing decomposition; and secondly, that the products of this decomposition are *ceten* and *athalic acid;* which fact serves to substantiate the correctness of the formula already taken for spermaceti, thus:

One atom ceten	$C_{32}\ H_{32}$
One atom hydrated athalic acid	$C_{32}\ H_{32}\ O_4$
One atom of spermaceti........................	$C_{64}\ H_{64}\ O_4$

Nature of Spermaceti.

From the foregoing researches I feel somewhat prepared to speculate upon the true nature of spermaceti; for although it may be difficult to arrive at any positive conclusion with regard to it, still we should not be deterred from forming a judgment upon probabilities.

For many reasons spermaceti would appear not properly to belong to the class of fatty bodies, and consequently not composed of an acid and a base. The fats, properly speaking, are known to be composed of acids, more or less different in their nature, in combination with glycerine; and when Chevreul found athal, as in spermaceti, accompanied with an acid, he considered athal as the base in this case, as well as making it the great mark of distinction between spermaceti and the fats.

Before going on to state the reasons why spermaceti should not be considered a fat, it would be well to mention what I suppose to be its proper position among the organic bodies. *Spermaceti* ought properly to be classed with *cholesterine* and *athal*, although approaching nearer to the fats than either of these substances; and that both the athalic acid and athal resulting from the saponification are simply products of decomposition brought about by the action of an alkali, neither of them existing ready formed.

The first reason for so believing is based upon the extreme difficulty with which spermaceti is saponified, it requiring to be digested for a number of days in a strong solution of potash or soda, or to be fused with the same alkalies at a temperature of from 212° to 220° Fah. before this change takes place. Now, from the experiments of Dumas and others, it will be seen that the action of hydrated potash upon organic substances, at a temperature more or less elevated, is to decompose them by changing their molecular arrangement, and that among the products formed acids play the most conspicuous part. The atom of water in the alkali is often important in bringing about this change by furnishing oxygen, hydrogen gas being evolved; but the action of this water appears to be but a secondary thing, and its influence is only felt where oxygen does not exist in sufficient quantity in the substance acted upon by the alkali to furnish the products that are found with the quantity that they exact.

The above would appear to apply exactly to the case in question. The spermaceti contains oxygen enough, which, when combined with one half of its other elements, serves to give rise to an acid. It is quite possible that the action of the alkali, although not sufficiently strong at the temperature of 212° Fah. to determine the elements of the spermaceti, to appropriate the atom of water in the alkali to its complete conversion into *athalic acid* (I say complete conversion into athalic acid, for it will be shown that the action of an alkali at a high temperature is to convert spermaceti entirely into *athalic acid*), still it is of sufficient energy to disturb its atomic arrangement, most of its oxygen combining with one half of the other elements to form an acid which unites with the potash.

It may be said that if this explanation of the saponification of spermaceti be true, we should apply the same to the saponification of all fats, no longer considering them composed of acids and glycerine, but simply of carbon, hydrogen, and oxygen, in the proper proportions to form them. But there appears to me no necessity for forming such a conclusion, as the circumstances attending the saponification of spermaceti and that of the fats differ considerably; and if this difference be taken into consideration with what follows, there is no doubt that the justice of this explanation will be seen.

Another reason for supposing that spermaceti does not consist of an acid and a base, or rather that athal does not exist in it ready formed, is that in the products afforded by the distillation of spermaceti no trace of athal is to be found. This fact is one that should be considered of great value in establishing the nature of spermaceti, for there is no way of explaining the non-existence of athal among the products of the distillation, except by admitting that the substance distilled did not contain it, for athal is a body easily volatilized without decomposition.

If, on the contrary, we remark the action of a strong solution of potash upon spermaceti at 100°, we find athal to be volatilized during the process—an evidence of the ease with which this substance is volatilized, as well as the necessity of an alkali for its formation.

Let us compare with this the action of heat upon the fats, with reference to the change that the glycerine undergoes. We find that if a fat be distilled, a portion of the glycerine is decomposed, giving rise to acroleine (a mixture of acetic acid, etc.), and another portion passes over undecomposed; whereas in the distillation of spermaceti its athal (supposing it to contain it) undergoes *complete decomposition*, although athal distilled by itself does not undergo *the least decomposition.*

This second reason then serves to increase the difference between the nature of spermaceti and that of the fats; but I am able to advance another fact stronger than either of the above augmenting this difference.

Dumas and Stass have shown that if athal be acted upon by potash at a temperature of from 410° to 428° Fah., an acid is the result, which acid they called *athalic acid*—the same that has been shown to result from the saponification of spermaceti, where the same alkali was employed, but at a much lower temperature. The action then of potash upon spermaceti, assisted by the proper temperatures, is to produce but one body, *athalic acid*, which circumstance would hardly take place were spermaceti composed of two or more proximate principles. We have no similar example among the fats.

Although cholesterine does not undergo any change by the action of a solution of potash at 212° Fah., still the analogy between it and the spermaceti may exist, for it must be observed that cholesterine, having an atomic weight of more than one half that of spermaceti, contains only one atom of oxygen, and not sufficient to give rise to an acid without the aid of an additional quantity; and it is probable that if cholesterine be treated with an alkali at a high temperature, that an acid similar to athalic acid would be the result, for then oxygen would be furnished from the water of the hydrated alkali.

For the above reasons spermaceti should be considered a simple organic substance, having, as already shown, for its composition $C_{64} H_{64} O_4$. The action of an alkali upon it produces a decomposition, which may be represented thus:

Anhydrous athalic acid in combination with the alkali..........	$C_{32} H_{31} O_3$
And hydrated athal, its atom of water being obtained from the alkali in combination with the athalic acid..........	$C_{32} H_{33} O + HO$
Spermaceti, plus one atom of water..........	$C_{64} H_{64} O_4 + HO$

Under the head of the distillation of spermaceti the decomposition brought about by the action of heat was shown to be represented thus:

Hydrated athalic acid..........	$C_{32} H_{32} O_4$
Ceten..........	$C_{32} H_{32}$
Spermaceti..........	$C_{64} H_{64} O_4$

OXIDATION OF SPERMACETI.

Having made mention of the oxidation of spermaceti as one of the evidences of the non-existence of oleic acid in this substance, I shall give a short statement of what has been done under this head, although but little, owing to the difficulty of isolating the products that are formed.

When nitric acid and spermaceti are heated together a gentle action takes place, and nitrous-acid fumes are given off; at the end of three or four days the spermaceti still floats upon the surface of the acid, but considerably changed in its nature, having nearly the consistency of hog's lard and an odor of rancid butter, owing probably to the presence of phocenic or butyric acid, but I am more inclined to believe phocenic acid, as this acid is found in the oil, in connection with which spermaceti is found in its natural state, and the spermaceti may no doubt play some part in its formation. This fact is interesting and worthy of future examination.

The action of the acid being continued (renewing it as it evaporates), in about ten days the spermaceti is in complete solution when the liquid is hot, and at the expiration of eighteen or twenty days the oxidation is completed, and if the solution be concentrated a crystalline deposit takes place.

The examination of the products formed is as yet imperfect; the following is all that has been done that can be relied upon as accurate.

After the completion of the oxidation the mass was thrown upon a funnel containing in its neck a bit of asbestus; the fluid was thus separated from the crystalline deposit, which was washed with strong nitric acid. The fluid that passed through upon concentration furnished more of the same crystals.

The crystalline mass in the funnel gave upon examination no traces of suberic acid, but when dissolved in warm water and allowed to cool a deposit slowly took place having the form of little grains and the appearance of starch. Its reaction is strongly acid, and when crystallized several times from its aqueous solution, and dried at 212° Fah., it has a melting-point of 298° Fah. It sublimes easily in feather-formed crystals; its ammoniacal salt does not precipitate the chlorides of lime, of baryta, or of strontia, the sulphate of copper, sulphate of zinc, or neutral acetate of lead. With the basic acetate of lead a precipitate is formed, which is soluble in an excess of the lead-salt.

0.3645 gramme of this acid burnt with the bioxide of copper gave 0.666 carbonic acid and 0.230 water. In 100 parts

Carbon	50.20
Hydrogen	7.00
Oxygen	42.80
	100.00

The silver-salt is easily formed by double decomposition with the ammoniacal salt and nitrate of silver. It is slightly soluble in water and not easily altered by the action of light.

Exp. 1.—0.525 gramme of this salt when burnt gave 0.294 gramme silver.

Exp. 2.—0.612 gramme of this salt when burnt gave 0.343 gramme silver.

These give in 100 parts

1.—56.01 silver, or 60.00 oxide of silver.
2.—56.05 ·· 60.19 ·· ··

Burnt with the bioxide of copper:

Exp. 1.—0.708 gramme of silver-salt gave 0.582 gramme carbonic acid and 0.174 gramme water.

Exp. 2.—0.787 gramme of silver-salt gave 0.6465 gramme carbonic acid and 0.190 gramme water.

These experiments give the following percentage:

	1	2	Mean.
Carbon	22.56	22.60	22.58
Hydrogen	2.68	2.68	2.68
Oxygen	14.67	14.63	14.65
Oxide of silver	60.09	60.09	60.09
	100.00	100.00	100.00

Out of this the following formula of a bibasic salt is calculated:

		Atomic weight.	In 100 parts. Calculated.	Found.
14 atoms	Carbon	84.95	22.18	22.58
10 ··	Hydrogen	10.00	2.66	2.68
7 ··	Oxygen	56.00	14.65	14.65
2 ··	Oxide of silver	232.25	60.51	60.09
		383.20	100.00	100.00

This formula agrees with that of adipinate of silver, as made out by Bromeis, with the unimportant difference of one atom of hydrogen, and its physical properties and reactions are the same as adipinic acid. I consider it as such.

None of the other acids afforded by the oxidation of spermaceti have been obtained in a state of sufficient purity to be examined. There is, however, one among them whose copper and zinc salts are more soluble in cold than in warm water, and if a solution of either of them be heated a precipitate is formed, which redissolves upon cooling. This phenomenon is most striking in the zinc-salt. Those portions of the examination of this subject that are as yet incomplete I propose finishing at some future time.

THE CALCARIMETER:

A NEW INSTRUMENT FOR ESTIMATING THE QUANTITY OF CARBONATE OF LIME PRESENT IN CALCAREOUS SUBSTANCES.

Among the most ready methods used for the purpose of estimating the quantity of carbonate of lime contained in calcareous substances are Davy's pneumatic and Rogers's methods, the one estimating it from the bulk of carbonic acid, and the other by the weight of the carbonic acid afforded by the action of an acid. The principal objection to the former is the complication of the apparatus, and for the latter it is necessary to be furnished with a more than ordinary pair of balances, and a set of accurate weights; whereas the instrument about to be described is free from both these objections, with the additional advantage of affording more accurate results.

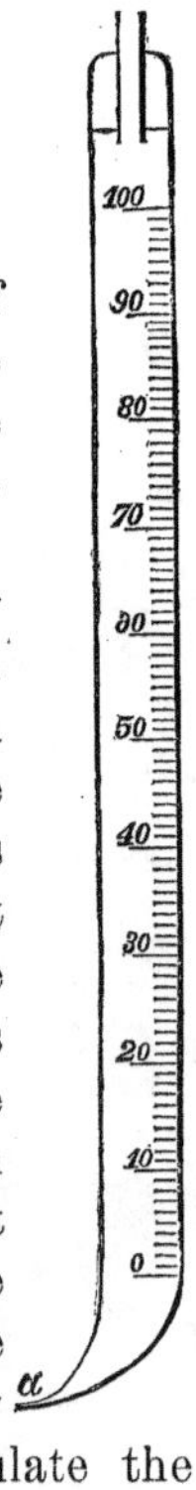

It appeared at first that by taking a certain quantity of the substance to be examined, and letting fall upon it by degrees a solution of acid, the strength of which we know, that it might be possible to estimate the quantity of carbonate of lime in the same manner as the carbonates of the fixed alkalies are estimated. But for this to succeed it is necessary that the substance should be finely pulverized, and free from any materials soluble in the acid used; but as it is not common to be furnished with these two conditions, another method had to be adopted, the principle of which is to treat the calcareous substance with an *excess* of acid, the strength of which is known, and then to find out the amount of this excess, thereby knowing the quantity of acid taken up, from which we can easily calculate the quantity of carbonate of lime present. In the application of this principle it will be found that any thing like difficult manipulation is avoided, and that there is no calculation required.

The first thing to be furnished with is an instrument, which consists simply of a tube about half an inch in diameter and ten inches long, having the principal part of it graduated in one hundred parts. The simplest form to be given to this tube is such as is represented in figure 1, the extremity *a* being drawn out and bent downward, leaving an opening so small as to allow a liquid to flow but slowly from the tube. To the upper part, for convenience' sake, is adapted a perforated cork, with a small tube. This is placed for the purpose of regulating the flow of the fluid, by placing upon it and withdrawing from it the finger, as we may wish to arrest or allow the liquid to flow from the extremity *a*. With this instrument, that I propose calling the *Calcarimeter* from its use, we must be furnished with two fluids, a solution of muriatic or nitric acid and a solution of ammonia, both of which are prepared of a certain strength.*

Preparation of the acid solution.—This solution is prepared as follows: weigh out fifty grains of dry, finely-powdered pure carbonate of lime, or what is better, carbonate of lime precipitated from any of its solutions by carbonate of potash or soda. Place this in a cupsule or other convenient vessel; add to it about an ounce of water (this is done simply for the purpose of moderating the action of the acid). Then take the muriatic or nitric acid of commerce, dilute it with one part of water. With this liquid fill the instrument to the 100 point; then let the acid fall gently upon the carbonate of lime, so as not to create a too great effervescence; and by proceeding carefully with the aid of a piece of litmus-paper we can find the exact point at which the carbonate of lime is all taken up by the solution having an acid reaction. When we see that nearly all the lime is taken up we proceed very cautiously, by adding but a few drops of the acid at a time, and agitating the mix-

*The capacity of the instrument from 0 to 100 is 30 c. c. m., and the length of the graduation had better be from eight to ten inches. Of course this will vary with the diameter of the tube. As they are all to be of the same capacity, the graduation may be made upon the tube itself, or upon a piece of paper and pasted on, then varnished, first with a solution of gum arabic, and afterward with copal varnish.

ture considerably for the purpose of bringing the insoluble carbonate well in contact with the different parts of the fluid. When the acid reaction commences the acid is no longer added, and the point at which the acid now stands in the tube is marked, and by subtracting that from 100 we have the number of degrees of acid used to dissolve fifty grains of carbonate of lime; but as it is desired that the liquid should be so made as to require 50° of it to dissolve fifty grains of the carbonate, it is diluted with the proper quantity of water. For example, suppose the fluid marked 65° after the experiment; this indicates that 35° of the acid solution were required to dissolve the 50 grains. Now instead of 35° we require it to take 50° to dissolve the same quantity, so that by making up the difference between the thirty-five and fifty with water the solution is prepared; that is to say, to every thirty-five parts of the acid experimented with fifteen parts of water are added. The solution can be again tested if necessary, and slight modifications made.

Preparation of the alkaline solution.—The alkaline solution is now prepared with ease. Let fall 50° of the acid into a vessel, then make a mixture of equal parts of ammonia and water, fill the instrument to the 100°, and let it flow upon the acid, and mark the point at which the acid is neutralized. Suppose it to be twenty, then 80° have been used for that purpose; but it must be so made as that it will require 100°; therefore to every eighty parts of the solution experimented with add twenty parts of water. In making either of these solutions one gallon can be made with the same ease as one ounce, and moreover, when they are once made, there is never any necessity of recurring to the carbonate of lime, as the acid may now be prepared with the aid of the ammonia.

a

b

c

Thus then 50° of acid dissolves exactly fifty grains of pure carbonate of lime, and 100° of the ammonia neutralizes fifty of the acid.

As using the same tube for both acid and alkali is attended with some inconvenience, having to wash it out after using one before introducing the other, I have used an additional tube (fig. 2), about the same diameter and a little more than half as long as the calcarimeter, for the acid. It has simply three marks upon it. The capacity of the tube from the point marked *a* to the lower extremity is equal to the capacity of 50° of the other tube, and the other two marks correspond to ten and five. The use that is made of these will be hereafter explained.

Manner of performing the analysis.—Being furnished with the two tubes, the two fluids, a capsule or other convenient vessel, a small piece of glass rod a few inches long, a wine-glass, and a piece of litmus-paper, a portion of which has been reddened by an acid, we proceed as follows: Weigh out fifty grains of the substance to be examined, place it in the capsule, and add to it about one ounce of water; fill the instrument last described up to the highest mark upon the stem with the *acid.* This is done by holding it between the thumb and fore-finger, having the little finger applied to the lower opening. After the acid is poured in, before withdrawing the finger, introduce the cork, and place the fore-finger of the other hand upon the opening of the tube on the cork, for the purpose of preventing the liquid flowing out when the lower opening is left unprotected. After seeing that the acid stands exactly at the mark it is allowed to flow gradually upon the substance. After all the action has ceased, stirring it toward the end to insure this result, we fill the graduated tube with the solution of ammonia, in the same manner as we did the last, and let it fall gradually upon the mixture of acid and calcareous substance, arresting at will the progress of the flow by simply placing the finger upon the tube in the cork. This instrument should always be transferred to the left hand and held in an inclined position. During the addition of ammonia the mixture should be well agitated with the glass rod, and occasionally tested by bringing a little of it upon the extremity of the rod in contact with the litmus-paper, and as soon as it ceases to turn this paper red, or begins to turn the red part of it blue, the experiment is completed, and we now look at what number of degrees the fluid stands in the tube, and we are furnished with the percentage of carbonate of lime contained in the calcareous substance examined.* We may be saved the trouble of testing

* If magnesia happens to be present it will be estimated as lime; but this will very seldom be a cause of error, as it exists very rarely in calcareous manures, for which this instrument is particularly intended.

too often by paying attention to the strength of the reaction of the fluid upon the litmus-paper.

In most marls which have served as the subjects of my experiments more or less alumina is to be found, a part of which is dissolved by the acid, of which part a very good use can be made. While adding the ammonia the alumina immediately around where the ammonia falls is thrown out of solution; and if we stir the liquid, the alumina will be redissolved so long as there is any free acid; so that when the flocks of alumina are no longer taken up we are furnished with an assurance that the process is nearly completed. The acid that the alumina and iron take up is acted upon by the ammonia with almost the same readiness as if free, so that no cause of error is to be apprehended from that source.

It may sometimes happen from oversight that too much ammonia is added. Notwithstanding this the analysis need not be lost. Still holding the instrument in the left hand over the cup, having of course arrested the flow of the fluid, we pour some of the acid solution into a wine-glass, introduce the small end of the acid instrument into it, and allow it to rise on the inside to either of the small marks, and add this acid to the liquid, and go on as before with the experiment, and at the conclusion read off what is indicated, and to it add 10 or 20 according as we may have added the acid measured by the first or second mark.

After what has been said a few words will suffice to explain how the instrument operates.

It takes 50° of acid to dissolve fifty grains of carbonate of lime, or 1° to dissolve one grain; and it takes 2° of the ammonia solution to neutralize one of the acid; and therefore in treating a substance consisting in part of carbonate of lime, for every grain that is present one degree of the acid is taken up, so that when we come to add the ammonia we know how much of the acid is taken up by the quantity of ammonia left behind, thereby knowing the number of grains of carbonate of lime, which we multiply by two (as fifty grains of the substance was used) to arrive at the percentage. This multiplication is not actually performed, as the instrument is so graduated as to dispense with it.

Were it at all necessary to give any evidence of its easy

application, I might state that it, along with the fluid, has been placed in the hands of persons entirely unacquainted with chemistry, and even with the principle of the instrument, and they have, with some little instruction in the manipulations necessary, obtained results only one or two per cent. out of the way in their *first* examination. The instrument is designed specially for examining calcareous manures.

ACTIONS OF NITRIC AND OXALIC ACIDS.

---

1. Action of Nitric Acid on the Chlorides of Potassium and Sodium. 2. Action of Oxalic Acid on the Nitrates and Chlorides of the same, with a ready Method of converting them into the Carbonates; Oxalic Acid enabling Zinc to decompose Water.

This note is intended as an appendix to my researches for determining the alkalies in insoluble silicates.

During that investigation many novel and interesting reactions were observed, several of which have already been alluded to. I present here one or two others of some interest.

It is well known that if nitric acid be added to a chloride, or hydrochloric acid to a nitrate, more or less of a decomposition will in either case ensue; but I believe it is not generally known how ready and complete the replacement is when nitric acid is heated with chloride of potassium or of sodium.

Among the experiments made, forty grammes of nitric acid were boiled gently with six grammes of chloride of potassium, and in twenty minutes no trace of chlorine could be found in the liquid. The same is true when the chloride of sodium is used. The operations appear to depend on the oxidizing property of the nitric acid, with the liberation of chlorine that combines with some of the elements of nitric acid to form the chloronitric acid that readily passes off. The decomposition of the nitrates of the alkalies by hydrochloric acid does not readily take place, it not being complete even after repeated evaporations to dryness with a large excess of hydrochloric acid.

Before settling on the plan I now adopt, an easy method was sought for separating the alkalies from magnesia by converting the two into carbonates—a plan that had previously been adopted; but the question with me was to change the nitrates to carbonates. The idea suggested itself of heating the nitrates with an excess of oxalic acid to the temperature at which the latter undergoes decomposition, when the nascent oxide of carbon might break up the constitution of the nitric acid, and the carbonic acid formed combine with the bases.

On making the experiment I was surprised to see an abundant evolution of nitrous-acid vapors at a temperature considerably below 212°. It was clear that the oxalic acid decomposed the nitrate, liberating the nitric acid, which reacting on the excess of oxalic acid gave rise to the nitrous-acid vapors. If crystallized oxalic acid and the nitrate of potash or soda, the former in excess, be placed together in a flask and heated over a water-bath, the mass soon enters into watery fusion, and at the temperature of from 130° to 140° bubbles of gas are evolved, consisting of nitrous acid and carbonic acid. At 212° the evolution is vigorous; and if after evaporation to dryness the water be renewed several times, the nitric acid will be completely expelled from the niter, there remaining the excess of oxalic acid and the oxalate of the alkalies.

It was natural to conclude from the above result that oxalic acid would likewise decompose the chlorides of the alkalies, and on experiment the conclusion proved to be correct. If an excess of oxalic acid be mixed with either the chloride of potassium or of sodium, and the whole warmed gently, abundant vapors of hydrochloric acid are evolved, and by careful manipulation all the chlorine may be driven off under this form.

If heat be applied to the mass resulting from the action of oxalic acid on either the nitrates or the chlorides, all the oxalic acid will be expelled and the oxalates converted into carbonates. A small amount of chloride of sodium can in this way be converted in a few moments into carbonate of soda; not, however, without some trace of the chloride being present. It is not my object to point to any special application of this decomposition, but it is one that may come into play in certain operations in analytical chemistry.

Experiments were made with the sulphates of the alkalies to see if the oxalic acid had any decomposing action on them, expecting to test for free sulphuric acid by the action of the solution of the mass on zinc or iron, taking for granted that the presence of oxalic acid alone would not cause the evolution of hydrogen gas. Experiment, however, showed that this manner of testing the question was fallacious, and no other method suggesting itself it was impossible to decide the question positively. Sufficient was ascertained to show that if the sulphate was decomposed it was only to a very minute extent.

In connection with this last experiment it is proved that zinc decomposes water readily in presence of oxalic acid, hydrogen gas being evolved. The action ceases in a short time from the formation of insoluble oxalate of zinc. With iron the action is very feeble even when the solution is heated.

The decomposing action of oxalic acid on the nitrates and chlorides of alkalies appears to be due simply to the fact of a more stable acid being able to replace a more volatile one, and in no way measures the relative strengths of the acids; it being a well-established fact that the physical as well as chemical properties of acids have much to do with their capability of replacing each other, a mere change of circumstances often reversing their relative action.

CHROMATE OF POTASSA:

A RE-AGENT FOR DISTINGUISHING BETWEEN THE SALTS OF BARYTA AND STRONTIA.

Having had occasion some months since to examine a specimen of fibrous celestine from Niagara, I was led to suspect from its specific gravity that baryta was present.

With this supposition I examined for baryta, in the usual way, with fluo-silicic acid; in fact, the only certain method that I was aware of. The indication that this test gave of its presence was so unsatisfactory that it led me at once to search for a more decisive and more delicate distinguishing test, and the following was the result of my labor.

It will be needless to detail the various re-agents that I had recourse to in my experiments, but suffice it to say chromate of potassa satisfied my most sanguine wishes, for no re-agent with which I am acquainted acts so promptly upon any body as does this upon the salts of baryta; and moreover, so delicate is this test that in one of my experiments, in which a grain of chloride of barium was dissolved in one gallon of water, it gave immediate indication of the presence of baryta, although sulphuric acid failed to do so; in fact, it will affect perceptibly a solution that contains less than one hundred-thousandth part of baryta.

When a strong solution of chromate of potassa is poured upon a strong solution of a salt of strontia a precipitate (similar to that which is produced when a salt of baryta is used) will take place. Solutions of these two salts of ordinary strength will not affect each other.

Lest this fact should, under any circumstance, cause erroneous conclusions, I sought for some acid which would dissolve the one precipitate and not the other. Acetic acid is the only acid among the many that I have tried which answered this end. If a small quantity of dilute acetic acid (common acetic acid diluted with five times its weight of water was used) be poured upon the precipitate produced in the case of strontia, it will be completely dissolved; whereas no impression is made on that from the salts of baryta.

Acetic acid, so concentrated as to crystallize when its temperature was below 50°, was poured on the precipitated chromate of baryta, and a portion of it taken up, but in no instance did any quantity of the acid dissolve the entire precipitate.

With the above means there need not now remain the least doubt in ascertaining promptly the presence of baryta in a salt of strontia supposed to contain it; for all that is necessary to be done is to add to a solution of the salt a solution of chromate of potassa, which, if baryta be present, will produce a light-yellow precipitate insoluble in acetic acid. This re-agent will also serve to distinguish baryta from lime.

BISULPHATE OF SODA AS A SUBSTITUTE FOR THE BISULPHATE OF POTASH

IN THE DECOMPOSITION OF MINERALS, ESPECIALLY THE ALUMINOUS MINERALS.

In referring to the more recent works on analytical chemistry I perceive that the bisulphate of potash is still used to the almost utter exclusion of bisulphate of soda in rendering certain minerals soluble: and it is still recommended as the proper agent to fuse with aluminous minerals, as corundum, emery, etc.

This subject occupied my attention to a considerable extent when engaged in the preparation of two memoirs on the geology and mineralogy of emery, presented to the French Academy of Science in 1850, as well as in some investigations I am now making on the emery from Chester, Mass. In the above researches I had a large number of corundums and emeries to analyze. The powdered minerals were fused with the bisulphate of potash in the usual way, and I found no difficulty in decomposing the minerals; but unfortunately during the operation a double salt of potash and alumina is formed which is almost insoluble in water or in the acids, and it is only by a solution of potash that it is first decomposed and afterward redissolved. There are many disadvantages and delays attendant upon this method which experience soon exhibits, as the constant deposition of alum if the solution is not kept quite dilute. I therefore experimented with the bisulphate of soda, knowing that the double salt of alumina and soda was quite soluble, and my results were every thing that could be desired; for while the soda-salt gives a decomposition at least as complete as the potash-salt, the melted mass is very soluble in water, and in the future operations of the analyses there is no embarrassment from a deposit of alum. The manner of employing the bisulphate of soda in the analysis of emery is referred to in the article on the emery of Chester, Mass.

PREPARATION OF THE BISULPHATE OF SODA.

The ordinary commercial article is not sufficiently pure for use, and I prepare it from pure carbonate of soda or sulphate of soda that has been purified by recrystallization. In either instance pure sulphuric acid is added in excess to the salt in a large platinum capsule, and heated over a flame until the melted mass, when taken up on the end of a glass rod, solidifies quite firmly. The mass is then allowed to cool; moving it over the sides of the capsule will facilitate this operation. When cool it is readily detached from the capsule, then broken up, and put into a glass-stoppered bottle. So far as my experience has yet gone, in almost every instance where we have been in the habit of using bisulphate of potash, the bisulphate of soda can be substituted.

ACTION OF POTASH UPON CHOLESTERINE.

For some reasons we would be induced to place cholesterine among the fatty bodies, but from many of its characters it would appear certainly not to belong to this class of bodies. The most important distinctions between these two bodies are, first, the want of action of a solution of potash upon cholesterine; and secondly, its high point of fusion, which is 298° Fah.*

Another difference which I am able to point out is that cholesterine is heavier than water, whereas the fats are lighter. It will be found in works on chemistry that cholesterine is lighter than water, and I attribute this to the fact that the substance, as it crystallizes out of alcohol, was found to float on the surface of water; but this is owing to the air adhering to the crystals. To show that it is heavier all that is necessary to be done is to throw a small piece of fused cholesterine into a vessel containing water, that must afterward be made to boil (this is done to drive away the air adhering to the surface of the body); after which it will be found to sink, and remain at the bottom of the vessel even when the water is cold. I dwell thus much upon this because I feel confident that there are other organic bodies that are said to be lighter than water, but which are actually heavier; for, owing to the looseness of their structure, air insinuates itself between the molecules, and is afterward held so firmly that it is impossible to drive it away by the ordinary means. I now return to the first distinguishing character between cholesterine and the fats—the difference of the action of potash upon the two bodies.

Chevreul and others have shown that if cholesterine be digested a great length of time in a boiling solution of potash no change takes place; but here the cholesterine is not subjected to the action of the potash under the same circumstances

* The melting-point of most of the fats is below 140°.

as the fats; for in the case of the latter, the point of fusion being considerably below that of boiling water, the force of aggregation is in a great degree destroyed, and consequently does not oppose itself to the chemical action; whereas in the case of cholesterine, its point of fusion being much higher than that of boiling water, it remains solid, and therefore its force of aggregation opposes itself strongly to the action of potash (supposing one to exist). So then the difference of the action of a solution of potash upon these substances is not such a strong mark of distinction as it would at first sight appear to be, as it is impossible to subject them to this action under similar circumstances.

This fact is mentioned not to show that cholesterine may be a species of fat. Far from it. It is simply to attempt to exhibit that there is no stronger reason for supposing that cholesterine is not a fat, because a boiling solution of an alkali does not act upon it, than there is for considering spermaceti a fat, because it is acted upon; as here the spermaceti is in a state of fusion, one that is favorable to this action; and the cholesterine solid, a state opposing this action.

In an article on spermaceti I stated my reasons at large for not believing this body to be a fat, properly speaking, and at the same time explained how I supposed an alkali to react upon it. It was there ranked with athal and cholesterine. I then also stated that although a boiling solution of an alkali might not react upon cholesterine, still I had no doubt that the alkali by itself, aided with a high temperature, would react upon it in a manner similar to that which it did upon spermaceti. From the kindness of M. Pelouze, who furnished me with a small quantity of cholesterine, I have been able to examine into the truth of this supposition.

The first circumstance necessary to be observed in the examination of this reaction is to have the cholesterine intimately in contact with the potash, and this is done by rubbing together equal parts of the two substances in a mortar. The mixture was placed in a watch-glass, and spread out so as to expose a large surface to the air; the watch-glass was placed on a support in a copper vessel (the air contained in this vessel could be brought to any required temperature). The experiment being thus disposed, the vessel was heated, and by the time that the air in the interior arrived at 248° Fah. a change began to take place in the mixture, and at 266° Fah. it was of a dark-brown color.

This was now treated with cold ether, which dissolved the unaltered cholesterine, and also a matter of a resinous character, which when dissolved in alcohol, and the alcohol allowed to evaporate spontaneously, is deposited in the form of little round concretions entirely devoid of crystalline structure. It is not soluble in any of the alkalies. What remains after the treatment by ether is of a brown color and completely soluble in water. If hydrochloric acid be added to this solution it is decomposed, and a yellowish substance arises to the surface. This substance is soluble in ether, alcohol, potash, soda, and ammonia, as well as their carbonates. It does not crystallize. Its alcoholic solution reacts slightly acid upon litmus-paper. In fact it is an acid of a resinous character. Its combinations with alkalies have the character of soaps. Its silver-salt is of a yellow color, but soon becomes black by exposure to the light.

From the small quantity of cholesterine that was at my disposal I have not been able to obtain sufficient of the acid to examine its composition, but I have no doubt that it is a new one.

If the mixture when heated be not well exposed to the air, very little of this acid is formed, even if we elevate the temperature as high as 300° Fah.; but, on the contrary, a considerable quantity of the resin before mentioned (soluble in ether) is formed. This though is capable of being converted into the acid by the action of potash, a high temperature, and free access of air. Thus then it will be seen that the action of potash, instead of being a means of showing that spermaceti and cholesterine are two substances of entirely different natures, affords strong evidence of their being similar bodies. Further, the action of potash upon spermaceti is to produce athalic acid and athal, the former capable of forming soaps with the alkalies, and the latter of being converted into the former by an alkali and a high temperature.

The action upon cholesterine is to form an acid (which it is impossible for me as yet to name) and a basic resin. The former forms soaps with alkalies, and the latter by the action of potash at a high temperature is converted into the former.

This article is meant as an appendix to the one on spermaceti, and as an additional proof of the analogy that exists between that body and cholesterine, they being two of a class of bodies which will no doubt be found to be tolerably numerous, and which class I propose to call *pseudo-gras*. Among them may be mentioned spermaceti, cholesterine, athal, ambreine, and probably stearérine and elaïérine, two fatty substances found in linseed-oil, and which M. Chevreul brought to the notice of the Academy of Sciences not long since. This class of bodies would appear to be a link between the fats and resins.

NEUTRAL ALKALINE PHOSPHATES.

ACTION OF THE NEUTRAL PHOSPHATES OF THE ALKALIES UPON CARBONATE OF LIME.

It is a fact that, notwithstanding the advanced state of the science of chemistry, we are ignorant of some of the laws that govern the relative affinities of acids for bases, and the action of neutral salts upon each other. It is true such and such acids are ranked according to what is termed their strength, and such bases are said to be more powerful than others; still from time to time facts are developing themselves that contradict these established rules. The decomposition of the sulphate of lead by certain neutral alkaline salts (Am. Jour., xlvii, 81) I thought could be explained upon a known law, that when there existed two acids and two bases in solution (the sulphate of lead being dissolved by the salts used) the stronger acid sought the stronger base, and the feebler acid had to combine with the feebler base, notwithstanding being originally in combination with an alkali. But how are we to explain the fact about to be mentioned, which, so far as my information goes, has not been previously observed? It is that the feeblest solution of the *neutral phosphate of soda or potash* will decompose the *carbonate of lime* in the cold, giving rise to carbonate of soda and phosphate of lime.

This fact was first observed while analyzing the ashes of a plant, which was fused with carbonate of soda, for the purpose of estimating the phosphoric acid. The fused mass was thrown into about four ounces of water, and digested at about 180° Fah. for a couple of hours. The insoluble portion was separated and treated with an acid, when to my astonishment it dissolved with but a *very slight* effervescence; in fact, with the escape of only a bubble or two of gas, the carbonate of lime expected not being present. It was known that this circumstance could not arise from a want of decomposition of the original matter, as it was kept fused for half an hour with four times its weight of carbonate of soda; therefore the only rational conclusion was that the phosphate of lime was in the first case decomposed by the soda, but was subsequently reformed upon treating the fused mass with water. This has been verified by direct experiment.

Twelve grains of neutral phosphate of soda and six of carbonate of lime were digested for two hours in four ounces of water at 180° Fah., when the carbonate of lime was found almost completely decomposed, and the clear solution upon evaporation furnished carbonate of soda.

Six grains of precipitated carbonate of lime added to a solution of twenty grains of phosphate of soda (equivalent proportions of each), in one ounce of water, were kept in a vial for one month, the temperature never exceeding 65° Fah. At the expiration of this time the insoluble portion contained three and a half grains of phosphate of lime, corresponding to a decomposition of about two and a half grains of the carbonate of lime. The soluble portion indicated a corresponding portion of carbonate of soda.

Other insoluble carbonates were experimented with, as the carbonates of magnesia, strontia, baryta, and lead. The results were the same, differing only in degree. Even hydrated alumina decomposes slightly the phosphate of soda when boiled with it for a length of time.

I tried two other neutral salts, the acids of which produce insoluble salts with lime, to see if they would act in the same way. The chromate and the tartrate of potash were digested a length of time upon the carbonate of lime, but no decomposition ensued.

I shall not attempt to seek for an explanation of this at present, but shall go on collecting facts of a similar character, to endeavor to find out some general principle that may operate in this and in other cases. This fact itself would not be published at the present time if it were not of the greatest importance to put analytical chemists upon their guard; for but a few days ago an individual wrote to me that he was estimating the phosphate of lime in a certain class of bodies by fusing them with carbonate of soda, which will certainly be productive of some error; and although it is to be regretted that our methods of arriving at phosphoric acid in analysis may be diminished by this fact, still it will only stimulate us to find out some other to solve this, one of the most difficult and annoying problems in analytical chemistry.

REMOVAL OF SAL AMMONIAC IN MINERAL ANALYSIS.

About twenty years ago, in a publication made upon the analysis of the natural silicates, I gave the details of some interesting experiments made upon the removal of sal ammoniac, which so commonly accumulates in these analyses.

The method of accomplishing the removal of this salt, being embodied in a lengthy paper embracing many other and more important points, has been to a great extent overlooked by analytical chemists. I have been frequently asked for details in connection with the removal of this salt, and some recent investigations have given me renewed appreciation of the invaluable nature of the process, where very large quantities of sal ammoniac had accumulated and remained associated with a very minute quantity of material that formed the subject of research.

It may be of interest to bring this process more clearly to the attention of chemists. The manner of proceeding is as follows: The solution containing the sal ammoniac is concentrated in a capsule, best over a water-bath or in a glass flask; pure nitric acid is added, about three grammes of it to every gramme of sal ammoniac supposed to exist in the liquid; a little habit will suffice to guide one in adding the nitric acid, as even a large excess has no effect on the accuracy of the analysis.

The flask or capsule is now warmed very gently, and before it reaches the boiling-point of water a gaseous decomposition will take place with great rapidity. This is caused by the decomposition of the sal ammoniac. It is no advantage to push the decomposition with too great rapidity; a moderately warm place on the sand-bath is well adapted for this purpose. I, however, prefer a porcelain capsule of about three and a half to four inches diameter (in the ordinary operations in mineral analysis), inverting a clean funnel of smaller diameter over it, and evaporating to dryness over the water-bath; at the end of the operation the heat can be increased to four or five hundred degrees.

By this operation, which requires no superintendence, one hundred grammes of sal ammoniac may be separated as easily and safely as one gramme from five milligrammes of alkalies, and no loss of the latter be experienced.

The following are some experiments made with given quantities of sal ammoniac and nitric acid, heated thus in a capsule over a water-bath:

	Nitric Acid.	Sal Ammoniac
5 grammes of sal ammoniac...	5 c. cent. left.	3.190 grammes.
5 " " " ...	7 " "	2.610 "
5 " " " ...	10 " "	.790 "
5 " " " ...	13 " "	.010 "

The decomposition commences before the temperature reaches 140° Fah. The results of the decomposition were fully explained in a note to an article of mine published in the Amer. Jour. of Science and Arts, March, 1853. It results principally in the formation of protoxide of nitrogen and chlorine, the former constituting over seven eighths of the gas formed.

MEMOIR ON METEORITES.

PART I.

A DESCRIPTION OF FIVE NEW METEORIC IRONS.

1. Meteoric Iron from Tazewell County, E. Tenn.

This meteorite was placed in my possession through the kindness of Prof. J. B. Mitchell, of Knoxville, in the month of August, 1853. It was found by a son of Mr. Rogers, living in that neighborhood, while engaged in plowing a hillside; his attention was drawn to it by its sonorous character. As it very often happens among the less informed, it was supposed to be silver, or to contain a large portion of that metal. With some difficulty the mass was procured by Prof. Mitchell and passed over to me. Nothing could be ascertained as to the time of its fall. It is stated among the people living near where the meteorite was found that a light has been often seen to emanate from and rest upon the hill—a belief that may have had its foundation in the observed fall of this body.

The weight of this meteorite was fifty-five pounds. It is of a flattened shape, with numerous conchoidal indentations, and three annular openings passing through the thickness of the mass near the outer edge. Two or three places on the surface are flattened, as if other portions were attached at one time, but had been rusted off by a process of oxidation that has made several fissures in the mass so as to allow portions to be detached by the hammer, although when the metal is sound the smallest fragment could not be thus detached, it being both hard and tough. Its dimensions are such that it will just lie in a box thirteen inches long, eleven inches broad, and five and a half inches deep. The accompanying figure gives a correct idea of the appearance of this meteorite

The exterior is covered with oxide of iron; in some places so thin as hardly to conceal the iron, in other places a quarter of an inch deep. Its hardness is so great that it is almost impossible to detach portions by means of a saw. Its color is

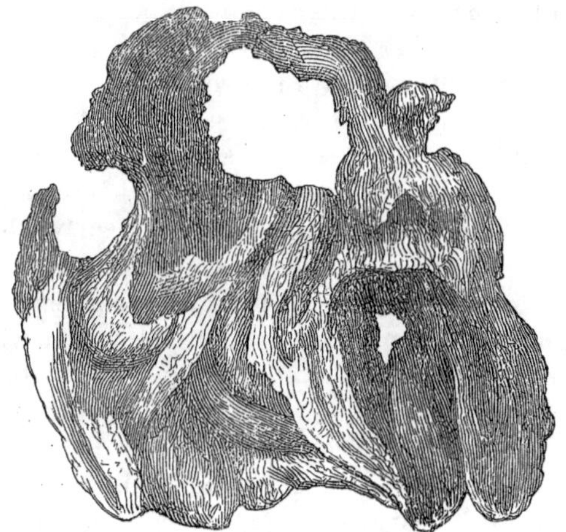

Fig. 1.

white, owing to the large amount of nickel present; and a polished surface, when acted on by hot nitric acid, displays in a most beautifully regular manner the Widmannstättian figures. The specific gravity taken on three fragments selected for their compactness and purity is from 7.88 to 7.91.

The following minerals have been found to constitute this meteorite: 1. *Nickeliferous iron*, forming nearly the entire mass. 2. *Protosulphuret of iron*, found in no inconsiderable quantity on several parts of the exterior of the mass. 3. *Schreibersite*, found more or less mixed with the pyrites and in the crevices of the iron, in pieces from the thickness of the blade of a penknife to that of the minutest particles. 4. *Olivine;* two or three very small pieces of this mineral have been found in the interior of the iron. 5. *Protochloride of iron;* this mineral has been found in this meteorite *in the solid state*, which I believe is the first observation of this fact; it was found in a crevice that had been opened by a sledge-hammer, and in the same crevice schreibersite was found. Chloride of iron is also found deliquescing on the surface; some portions of the surface are entirely free from it, while others again are covered with an abundance of rust arising from its decomposition.

Besides the above minerals two others were found—one a siliceous mineral, the other in minute rounded black particles; both, however, were in too small quantity for any thing like a correct idea to be formed of their composition.

The different minerals that admitted of it were examined chemically, and the following are the results:

1. *Nickeliferous Iron.*—The specific gravity of this iron is, as already stated, from 7.88 to 7.91. It is not readily acted on by any of the acids in the cold; nitric acid, either concentrated or dilute, has no action on it until heated to nearly 200° Fah., when the action commences, and continues with great vigor even after the withdrawal of heat. With reference to the action of sulphate of copper, it is *passive*, although when im mersed in a solution of sulphate of copper, and allowed to remain for several hours, the latter metal deposits itself in spots on the surface of the iron.

Thorough digestion in hot nitric acid dissolves the iron completely. When boiled with hydrochloric acid the iron dissolves with the liberation of hydrogen, leaving undissolved the schreibersite; but by long-continued action this latter is also dissolved with the evolution of phosphureted hydrogen.

The following ingredients were detected on analysis of two specimens:

	1	2
Iron	82.39	83.02
Nickel	15.02	14.62
Cobalt	.43	.50
Copper	.09	.06
Phosphorus	.16	.19
Chlorine	......	.02
Sulphur	......	.08
Silica	.46	.84
Magnesia	......	.24
	98.55	99.57

Tin and arsenic were looked for, but neither of those substances detected. The magnesia and silica are doubtless combined, probably in the form of olivine, and disseminated in minute particles through the iron. The phosphorus is in combination with a given portion of iron and nickel, forming schreibersite. The 0.16 per cent. of phosphorus corresponds to 1.15 of schreibersite; so the metal mass may be looked on as composed of nickeliferous iron 98.97, screibersite 1.03=100.00

The composition of the nickeliferous iron corresponds to five atoms of iron and one of nickel: iron, 5 atoms, 82.59; nickel, 1 atom, 17.41=100.00.

2. *Protosulphuret of Iron.*—This variety of sulphuret of iron found with meteorites is usually designated as magnetic pyrites, leaving it to be inferred that its composition is the same as the terrestrial variety. Without alluding to the doubt among some mineralogists as to the true composition of the terrestrial magnetic pyrites, I have only to say that most careful examination of the sulphuret detached from the meteorite in question proves it to be a protosulphuret—a conclusion to which Rammelsberg had already come with reference to the pyrites of the Seelasgen iron, which latter pyrites I have also examined, confirming the results of Rammelsberg.

This pyrites incrusts some portion of the iron, and in places is mixed with a little schreibersite. It presents no distinct crystalline structure, has a gray metallic luster, and a specific gravity of 4.75. The Seelasgen pyrites gave me for specific gravity 4.681. The specimen of pyrites in question gave, on analysis: iron, 62.38; sulphur, 35.67; nickel, 0.32; copper, trace; silica, 0.56; lime, 0.08=98.91. The formula $Fe\,S$ requires sulphur 36.36, iron 63.64. The magnetic property of this mineral is far inferior to that possessed by schreibersite.

3. *Schreibersite.*—It is found disseminated in small particles through the mass of the iron, and is made evident by the action of hydrochloric acid; it is also found in flakes of little size, inserted as it were into the iron; and owing to the fact that in many parts where it occurs chloride of iron also exists, this last has caused the iron to rust in crevices, and on opening these schreibersite was detached mechanically. This mineral as it exists in the meteorite in question so closely resembles magnetic pyrites that it can readily be mistaken for this latter substance, and I feel confident in asserting that a great deal of the so-called magnetic pyrites associated with various masses of meteoric iron will upon examination be found not to contain a trace of sulphur, and will, on the contrary, prove to be schreibersite, that can be easily recognized by the characters to be fully detailed a little farther on.

Its color is yellow or yellowish-white, sometimes with a greenish tinge; luster metallic; hardness 6; specific gravity

7.017. No regular crystalline form was detected; its fracture in one direction is conchoidal. It is attracted very readily by the magnet, even more so than magnetic oxide of iron; it acquires polarity and retains it. I have a piece three tenths of an inch long, two tenths of an inch broad, and one twentieth of an inch thick, which has retained its polarity over six months; unfortunately the polarity was not tested immediately when it was detached from the iron, and not until it had come in contact with a magnet, so that it can not be pronounced as originally polar.

Before the blowpipe it melts readily, little blisters forming on the surface from the escape of *chlorine*, and blackens. The magnet is a most ready means of distinguishing the schreibersite from the pyrites commonly found in meteoric irons; for, although the pyrites is attracted by the magnet, it is necessary that the latter should be brought quite near to it for the effect to be produced; whereas if the particles exposed to the magnet be schreibersite, they will be attracted with almost the readiness of iron filings.

Hydrochloric acid acts exceedingly slow on this mineral when pulverized with the formation of phosphureted hydrogen. Nitric acid acts more vigorously, and readily dissolves it when finely pulverized. The composition of this substance has in all cases but one been made out from the residue of meteoric iron, after having been acted on by hydrochloric acid, which accounts for the great variation in the statements of the proportion of its constituents.

Mr. Fisher examined pieces of schreibersite detached from the Braunau iron with the following results: iron, 55.430; nickel, 25.015; phosphorus, 11.722; chrome, 2.850; carbon, 1.156; silex, 0.985=98.158.

The results of my analyses do not differ very materially from this. They are as follows:

	1	2	3
Iron	57.22	56.04	56.53
Nickel	25.82	26.43	28.02
Cobalt	.32	.41	.28
Copper	trace.	not est.	......
Phosphorus	13.92	......	14.86
Silica	1.62	......	......
Alumina	1.63	......	......
Zinc	trace.	not est.	......
Chlorine	.13	......	......
	100.66		99.69

Nos. 1 and 2 were separated mechanically from the iron; No. 3 chemically. The silica, alumina, and lime were almost entirely absent from No. 3, and in the other specimen they are due to a siliceous mineral that I have found attached in small particles to the schreibersite. There is no essential difference in my results; yet in neither instance do I suppose the mineral was obtained perfectly pure, although enough so, it is believed, to furnish the correct chemical formula; and as from what has been previously said schreibersite will be found to exist in larger quantities than it was suspected, it will not be long before the question of the uniformity of its composition will be settled, a point of interest bearing upon the theoretical consideration of meteoric stones.

The formula of schreibersite I consider to be $Ni^2 Fe^4 P$.

	Atom.	Per cent.
Phosphorus	1	15.47
Nickel	2	29.17
Iron	4	55.36

This mineral, although not usually much dwelt upon when speaking of meteorites, is decidedly the most interesting one associated with this class of bodies, even more so than the nickeliferous iron. It has no representative in genus or species among terrestrial minerals, and is one possessed of highly interesting properties. Although among terrestrial minerals phosphates are found, not a single phosphuret is known to exist. So true is this that, with our present knowledge, if any one thing could convince me more strongly than another of the non-terrestrial origin of any natural body, it would be the presence of this or some similar phosphuret. It is commonly alluded to as a residue from the action of hydrochloric acid upon meteoric iron, when in fact it exists in plates and fragments of some size in almost all meteoric iron, and there is reason to believe that it is never absent from any of them in some form or other. What is meant by "some size" is that it is in pieces large enough to be seen by the naked eye, and to be detached mechanically.

In an examination of the meteoric specimens in the Yale College Cabinet more than half of them have been discovered to contain schreibersite, visible to the eye, that had been considered pyrites. Among them the large Texas meteorite was examined; and although none was visible on the surface, a small fragment of the same mass given me by Prof. Silliman contains a piece of schreibersite of over a grain weight.

The reason why it has not attracted more attention arises from its resemblance to pyrites. I will therefore state a ready manner of telling whether it be such or not.

Detach a small fragment, and hold a magnet capable of sustaining five or six ounces or more within half an inch or an inch of the fragment. If it be schreibersite it will be attracted with great readiness, the magnetic pyrites requiring a very close approximation of the magnet before attracted. This, with some little experience, becomes a ready method of separating the two. It is not, however, to be expected that this method alone is to satisfy us when other means can be appealed to for distinguishing this mineral. The following is one which is readily accomplished with the smallest fragment (half a milligramme). Melt in a small loop of platinum wire a little carbonate of soda; add the smallest fragment of nitrate of soda and the piece of mineral; hold the mixture in the flame of a lamp for two or three minutes; place the bead of soda in a watch-glass; add a little water, and filter. To the filtrate add a drop or two of acid to neutralize the excess of carbonate of soda; evaporate nearly to dryness; add a drop of ammonia, and then a drop of ammoniacal sulphate of magnesia, when the double phosphate of magnesia and ammonia will show itself, and the crystalline form will be recognized under the microscope. If the piece examined be several milligrammes in weight, the operation can be carried on in a small platinum capsule. This reaction can also be had by acting on the mineral, however small the piece, by aqua regia; evaporate until only a little of the liquid is left; add a little tartaric acid, then a drop or two of ammonia to supersaturate the acid; and lastly a little ammoniacal sulphate of magnesia, when the crystals of the double phosphate of magnesia and ammonia will appear.

4. *Protochloride of Iron.*—In breaking open one of the fissures of this meteoric iron a small amount of a green substance was obtained that was easily soluble in water; and although not analyzed quantitively, it left no doubt upon my mind as to its being protochloride of iron; and the manner of its occurrence gave strong evidence of its being an original constituent of the mass, and not formed since the fall of the mass. Chloride

of iron was apparent on various parts of the iron by its deliquescence on the surface.

2. Meteoric Iron from Campbell County, Tenn.

This meteorite was discovered in July, 1853, in Campbell County, Tenn., in Stinking Creek, which flows down one of the narrow valleys of the Cumberland Mountains. It was found by a Mr. Arnold in the channel of this stream; and having been obtained by Prof. Mitchell, of Knoxville, he kindly presented it to me. It is a small oval mass, two and a fourth inches long, one and three fourths broad, and three fourths thick, with an irregular surface and several cavities perforating the mass. It was covered with a thin coat of oxide; and on one half of it chloride of iron was deliquescing from the surface, while on another portion there was a thin siliceous coating.

The iron composing the mass was quite tough, highly crystalline, and exhibited small cavities on being broken, resembling very much in this respect, as well as in many other points, the Hommony Creek iron. A polished surface when etched exhibited distinct irregular Widmannstättian figures.

The weight is four and a third ounces; specific gravity 7.05. The lowness of the specific gravity is accounted for by its porous nature. Composition:

Iron	97.54
Nickel	.25
Cobalt	.60
Copper, too small to be estimated.	
Carbon	1.50
Phosphorus	.12
Silica	1.05
	100.52

Chlorine exists in some parts in minute proportion. The amount of nickel, it will be seen, is quite small; but its composition is nevertheless perfectly characteristic of its origin.

3. Meteoric Iron from Coahuila, Mexico.

This meteorite was brought to this country by Lieut. Gouch, of the United States army, he having obtained it at Saltillo. It was said to have come from the Sancha estate, some fifty or sixty miles from Santa Rosa, in the north of Coahuila. Various accounts were given of the precise locality, but none seemed very satisfactory. When first seen by Lieut. Gouch it was used as an anvil, and had been originally intended for the Society of Geography and Statistics in the City of Mexico. It is stated that where this mass was found there are many others of enormous size. These stones, however, it is well known, are to be received with many allowances. Mr. Weidner, of the mines of Freiberg, states that near the south-western edge of the Balson de Mapimi, on the route to the mines of Parral, there is a meteorite near the road of not less than a ton weight. Lieut. Gouch also states that the intelligent but almost unknown Dr. Berlandier writes in his journal of the commission of limits that at the hacienda of Venagas there was (1827) a piece of iron that would make a cylinder one yard in length, with a diameter of ten inches. It was said to have been brought from the mountains near the hacienda. It presented no crystalline structure, and was quite ductile.

The meteoric mass in question, which is at the Smithsonian Institution, is of the form represented in the figure, and one

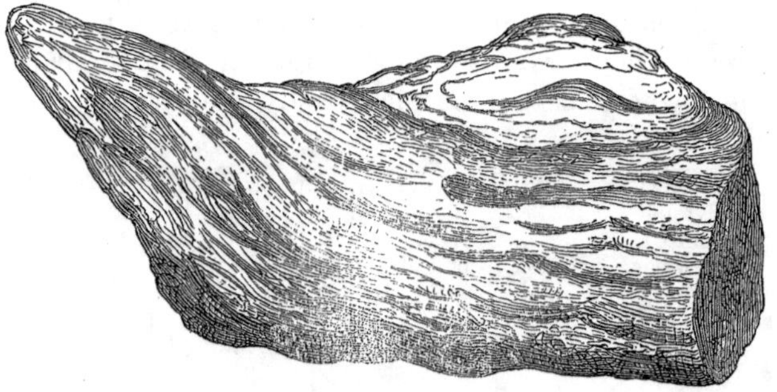

Fig. 2.

well adapted for an anvil. Its weight is two hundred and fifty-two pounds, and from several flattened places I am led to suppose that pieces have become detached. The surface, although irregular in some places, is rather smooth, with only here and there thin coatings of rust, and, as might be expected, but very feeble evidence of chlorine, and that only on one or two spots on the surface. Specific gravity 7.81. It is highly crystalline, quite malleable, and not difficult to cut with the saw. Its surface etched with nitric acid presents the Widmannstättian figures, with a finely-specked surface between the lines, resembling the representation we have of the etched surface of Hauptmannsdorf iron. Schreibersite is visible in the iron, but so inserted in the mass that it can not be readily detected by mechanical means. Hydrochloric acid leaves a residue of beautifully brilliant patches of this mineral. Subjected to analysis it was found to contain

Iron	95.82
Cobalt	.35
Nickel	3.18
Copper, minute quantity, not estimated.	
Phosphorus	.24
	99.59

Which corresponds to

Nickeliferous iron	98.45
Schreibersite	1.55
	100.00

The iron is remarkably free from other constituents. It is especially interesting as the largest mass of meteoric iron in this country next to the Texas meteorite at Yale College.

4. Meteoric Iron from Tucson, Mexico.

We have had several accounts of meteoric masses which exist at Tucson, Dr. J. L. LeConte having made them known some few years ago. Since that time Mr. Bartlett, of the Boundary Commission, has seen them and made a drawing of one which he has kindly allowed me the use of, as well as the manuscript notice of them, which is, however, quite brief. This mass is used for an anvil, resembles native iron, and weighs about six hundred pounds. Its greatest length is five feet. Its exterior is quite smooth, while the lower part which projects from the larger leg is very jagged and rough. It was found about twenty miles distant toward Tubac, and about eight miles from the road, where we are told are many larger masses. The following figure (3) represents the appearance of that meteorite.

Since my communication last April I have obtained fragments of the meteorite from Lieut John G. Parke, of the U. S. Topographical Engineers, who cut them from the mass at Tucson, and to whose kindness I feel much indebted.

Some of the fragments were entirely covered with rust, and

in some parts little blisters existed, arising from chloride of iron. Portions of the broken surface retain their metallic luster untarnished. The Widmannstättian figures are very imperfectly developed, owing to the porous nature of the iron, the pores of which are filled with a stony mineral. The specific gravity taken on three specimens were 6.52, 6.91, 7.13. The last was the most compact and free from stony particles that could be found, and upon that the chemical examination was made.*

Fig. 3.

On examination it is seen to consist of two distinct parts, metallic and stony. The latter was only in minute particles, yet it was impossible, among the specimens at my disposal, to find a piece that was without it. On analysis the following ingredients were found:

Iron	85.54
Nickel	8.55
Cobalt	.61
Copper	.03
Phosphorus	.12
Chromic oxide	.21
Magnesia	2.04
Silica	3.02
Alumina	trace
	100.12

Which represent the following minerals:

Nickeliferous iron	93.81
Chrome iron	.41
Schreibersite	.84
Olivine	5.06
	100.12

* This iron is now in the Smithsonian Institution, as has been for several years. April, 1873.

Some few particles of olivine were separated mechanically, and readily recognized as such under the magnifying glass in connection with the action of acids, which readily decompose it, furnishing silica and magnesia. Some of the olivine is in a pulverulent condition, resembling that of the Atacama iron. The nickeliferous iron of this Tucson meteorite also resembles that of the Atacama iron. Calculated from the above results it consists of: Iron, 90.91; nickel, 8.46; cobalt, 63; copper, trace=100.00.

This meteorite is one of much interest, and it is to be hoped that some of our enterprising U. S. Topographical Engineers will yet be able to persuade the owners to part with it, and bring it to this country.

5. Meteoric Iron of Chihuahua, Mexico.

For the description of this meteorite I am indebted to the manuscript of Mr. Bartlett, and had hoped to have obtained a

Fig. 4.

fragment of it for examination from Dr. Webb, who detached pieces from the mass; but when applied to they were no longer in his possession. It exists at the *Hacienda de Conception*, about ten miles from Zapata. "The form is irregular. Its greatest height is forty-six inches; greatest breadth thirty-seven inches; circumference in thickest part eight feet three inches. Its weight, as given by Senor Urquida, is about three thousand eight hundred and fifty-three pounds. It is irregular in form, as seen by the figure; and one side is filled with deep cavities, generally round and of various dimensions. At its lower part, as it now stands, is a projecting leg, quite similar to the one on the meteorite at Tucson. The back or broadest part is less jagged than the other portions, and contains fewer cavities, yet, like the rest, is very irregular."

PART II.

THEORETICAL CONSIDERATIONS.

Under this head no mention will be made of the phenomena accompanying the fall of meteorites, as their light, noise, bursting, and their black coating, which arise after the bodies have entered the atmosphere, and are brought about by its agency. This omission will affect in no way the theoretical views under consideration, and the introduction of these particulars would uselessly increase the length of this memoir.

The lessons to be learned from meteorites, both stony and metallic, are probably not as much appreciated as they ought to be. We are usually satisfied with an analysis of them and surmises as to their origin, without due consideration of their physical and chemical characters.

The great end of science is to generalize facts that are observed. Thus terrestrial gravitation has been extended to the solar system, and in fact to the whole universe. The astronomer by his discoveries only proves the universality of this one law of nature operating on matter. He has found no evidence that any other force pertaining to terrestrial matter displays itself in a similar manner in other spheres. However true and self-evident it may appear that all matter in space is under the same laws, be they those of gravitation, cohesion, chemical affinity, etc., it is none the less interesting to have the fact proved; and meteorites when looked upon as bringing these proofs acquire additional interest.

Meteorites studied in the way just mentioned lead us to the inference that the materials of the earth are exact representatives of the materials of our system, for up to the present time no element has been found in a meteorite that has not its counterpart on the earth; or, if we are not warranted in making such a broad assumption, we certainly have the proof, as far as we may ever expect to get it, that materials of other portions of the universe are identical with those of our earth.

Meteorites also show that the *laws of crystallization* in bodies foreign to the earth are the same as those affecting terrestrial matter, and in this connection we may instance pyroxene, olivine, and chrome iron, affording in their crystalline form angles identical with those of terrestrial origin.

But perhaps of all the interesting facts under this head developed by meteorites is the universality of the laws of chemical affinity, or the truth that all the laws of chemical combination and atomic constitution are to be equally well seen in extra-terrestrial and terrestrial matter; so that were Dalton or Berzelius to seek for the atomic weights of iron, silica, or magnesia, they might learn them as well from meteoric minerals as from those taken from the bowels of the earth. The atomic constitution of meteoric anorthite or of pyroxene is the same as that which exists in our own rocks.

Keeping in view then the physical and chemical characters of meteorites, I propose to offer some theoretical considerations which to be fully appreciated must be followed step by step. These views are not offered because they individually possess particular novelty. It is the manner in which they are combined to which especial attention is called.

Physical Characteristics to be noted in Meteorites.

The first physical characteristic to be noted is their form. No masses of rock, however rudely detached from a quarry, or blasted from the side of a mountain, or ejected from the mouth of a volcano, would present more diversity of form than meteoric stones. They are rounded, cubical, oblong, jagged, flattened, and in fine they present a great variety of fantastic shapes. Now the fact of form I conceive to be a most important point for consideration in regard to the origin of these bodies; as the form alone is strong proof that the individual meteorites have not always been cosmical bodies, for had they been their form must have been spherical or spheroidal. As this is not so, it is reasonable to suppose that at one time or another they must have constituted a part of some larger mass. But as this subject will be taken up again, I pass to another point; namely, the crystalline structure, more especially that of the iron, and the complete separation in nodules, in the interior of the iron, of sulphuret and phosphuret of the metals constituting the mass. When this is properly examined it is seen that these bodies must have been in a plastic state for a great length of time, for nothing else could have determined such crystallization as we see in the iron, and allow such perfect separation of sulphur and phosphorus from the great bulk of the metal, combining only with a limited portion to form particular minerals; and did we aim to imitate such separation by artificial processes, we could only hope to do it by retaining the iron in a plastic condition for a great length of time. Also, no other agent than fire can be conceived of by which this metal could be kept in the condition requisite for the separation.

If these facts with reference to the crystalline structure be admitted, the natural suggestion is that they could only have been thus heated while a part of some large body.

Another physical fact worthy of being noticed here is the manner in which the metallic iron and stony parts are often interlaced and mixed, as in the Pallas and Atacama irons, where nickeliferous iron and olivine in nearly equal portions (by bulk) are intimately mixed, so that when the olivine is detached the iron resembles a very coarse sponge. This is an additional fact in proof of the great heat to which the meteorites must have been submitted, for with our present knowledge of physical laws there is no other way in which we can conceive that such a mixture of iron and olivine could have been produced.

Other physical points might be noticed, but as they are familiar to all, and would add nothing to the theoretical considerations, they will be passed over.

Mineralogical and Chemical Points in Meteorites.

The rocks or minerals of meteorites are not of a sedimentary character, nor such as are produced by the action of water. This is obvious to any one who will examine these bodies. A mineralogist will also be struck with the thin dark-colored coating on the surface of the stony meteorites. The coating, in most if not in all instances, is of atmospheric origin, being acquired after the meteorite enters the atmosphere, and as such no further notice will be taken of it; but I will proceed at once to notice the most interesting peculiarities under this head. First of all, metallic iron, alloyed with more or less nickel and cobalt, is of constant occurrence in meteorites, with but three or four exceptions—in some instances constituting the entire mass, at other times disseminated in fine particles through stony matter. The existence of this highly oxidizable mineral in its metallic condition is a positive indication of a scarcity or total absence of oxygen (in its gaseous state or in the form of water) in the locality whence it came.

Another mineralogical character of significance is that the stony portions of the meteorites resemble the older igneous rocks, and in even a more striking manner the volcanic rocks belonging to various active and extinct volcanoes. It is useless to dwell on this fact, as it is one well known to all mineralogists who may have examined this matter; and none have given more especial attention to it than Rammelsberg, who in a paper published in 1849 details his examination of a great variety of lavas, and traced the perfect parallelism between them and stony meteorites. He showed that the Juvenas stone has the same constitution as the Thjorsa lava of Heckla, both consisting substantially of augite and anorthite, even in nearly the same relative proportions; while the Chateau Renard and Nordhausen stones have labradorite replacing the anorthite; and the Blansko, Chantonnay, and Utrecht stones have oligoclase as the feldspar, and resemble the lavas of Ætna, Stromboli, and the newer lavas of Heckla.

The inference to be drawn from the last character is very evident. It is highly significative of the igneous origin of these bodies, and of an igneous action similar to that now existing in our volcanoes.

Yet another point of resemblance to certain of our terrestrial igneous rocks is the presence of metallic iron, for lately Mr. Andrews has proved the existence of metallic iron in basaltic rocks; but this will not be insisted on, as the quantity of iron discovered in basaltic rocks is so minute as only to be detected by the most delicate means of investigation.

Ever since the labors of Howard in 1802 the chemical constitution of meteorites has attracted much attention, more especially the elements associated in the metallic portion; and although we find no new elements, still their association, so far as yet known, is peculiar to this class of bodies. Thus nickel is a constant associate of iron in meteorites (if we except the Oswego, N. Y., meteorite upon whose claims to meteoric origin there yet remains some doubt); and although cobalt and copper are mentioned only as occasional associates, in my examination of near thirty known meteorites (in more than one half of which these constituents were not mentioned) I have found both of the last-mentioned metals as constantly as the nickel. With our more recent method of separating cobalt from nickel very accurate and precise results can be obtained as relates to the cobalt. The copper exists always in so minute proportion that the most careful manipulation is required to separate it.

Another element frequently but not always mentioned as associated with the iron is phosphorous. Here again my testing of thirty specimens led me to a similar generalization concerning phosphorus; namely, that no meteoric iron is to be expected without it. My examination has extended as well to the metallic particles separated from the stony meteorites as to the meteoric irons proper. It may be even further stated that in most instances the phosphorus was traceable directly to the mineral schreibersite.

These four elements then—iron, nickel, cobalt, and phosphorus—I consider remarkably constant ingredients. First in the meteoric irons proper, and secondly in the metallic particles of the stony meteorites; there being only some three or four meteorites among hundreds that are known in which they are not recognized.

As regards the combination of these elements it is worthy of remark that no one of them is associated with oxygen, although all four of them have strong affinity for this element, and are never found (except copper) in the earth uncombined with it, except where some similar element (as sulphur, etc.) supplies its place.*

The inference of the absence of oxygen in a gaseous condition, or in water, is drawn from such substances as iron and nickel being in their metallic state, as has been just mentioned. But it must not be inferred that oxygen is absent in all forms at the place of origin of meteorites, for the silica, magnesia, protoxide of iron, etc., contain this element. The occurrence of one class of oxides and not another would indicate a limited supply of the element oxygen, the more oxidizable elements, as silicon, magnesium, etc., having appropriated it in preference to the iron.

Many other elements worthy of notice might be mentioned here, and some of them, for aught we know, may be constant ingredients; but in the absence of strong presumption at least on this head they will be passed over, as those already mentioned suffice for the support of all theoretical views to be advanced.

I can not, however, avoid calling attention to the presence of *carbon* in certain meteorites; for although its existence is denied by some chemists, it is nevertheless a fact that can be as easily established as the presence of the nickel. The interest to be attached to it is due to the fact that it is so commonly regarded in the light of an organic element. It serves to strengthen the notion that carbon can be of pure mineral origin, for no one would be likely to suppose that the carbon found its way into a meteorite either directly or indirectly from an organic source.

Having thus noted the predominant physical, mineralogical, and chemical characteristics of meteorites, I pass on to the next head.

Marked Points of similarity in the Constitution of Meteoric Stones.

Had this class of bodies not possessed certain properties distinguishing them from terrestrial minerals, much doubt would

---

*The traces of iron found in basaltic rock already alluded to form too insignificant an exception to be insisted on.

even now be entertained of their celestial origin, and various would be the explanations made even in those cases where the bodies were seen to fall, and afterward collected. Chemistry has entirely dissipated all doubts in the matter, and now an examination in the laboratory of the chemist is entitled to more credit than evidence from any other source in pronouncing on the meteoric origin of a body. No question need be asked as to whether it was seen to fall, or whether this or that rock or mineral exists in the neighborhood where it may have been collected. The reagents of the chemist alone are unerring indications that suffice to set aside all caviling in the matter.

It is the object of this part of the paper to explain more prominently perhaps than has yet been done how it is that chemistry pronounces with such unerring certainty on the celestial origin of certain bodies; and I propose to go even a step farther, and see if the chemical constitution of the meteorites can indicate from what part of the heavens they may have come.

When the mineralogical and chemical composition of these bodies are regarded, the most ordinary observer will be struck with the wonderful family likeness running through them all, however unlike at first sight. There will be seen to be three great divisions of meteoric bodies (omitting three or four); namely, metallic, stony with small particles of metal, and a mixture of metallic and stony, in which the former predominates, as in the Pallas and Atacama meteorites.

As regards external appearances, these three classes differ in a very marked manner from each other. The *meteoric iron* being ordinarily of a compact structure, more or less corroded externally, and when cut showing a dense structure with most of the peculiarities of pure iron, only a little harder in texture and whiter in color. The *stony meteorites* are usually of a gray or greenish-gray color, granular structure, readily broken by a blow of the hammer, and exteriorly are covered with a thin coating of fused material. The *mixed meteorite* presents characters of both of the above; a large portion of it is constituted of the kind of iron already mentioned, cellular in its character, and the cells filled up with stony materials, similar in appearance to those constituting the second class.

Although there are some instances of bodies of undoubted

meteoric origin not falling properly under either of the above three heads, still they will be seen upon close investigation not to interfere in any way with the general conclusions that are attempted to be arrived at; for these constituents are represented in the stony materials of the second class, from which their only essential difference consists in the absence of metallic particles.

If we now examine chemically the three classes mentioned, we find them all possessed of certain common characteristics that link them together, and at the same time separate them from every thing terrestrial. Take first the metallic masses; and in very many instances, in some fissure or cavity exposed by sawing or otherwise, stony materials will frequently be found, and a stony crystal is sometimes exposed; now examine the composition of these, and then compare the results with what may be known of the stony meteorites, and in every instance it will agree with some mineral or minerals found in this latter class, as olivine or pyroxene—most commonly the former; but in no instance is it a mineral not found in the stony meteorites. If these last in their turn be examined, differing vastly in their appearance from the metallic meteorites, they will with but two or three exceptions be found to contain a malleable metal identical in composition with the metal constituting the metallic meteorites.

As to those mixed meteorites in which the metallic and stony portions seem to be equally distributed; their two elements are but representatives of the two classes just described. Examined in this way, there will be no difficulty in tracing the same signature on them all, indorsing the above as their true character, and almost serving to tell us whence they came. They may emphatically be said to have been linked in their origin by a chain of iron.

There is one mineral which there is every reason to believe constantly accompanies the metallic portions, and which may be regarded as a most peculiar mark of difference between meteorites and terrestrial bodies. It is the mineral screibersite (see first part of this memoir), to which the constant presence of phosphorus in meteoric iron is due. This mineral, as already remarked, has no parallel on the face of the globe, whether we consider its specific or generic character, there being no such thing as phosphuret of iron and nickel or any other phosphuret found among minerals. These facts render the consideration of schreibersite one of much interest, running as it probably does through all meteorites, and forming another point of separation between meteorites and terrestrial objects.

Another striking similarity in the composition of meteorites is the limited action of oxygen on them. In the case of the purely metallic meteorites we trace an almost total absence of this element. In the stony meteorites the oxygen is in combination with silicon, magnesium, etc., forming silica, magnesia, etc., that combine with small portions of other substances to form the predominant earthy minerals of meteorites. When iron is found in combination with oxygen it is found in its lowest state of oxidation, as in the protoxide of the olivine and chrome iron, and as magnetic oxide.

Without going further into detail as regards the similarity of composition of meteorites, they will be seen to have as strongly marked points of resemblance as minerals coming from the same mountain—I might almost say from the same mine—and it is not asking much to admit their having a *common center of origin*, and that, whatever the body from which they originate, it must contain no uncombined oxygen, and, I might even add, none in the form of water.

What is this center of origin? Physics does not point it out; and although the chemist can not explore the elementary constitution of any of the other great celestial bodies than the earth, he can examine those smaller celestial masses which come to the earth, and from his results stand on a firmer basis for theoretical conclusions.

Origin of Meteoric Stones.

In taking up the theoretical considerations of the origin of meteoric stones, it is of the utmost importance to reflect well before we confound shooting-stars and meteoric stones as all belonging to the same class of bodies—a view entertained by many distinguished observers. It is doubtless owing to the fact of their having been confounded that but little advance has been made in settling upon the origin of these bodies; in fact, owing to this manner of viewing the subject, observers such as Arago, Bissel, Olbers, and others have turned away from the original conception of the origin of meteoric stones to views of a different character, based on observations of the shooting-stars.

It may be a broad assumption to start with, that there is not a single evidence of the identity of shooting-stars (as exemplified by the periodical meteors of August and November) and these meteors which give rise to meteoric stones, and this conclusion is one arrived at by as full an examination of the subject as I am capable of making.* Some of the prominent reasons for such a conclusion will be mentioned.

Were shooting-stars and meteoric stones the same class of bodies, it is natural to suppose that the fall of the latter would be most abundant when the former are most numerous. In other words, these periodic occurrences of shooting-stars in August and November, and more particularly those immense showers that have been sometimes seen, ought to have been attended with the falling of one or more meteoric stones; whereas there is not a single instance on record where these showers have been accompanied with the falling of a meteoric stone. Again, in all instances where a meteoric body has been seen to fall, and has been observed even from its very commencement, it has been alone and not accompanied by other meteors. Very little reflection will serve to convince any one that an objection to the identity of the two classes of bodies, based upon the above fact, is of great weight.

Another strong objection to considering the bodies of the same nature is based on the want of proof of their velocities being the same. It is a pretty well-established fact that the average velocity of shooting-stars is sixteen and a half miles a second—a result arrived at by different observers, and doubtless a close approximation to the truth, as from the constant occurrence of shooting-stars thousands of observations may be made with comparative ease by different observers noting the same stars. Not so with meteoric stones, these occurrences being rare, sudden, and unexpected, and no two observers being ever

* Prof. D. Olmsted, in a most interesting article on the subject of meteors, to be found in the 26th volume of Amer. Jour. of Science and Arts, p. 132, insists upon the difference between shooting-stars and meteorites, and the time and attention he has devoted to the phenomena of meteors give weight to his opinion.

prepared to note the data requisite for calculating their velocities; besides, I am prepared to prove that the two or three cases of supposed determination of velocities of meteoric stones can not be considered even gross approximation to the truth; in fact, the difficulties in the way are so great that we probably never shall arrive at a knowledge of their velocities.* Not even their effect on striking the earth will furnish any data whereby to calculate their velocities before entering the atmosphere, for this medium must offer such enormous resistance to bodies penetrating at great velocities that these velocities must be reduced to but a fraction of what they originally were; and it is a question whether a body entering our atmosphere at ten miles a second would penetrate the soil to a much greater depth than one entering it at five miles a second, for the increased velocity of the former would cause an increased resistance in the atmosphere, and therefore have received proportionally a greater check before striking the earth.

Another fact tending to prove a dissimilarity between shooting-stars and meteoric stones is that the velocity of no one of the shooting-stars has been observed to be so low as to allow of their being considered satellites to the earth; their average velocity is sixteen and a half miles a second, and it requires a reduction to less than six miles a second for them to assume a path around the earth. Now assume what we may as to the original orbit of the meteoric stones, and as to their original velocity—let

---

* Under this head I will merely note what is considered one of the best established cases of the determination of velocity of a meteoric stone; namely, that of the Weston meteorite, the velocity of which Dr. Bowditch estimated to "*exceed three miles a second.*" Mr. Herrick considers the velocity very much greater, and writes among other things what follows: "The length of its path, from the observations made at Rutland, Vt., and at Weston, was at least one hundred and seven miles. This space being divided by the duration of the flight as estimated by two observers—viz., thirty seconds—we have for the meteor's relative velocity about *three and a half miles a second.* The observations made at Wenham, Mass., are probably less exact in this respect, and need not be mentioned here. An experienced observer, however intelligent, will give the time ten or even twenty-fold too large. One not unversed in science who saw the meteor is confident it could not have been in sight as long as ten seconds." The above is given as a specimen of the uncertain data we are to proceed upon in estimating the velocity of meteoric stones.

their orbit be around the sun and their velocity sixteen miles a second — there is one thing we know; namely, that these bodies do enter our atmosphere, and, it is but right to assume, often pass through the atmosphere without falling to the earth, sometimes passing through the very uppermost portion of that medium, at other times lower. What becomes of their original assumed velocity after this passage? As they can be so checked as to be drawn to the earth's surface, and thus stopped altogether in their passage, their velocities may be changed to any velocity from sixteen miles a second to zero, according to the amount of resistance they meet with; and what is equally true in this connection is that when the velocity falls below six miles a second (or thereabout) they can no longer escape from the attraction of the earth and resume their solar orbit, but must revolve as a satellite around the earth until ultimately brought to its surface by repeated disturbances.

The deduction from the above fact is as follows: that as the most correct observations have never given a velocity of less than nine miles a second to a shooting-star, it is reasonable to suppose that none have ever entered our atmosphere, or, what is perhaps still more reasonable, that the matter of which they are composed is as subtile as that of Encke's comet, and any contact with even the uppermost limit of the atmosphere destroys their velocity and disperses the matter of which they are composed. Other grounds might be mentioned for supposing a difference between shooting-stars and meteoric stones, and I have dwelt on it thus much because it is conceived of prime importance in pursuing the correct path that is to lead to the discovery (if it can be made) of their origin. It is also of no small value to the beautiful and probable theory of shooting-stars that we should separate every thing from it that may tend to affect its plausibility.

Various theories have been devised to account for their origin. One is that they are small planetary bodies revolving around the sun, and that at times they become entangled in the atmosphere, lose their orbital velocity by the resistance of the atmosphere, and are finally attracted to the earth. They are also supposed to have been ejected from the volcanoes of the moon; and lastly, they are considered as formed from particles floating in the atmosphere. The exact nature of this last theory is understood by reading the views of Prof. C. U. Shepard, as expressed in an interesting report on meteorites published in 1848. The author* says: "The extra-terrestrial origin of meteoric stones and iron masses seems likely to be more and more called in question with the advance of knowledge respecting such substances, and as additions continue to be made to the connected sciences. Great electrical excitation is known to accompany volcanic eruptions, which may reasonably be supposed to occasion some chemical changes in the volcanic ashes ejected; these being wafted by the ascensional force of the eruption into the regions of the magneto-polar influence, may there undergo a species of magnetic analysis. The most highly magnetic elements (iron, nickel, cobalt, chromium, etc.), or compounds in which these predominate, would thereby be separated, and become suspended in the form of metallic dust, forming those columnar clouds so often illuminated in auroral displays, and whose position conforms to the direction of the dipping-needle. While certain of the diamagnetic elements, or combinations of them, on the other hand, may under the control of the same force be collected into different masses, taking up a position at right-angles to the former (which Faraday has shown to be the fact in respect to such bodies), and thus produce those more or less regular arches, transverse to the magnetic meridian, that are often recognized in the phenomena of the aurora borealis. Any great disturbance of the forces maintaining these clouds of meteor-dust, like that produced by a magnetic storm, might lead to the precipitation of portions of the matter thus suspended. If the disturbance was confined to the magnetic dust, iron masses would fall; if to the diamagnetic dust, a non-ferruginous stone. If it should extend to both classes simultaneously, a blending of the two characters would ensue in the precipitate, and a rain of ordinary meteoric stones would take place. The occasional raining of meteorites might therefore on such a theory be as much expected as the ordinary deposition of moisture from the atmosphere. The former would originate in a mechanical

*I must, in justice to Prof. Shepard, say that since his paper was written he has informed me that he no longer entertains these views; and I would now omit the criticism of them did they not exist in his memoir uncontradicted, and also were they not views still entertained by some.

elevation of volcanic ashes and in matter swept into the air by tornadoes, the latter from simple evaporation. In the one case the matter is upheld by magneto-electric force; in the other by the law of diffusion which regulates the blending of vapors and gases, and by temperature. A precipitation of metallic and earthy matter would happen on any reduction of the magnetic tension; one of rain, hail, or snow on a fall of temperature. The materials of both originate in our earth. In the one instance they are elevated but to a short distance from its surface, while in the other they appear to penetrate beyond its farthest limits, and possibly to enter the interplanetary system. In both cases, however, they are destined, through the operation of invariable laws, to return to their original repository."

This theory, coming as it does from one who is justly entitled to high consideration from the fact of the special attention he has given to the subject of meteorites, may mislead; and for that reason objections will be advanced which will doubtless entirely set aside this notion of terrestrial origin, and to this end I would consider two fundamental principles of it. First of all it must be proved that terrestrial volcanoes contain all the varieties of matter found in the composition of the meteoric bodies. There is no doubt that many of the varieties are ejected from volcanoes, as olivine, etc.; but then the principal one, nickeliferous iron, has never in a single instance been found in the lava or other matter coming from volcanoes, although frequently sought for.

But the physical obstacles are a still more insuperable difficulty in the way of adopting this theory. In the first place it is considered a physical impossibility for tornadoes or other currents of air to waft matter, however impalpable, "beyond the farthest limits of the earth, and possibly into interplanetary space." Again, if magnetic and diamagnetic forces cause the particles to coalesce and form solid masses, by the cessation of those forces the bodies would crumble into powder. Another strong physical objection to the theory is, that as the consolidation of these masses is expected to take place in "magneto-polar" regions, their fall should only be in those portions of the earth; for, like rain and hail (to which the consolidation of these bodies are assimilated in this theory), they should fall perpendicularly, or nearly so, from their points of condensation. And lastly, under the head of physical objections, how can bodies so formed be precipitated in such very oblique directions as many are known to have, and that too from east to west and not from the north?

We pass on to a concise statement of some of the chemical objections to this theory of atmospheric origin, and, if possible, they are more insuperable than the last mentioned. Contemplate for a moment the first meteorite described in this paper. Here is a mass of iron, of about sixty pounds, of a most solid structure, highly crystalline, composed of nickel and iron chemically united, containing in its center a crystalline phosphuret of iron and nickel, and on its exterior surface a compound of sulphur and iron also in atomic proportions, and then see if the mind can be satisfied in supposing that the dust wafted from the crater of a volcano into the higher regions of the atmosphere could *in a few moments of time* be brought together by any known forces so as to create the body in question. However finely divided this volcanic dust might be, it can never be subdivided into atoms, a state of things that must exist to form bodies in atomic proportions where no agency is present to dissolve or fuse the particles concerned. One other objection and I am done with this theory.

The particles of iron and nickel supposed to be ejected from the volcano must pass from the heated mouth of a crater, ascend through the oxygen of the atmosphere without undergoing the slightest oxidation; for if there be any one thing which marks the meteorites more strongly than any other it is the freedom of the masses of iron from oxidation except on the surface; but a still more remarkable abstinence from oxidation would be the ascent of the particles of phosphorus to form the schreibersite traceable in so many meteorites. Having noticed the prominent objections to this theory, I pass on to consider, in as few words as possible, the other two theories.

A very commonly-adopted theory of the origin of meteoric bodies is that they are small planetary bodies revolving around the sun, one portion of their orbit approaching or crossing that of the earth; and from the various disturbing causes to which these small bodies must necessarily be subjected their orbits are constantly undergoing more or less variation until intersected by our atmosphere, when they meet with the most serious derangement, and fall to the earth's surface in whole or in part. This may not occur in their first passage through the atmosphere, but repeated obstructions in this medium at different times must ultimately bring about the result. In this theory their origin is supposed to be the same as that of other planetary bodies, and they are regarded as always having had an individual cosmical existence. Now, however reasonable the admission of this orbital motion immediately before and for some time previous to their contact with the earth, the assumption of their original cosmical origin would appear to have no support in the many characteristics of meteoric bodies as enumerated some pages back. The form alone of these bodies, is any thing but what ought to be expected from a gradual condensation and consolidation. All the chemical and mineralogical characters are opposed to this supposition. If the advocates of this theory do not insist on the last feature of it, then the theory amounts to but little else than a statement that meteoric stones fall to us from space while having an orbital motion. In order to entitle this planetary theory to any weight it must be shown how bodies formed and constructed as these are could be other than fragments of some very much larger mass.

As to the existence of meteoric stones in space, traveling in a special orbit prior to their fall, there can be but little doubt when we consider their direction and velocity, their composition proving them to be of extra-terrestrial origin. This, however, only conducts in part to their origin; and those who will examine them chemically will feel convinced that the earth is not the first great mass that meteoric stones have been in contact with, and this conviction is strengthened when we reflect on the strong marks of community of origin so fully dwelt upon. It is then in consideration of what was the connection of these bodies prior to their having an independent motion of their own that this memoir will be concluded.

LUNAR ORIGIN OF METEORIC STONES.

It only remains to bring forward the facts already developed to prove the plausibility of this origin of meteorites.

It is a theory that was proposed as early as 1660 by an Italian philosopher, Terzago, and advanced by Olbers, in 1795, without any knowledge of its having been before proposed. It was sustained by Laplace with all his mathematical skill from the time of its adoption to his death. It was also advocated on chemical grounds by Berzelius, whom I have no reason to believe ever changed his views in this matter; and to these we have to add the following distinguished mathematicians and philosophers: Biot, Brandes, Poisson, Quetelet, Arago, and Benzenberg, who have at one time or another advocated the lunar origin of meteorites.

Some of the above astronomers abandoned the theory, among them Olbers and Arago; but they did not do so from any supposed defect in it, but from adopting the assumption that shooting-stars and meteorites were the same; and on studying the former, and applying the phenomena attendant upon them to meteorites, the supposed lunar origin was no longer possible.

On referring to the able researches of Sears C. Walker on the periodical meteors of August and November (Trans. Am. Phil. Soc., Jan., 1841), that astronomer makes the following remarks about Olbers's change of views: "In 1836 Olbers, the original proposer of the theory of 1795, being firmly convinced of the correctness of Brandes's estimate of the relative velocity of meteors, renounces his *selenic theory*, and adopts the *cosmical* theory as the only one which is adequate to explain the established facts before the public."

For reasons already stated it appears wrong to assume the identity of meteorites and shooting-stars; so that whatever difficulty the phenomena of shooting-stars may have interposed in conceiving this or that to have been the origin of meteoric stones, it now no longer exists; and we are fully authorized in forming our conclusions concerning them to the utter disregard of the phenomena of shooting-stars. Had Olbers viewed the matter in this light he would doubtless have retained his original convictions, to which no material objection appears to have occurred to him for forty years.

It is not my object to enter upon all the points of plausibility of this assumed origin, or to meet all the objections which have been urged to it, for most of them have already been ably treated of. The object now is simply to urge such points developed in this memoir as appear to give strength to the lunar theory. They may be summed up under the following heads:

1. That all meteoric masses have a community of origin.
2. At one period they formed parts of some large body.
3. They have all been subject to a more or less prolonged igneous action, corresponding to that of terrestrial volcanoes.
4. That their source must be deficient in oxygen.
5. That their average specific gravity is about that of the moon.

From what has been said under the head of common characters of meteorites, it would appear far more singular that these bodies should have been formed separately from each other than that they should have at one time or another constituted parts of the same body; and from the character of their formation that body should have been of great dimensions. Let us suppose all the known meteorites assembled in one mass, and regarded by the philosopher, mindful of our knowledge of chemical and physical laws. Would it be considered more rational to view them as the great representatives of some one body that had been broken into fragments, or as small specks of some vast body in space that at one period or another has cast them forth? The latter, it seems to me, is the only opinion that can be entertained in reviewing the facts of the case.

As regards the igneous character of the minerals composing meteorites, nothing remains to be added to what has already been said. In fact, no mineralogist can dispute the great resemblance of these minerals to those of terrestrial volcanoes, they having only sufficient difference in association to establish that, although igneous, they are extra-terrestrial. The source must also be deficient in oxygen, either in a gaseous condition or combined as in water. The reasons for so thinking have been clearly stated as dependent upon the existence of *metallic iron* in meteorites; a metal so oxidizable that in its terrestrial associations it is almost always found combined with oxygen, and never in its metallic state.

What then is that body which is to claim common parentage of these celestial messengers that visit us from time to time? Are we to look at them as fragments of some shattered planet, whose great representatives are the numerous asteroids between Mars and Jupiter, and that they are "minute outriders of the asteroids" (to use the language of R. P. Greg, jr., in a late communcation to the British Association) which have been ultimately drawn from their path by the attraction of the earth? For more reasons than one this view is not tenable. Many of our most distinguished astronomers do not regard the asteroids as fragments of a shattered planet; and it is hard to believe if they were, and the meteorites the smaller fragments, that these latter should resemble each other so closely in their composition, a circumstance that would not be realized if our earth was shattered into a million of masses large and small.

If then we leave the asteroids and look to the other planets, we find nothing in their constitution or the circumstances attending them to lead to any rational supposition as to their being the original habitation of the class of bodies in question. This leaves us then but the *moon* to look to as the parent of meteorites; and the more I contemplate that body the stronger does the conviction grow that to it all these bodies originally belonged.

It can not be doubted from what we know of the moon that it is in all likelihood constituted of such matter as compose meteoric stones; and that its appearances indicate volcanic action, which, when compared with the combined volcanic action on the face of the globe, is like contrasting Ætna with an ordinary forge, so great is the difference. The results of volcanic throws and outbursts of lava are seen for which we seek in vain any thing but a faint picture on the surface of our earth. Again, in the support of the present view it is clearly established that there is neither atmosphere nor water on the surface of that body, and consequently no oxygen in those conditions which would preclude the existence of metallic iron.

Another ground in support of this view is based on the specific gravity of meteorites, a circumstance that has not been insisted on; and although of itself possessing no great value, yet in conjunction with the other facts it has some weight.

In viewing the cosmical bodies of our system with relation to their densities, they are divided into two great classes—planetary and cometary bodies (these last embracing comets proper and shooting-stars)—the former being of dense and the latter of very attenuated matter; and so far as our knowledge extends there is no reason to believe that the density of any comet approaches that of any of the planets. This fact gives some grounds for connecting meteorites with the planets. Among the planets there is also a difference, and a very marked one, in their respective densities; Saturn having a density of 0.77 to 0.75, water being 1.0; Jupiter, 2.00–2.25; Mars, 3.5–4.1; Venus, 4.8–5.4; Mercury between 7 and 36; Uranus, 0.8–2.9; that of the Earth being 5.67.* If then from specific gravity we are to connect meteorites to the planets, as their mean density is usually considered about 3.0,† they must come within the planetary range of Mars, Earth, and Venus. In the cases of the first and last we can trace no connection, from our ignorance of their nature and of the causes that could have detached them.

This reduces us then to our own planet, consisting of two parts, the planet proper, with a density of 5.76, and the moon, with a density of about 3.62.‡ On viewing this, we are at once struck with the relation that these bear to the density of meteorites, a relation that even the planets do not bear to each other.

As before remarked, I lay no great weight on this view of the density, but call attention to it as agreeing with conclusions arrived at on other grounds.

The chemical composition is also another strong ground in favor of their lunar origin. This has been so ably insisted on by Berzelius and others that it would be superfluous to attempt to argue the matter any further here; but I will simply make a comment on the disregard that astronomers usually have for this argument. In the memoir on the periodic meteors by Sears C. Walker, already quoted from, it is stated, "The chemical objection is not very weighty, for we may as well suppose a uniformity of constituents in cosmical as in lunar substances." From this conclusion it is reasonable to dissent, for as yet we are acquainted with the materials of but two bodies, those of

* For these estimates of the density of the planets the author is indebted to Prof. Peirce.

† Although the average specific gravity of the metallic and stony meteorites is greater, yet the latter exceeding the former in quantity, the number 3.0 is doubtless as nearly correct as can be ascertained.

‡ Although the densities of the earth and moon differ, these two bodies may consist of similar materials, for the numbers given represent the density of bodies as wholes. The solid crust of the earth for a mile in depth can not average a density of 3.0.

the earth and those of meteorites; and their very dissimilarity of constitution is the strongest argument of their belonging to different spheres. In further refutation of this idea it may be asked, Is it to be expected that a mass of matter detached from Jupiter (a planet but little heavier than water) or from Saturn (one nearly as light as cork) or from Encke's comet (thinner than air) would at all accord with each other or with those of the earth? It is far more rational to suppose that every cosmical body, without necessarily possessing elements different from each other, yet are so constituted that they may be known by their fragments. With this view of the matter our specimens of meteorites are but multiplied samples of the same body, and that body, with the light we now have, appears to have been the moon.

This theory is not usually opposed on the ground that the moon is not able to supply such bodies as the meteoric iron and stone. It is more commonly objected to from the difficulty that there appears to be in the way of this body's projecting masses of matter beyond the central point of attraction between the earth and the moon. Suffice it to say that Laplace, with all his mathematical acumen, saw no difficulty in the way of this taking place, although we know that he gave special attention to it at three different times during a period of thirty years, and died without discovering any physical difficulty in the way. Also for a period of forty years Olbers was of the same opinion, and changed his views, as already stated, for reasons of a different character. And to these two we add Hutton, Biot, Poisson, and others whose names have been already mentioned.

Laplace's view of the matter was connected with present volcanic action in the moon; but there is every reason to believe that all such action has long since ceased in the moon. This, however, does not invalidate this theory in the least, for the force of projection and modified attraction to which the detached masses were subjected only gave them new and independent orbits around the earth, that may endure for a great length of time before coming in contact with the earth.

The various astronomers cited concur in the opinion that a body projected from the moon with a velocity of about eight thousand feet per second would go beyond the mutual point of attraction between the earth and moon, and already having an orbital velocity, may become a satellite of the earth with a modified orbit.

The important question then for consideration is the force requisite to produce this velocity. The force exercised in terrestrial volcanoes varies. According to Dr. Peters, who made observations on Ætna, the velocity of some of the stones was 1,250 feet a second, and observations made on the peak of Teneriffe gave 3,000 feet a second. Assuming, however, the former velocity to be the maximum of terrestrial volcanic effects, the velocity with which the bodies started (stones with a specific gravity of about 3.00) must have exceeded 2,000 feet a second to permit of an observed velocity of 1,250 feet through the denser portions of our atmosphere.

Doubtless the projectile force of lunar volcanoes far exceeds that of the terrestrial; and this we infer from the enormous craters of elevation to be seen upon its surface, and their great elevation above the general surface of the moon, with their borders thousands of feet above their center; all of which point to the immense internal force required to elevate the melted lava that must have at one time poured from their sides. I know that Prof. Dana, in a learned paper on the subject of lunar volcanoes, argues that the great breadth of the craters is no evidence of great projectile force, the pits being regarded as boiling craters, where force for lofty projection could not accumulate. Although his hypothesis is ingeniously sustained, still, until stronger proof is urged, we are justified, I think, in assuming the contrary to be true, for we must not measure the convulsive throes of nature at all periods by what our limited experience has enabled us to witness.

As regards the existence of volcanic action in the moon, without air or water, I have nothing at present to do, particularly as those who have studied volcanic action concede that neither of these agents is absolutely required to produce it. Moreover, the surface of the moon is the strongest evidence we have in favor of its occurring under those circumstances.

But it may be very reasonably asked, Why consider the moon the source of these fragmentary masses called meteorites? May not smaller bodies, either planets or satellites, as they pass by the earth and through our atmosphere, have portions detached by the mechanical and chemical action to which they are subjected? To this I will assent as soon as the existence of that body or those bodies is proved. Are we to suppose that each meteorite falling to the earth is thrown off from a different sphere which becomes entangled in the atmosphere? If so, how great the wonder that the earth has never intercepted one of those spheres, and that all should have struck the stratum of air surrounding our globe (some fifty miles in height), and escaped the body of the globe, 8,000 miles in diameter. It is said that the earth has never intercepted one of these spheres; for if we collect together all the known meteorites, in and out of cabinets, they would hardly cover the surface of a good-sized room, and no one of them could be looked upon as the maternal mass upon which we might suppose the others to have been grafted. And this would appear equally true if we consider the known meteorites as representing not more than a hundredth part of those which have fallen.

If it be conceived that the same body has given rise to them, and is still wending its path through space, only seeming by its repeated shocks with our atmosphere to acquire new vigor for a new encounter with that medium, the wonder will be greater that it has not long since encountered the solid part of the globe; but still more strange that its velocity has not been long since destroyed by the resistance of the atmosphere, through which it must have made repeated crossings of over one thousand miles in extent.

But it may be said that facts are stronger than arguments, and that bodies of great dimensions (even over one mile in diameter) have been seen traversing the atmosphere, and have also been seen to project fragments and pass on. Now of the few instances of the supposed large bodies I will only analyze the value of the data upon which the Wilton and Weston meteorites were calculated; and they are selected because the details connected with them are more accessible. The calculations concerning the latter were made by Dr. Bowditch; but his able calculations were based on deceptive data; and this is stated without hesitation, knowing the difficulty, admitted by all, of making correct observation as to size of luminous bodies passing rapidly through the atmosphere. Experiments that would be considered superfluous have been instituted to prove the perfect fallacy of making any but a most erroneous estimate of the size of luminous bodies by their apparent size, *even when their distance from the observer and the true size of the object are known.* How much more fallacious then any estimate of size made where the observer does not know the true size of the body, and not even his distance very accurately.

In my experiments three solid bodies in a state of vigorous incandescence were used: first, charcoal points transmitting electricity; second, lime heated by the oxyhydrogen blowpipe; third, steel in a state of incandescence in a stream of oxygen gas. They were observed on a clear night at different distances, and the body of light (without the bordering rays) compared with the disk of the moon, then nearly full and 45° above the horizon. Without going into details of the experiment the results will be tabulated:

	Actual diam. as seen at 10 in.	Apparent diameter at 200 yds.	Apparent diameter at ¼ mile.	Apparent diameter at ½ mile.
	Inch.	Diameter moon's disk.	Diameter moon's disk.	Diameter moon's disk.
Carbon points	$\frac{3}{10}$	½	3	3½
Lime light	$\frac{4}{10}$	⅓	2	2
Incandescent steel	$\frac{2}{10}$	¼	1	1

If then the apparent diameter of a luminous meteor at a given distance is to be accepted as a guide for calculating the real size of these bodies the

Charcoal* points would be 80 feet in diameter, instead of $\frac{3}{10}$ of an inch.
Lime " 50 " " " $\frac{4}{10}$ "
The steel globule " 25 " " " $\frac{2}{10}$ "

It is not in place to enter into any explanation of these deceptive appearances, for they are well-known facts, and were tried in the present form only to give precision to the criticism on the supposed size of these bodies. Comments on them are also unnecessary, as they speak for themselves. But to return to the two meteorites under review.

* Estimate made according to a table given by Prof. Olmsted (Amer. Jour. of Science and Arts, vol. xxvi, p. 155) for estimating the diameter of meteors on comparison with the moon.

That of Wilton was estimated by Mr. Edward C. Herrick (Amer. Jour. of Science and Arts, vol. xxxvii, p. 130) to be about one hundred and fifty feet in diameter. It appeared to increase gradually in size until *just before the explosion*, when it was at its largest apparent magnitude of one fourth the moon's disk; exploded twenty-five to thirty degrees above the horizon with a heavy report that was heard about thirty seconds after the explosion was seen. One or more of the observers saw luminous fragments descend toward the ground. When it exploded it was three or four miles above the surface of the earth; immediately after the explosion it was no longer visible. The large size of the body is made out of the fact of its appearing one fourth the apparent disk of the moon at about six miles distant. After the experiments just recorded and easy of repetition, the uncertainty of such a conclusion must be evident; and it is insisted on as a fact easy of demonstration that a body in a state of incandescence (as the ferruginous portions of a stony meteorite) might exhibit the apparent diameter of the Wilton meteorite at six miles distance, and not be more than a few inches or a foot or two in diameter, according to the intensity of the incandescence.

Besides, if that body was so large, where did it go to after throwing off the supposed small fragments? The fragments were seen to fall, but the great ignited mass suddenly disappeared at thirty degrees above the horizon, four miles from the earth, when it could not have had less than six or seven hundred miles of atmosphere to traverse before it reached the limit of that medium; it has already acquired a state of ignition in its passage through the air prior to the explosion, and should have retained its luminous appearance consequent thereupon, at least while remaining in the atmosphere; but as this was not the case, and a sudden disappearance of the entire body took place in the very lowest portions of the atmosphere, and descending luminous fragments were seen, the natural conclusion appears to be that the whole meteorite was contained in the fragments that fell.

As to the Weston meteorite, it is stated that its direction was nearly parallel to the surface of the earth at an elevation of about eighteen miles; was one mile farther when it exploded; the length of its path from the time it was seen until it exploded was at least one hundred and seven miles; duration of flight estimated at about thirty seconds, and its relative velocity three and a half miles a second; it exploded; three heavy reports were heard; *the meteorite disppaeared at the time of the explosion.*

As to the value of the data upon which its size was estimated, the same objection is urged as in the case of the Wilton meteorite, and it is hazarding nothing to state that the apparent size may have been due to an incandescent body a foot or two in diameter. Also with reference to its disappearance there is the same inexplicable mystery. It is supposed from its enormous size that but minute fragments of it fell, yet it disappeared at the time that this took place, which, it is supposed, occurred nineteen miles above the earth—an estimate doubtless too great when we consider the heavy reports. Accepting this elevation, what do we have? A body one mile and a half in diameter in a state of incandescence, passing in a curve almost parallel to the earth, and while in the very densest stratum of air that it reaches with a vigorous reaction between the atmosphere and its surface, and a dense body of air in front of it, is totally eclipsed; while if it had a direction only tangential to the earth, instead of nearly parallel, it would at the height of nineteen miles have had upward of five hundred miles of air of variable density to traverse, which at the relative velocity of three and a half miles a second (that must have been constantly diminishing by the resistance) would have taken about one hundred and forty-three seconds. It seems most probable that if this body was such an enormous one, it should have been seen for more than ten minutes after the explosion, for the reasons above stated. The fact of its disappearance at the time of the explosion is strong proof that the mass itself was broken to fragments, and that these fragments fell to the earth, assuring us that the meteorite was not the huge body represented, but simply one of those irregular stony fragments which, by explosion from heat and great friction against the atmosphere, become shattered. I say irregular, because we have strong evidence of this irregularity in its motion, which was "scolloping"—a motion frequently observed in meteorites, and doubtless due to the resistance of the atmosphere upon the irregular mass, for a spherical body passing through a resisting medium at a great velocity would not show this. In fact, if almost any of the specimens of meteorites in our cabinets were discharged from a cannon, even in their limited flight the scalloping motion would be seen.

This then will conclude with what I have to say in contradiction to the supposition of large solid cosmical bodies passing through the atmosphere and dropping small portions of their mass. The contradiction is seen to be based, first, upon the fact that no meteorite is known of any great size—none larger than the granite balls to be found at the Dardanelles alongside of the pieces of ordnance from which they are discharged; secondly, on the fallacy of estimating the actual size of these bodies from their apparent size; and lastly, from its being opposed to all the laws of chance that these bodies should have been passing through an atmosphere for ages and none have yet encountered the body of the earth.

To sum up the theory of the lunar origin of meteorites, it may be stated *that the moon is the only large body in space of which we have any knowledge possessing the requisite conditions demanded by the physical and chemical properties of meteorites; and that they have been thrown off from that body by volcanic action (doubtless long since extinct), and, encountering no gaseous medium of resistance, reached such a distance as that the moon exercised no longer a preponderating attraction; the detached fragment, possessing an orbital motion and an orbital velocity, which it had in common with all parts of the moon, but now more or less modified by the projectile force and new condition of attraction in which it was placed with reference to the earth, acquired an independent orbit more or less elliptical. This orbit, necessarily subject to great disturbing influences, may sooner or later cross our atmosphere and be intercepted by the body of the globe.*

In concluding this lengthy examination I must say that a discussion of the phenomena accompanying the falling of meteorites has been avoided, as well as many points connected with their history. This has been done from its having no immediate connection with the object of this memoir, which is intended simply to present some new views and many old views in a new light, so as to awaken attention to the study of this most interesting class of bodies.

BISHOPVILLE METEORIC STONE:

CHLADNITE PROVED TO BE A MAGNESIAN PYROXENE.

In 1846 Professor C. U. Shepard published an account of an exceedingly interesting meteoric stone that fell at Bishopville, S. C., in 1843, differing in its external character from other meteoric stones; the fractured mass being exceedingly white, except where metallic iron and other associate minerals occur. I would refer the reader to Prof. Shepard's description of it in the American Journal of Science and Arts, September, 1846, page 381.

The composition of the snow-white mineral (constituting about ninety per cent. of the entire mass), as given by Prof. Shepard, is

		Oxygen.	Oxy. ratio.
Silica	70.41	35.205	3
Magnesia	28.25	11.300	1
Soda	1.39	.338	..

From the results of this analysis he considered it a *tersilicate of magnesia*, constituting a new species, to which he gave the name of *chladnite*.

Several years after this examination a fragment of this meteoric stone came into my possession, and separating a small portion of the mineral in question it was examined. The result of this incomplete examination justified the statement in a note to a memoir of mine on meteorites, presented to the American Scientific Association in April, 1854, and published in the Amer. Jour. of Science and Arts for March, 1855, page 162, "that from some investigations just made chladnite is likely to prove a pyroxene."

Since that announcement I have been placed in possession of other fragments of the meteorite, and have been able to separate the "chladnite" perfectly pure, and in sufficient quantity to submit it to a thorough analysis.

To render the chladnite soluble in acid it was fused with four times its weight of carbonate of soda and potash, with a small fragment of caustic potash placed on the top of the mixed powders in the crucible.* After fusion the analysis was proceeded with in the ordinary way. The results of two analyses were as follows:

	1	2
Silica	60.12	59.83
Magnesia	39.45	39.22
Peroxide of iron	.30	.50
Soda, with feeble potash and strong lithia reaction,	.74	.74
	100.61	100.29

The minute quantity of peroxide of iron came from exceedingly fine particles of iron diffused through the minerals, and could be seen by a magnifying-glass. One separate analysis was made for the soda. The constitution of the mineral, as made out from the numbers in analysis 1, is

	Oxygen.	Oxy. ratio.
Silica	31.22	2
Magnesia	15.51	1
Soda	.19	

corresponding to the formula $\dot{M}g^3 \dddot{S}i^2$, equivalent to the general formula of pyroxene, $\dot{R}^3 \dddot{S}i^2$.

The excess of silica obtained by Professor Shepard in his analysis is doubtless due to an imperfect fusion of the mineral with the carbonate of soda, an error easily made if the precautions I have already mentioned are not attended to.

"Chladnite" approaches those forms of pyroxene known as white augite, diopside, white coccolite, etc.; these last-named minerals having a part of the magnesia replaced by lime. It is identical in composition with *enstatite* of Kenngott, a pyroxenic mineral from Aloysthal, in Moravia.

From these observations it will be seen that the Bishopville meteoric stone, however different in external characteristics from other smaller bodies, is after all identical with the great family of pyroxenic meteoric stones.

* I would remark that I seldom or ever fuse a silicate with the alkaline carbonates without the addition of a small piece of caustic potash or soda, and never analyze a known or supposed pyroxene or hornblende without this precaution. I have no doubt that there are many minerals classified with hornblende which properly belong to pyroxene, the silica in the analyses being rated too high, an error arising from an imperfect fusion.

HARRISON COUNTY (IND.) METEORITES.

(Fell March 28, 1859.)

Having become acquainted with a remarkable phenomenon accompanied with a fall of stones that occurred in Harrison County, Indiana, I immediately made inquiries concerning it, expecting to visit the neighborhood on an early occasion; but I was fortunate enough to learn of some admirable observations made by Mr. E. S. Crosier, and in fact so complete were his examinations that I clearly saw no additional information could be elicited by my resorting to the spot. Mr. Crosier obtained for me the various stones that had been found, and also put himself to much trouble to obtain the information desired.

The stones fell on Monday the 28th of March, 1859, and Mr. Crosier visited the place on the Saturday following; in the mean time the following stones were discovered:

No. 1,	weighing	19 ounces,	discovered by	Goldsmith.
No. 2,	"	4½ "	"	Crawford.
No. 3,	"	420 grains,	"	Lamb.
No. 4,	"	167 "	"	Mrs. Kelly.

The following are the facts elicited by inquiry on the spot. The time at which it occurred (four o'clock in the afternoon) rendered the phenomenon of ready observation. The area of observation was about four miles square, and wherever persons were about in that area the stones were heard hissing in the air and then striking on the ground or among the trees. Hardly a single person in the immediate vicinity of the occurrence saw any flash or blaze, as was noticed by all who heard the report from a distance.

Three or four loud reports, like the bursting of bombshells, were the first intimations of any thing unusual. A number of smaller reports followed, resembling the bursting of stones in a lime-kiln. The stones were seen to fall after the first four loud explosions. Those who happened to be in the woods or near them heard the stones distinctly striking amongst the trees. In some places the noise of the falling stones in the woods alarmed the cattle and horses in the vicinity, so that they fled in terror. A peculiar hissing noise during the fall of the stones was heard for miles around. A very intelligent lady described it as very much like the sound produced by pouring water upon hot stones. The air seemed as if all at once it had become filled with thousands of serpents.

Mr. Crawford and his wife were standing in their yard at the time, and hearing a loud hissing sound overhead, on looking up a stone (No. 2) was seen to fall just before them, burying itself four inches in the ground; they dug it up immediately, but it did not possess any warmth; it had a sulphurous smell. Another, which they did not find, fell near them, when they thought it prudent to retire to the house.

Two sons of John Lamb were in the barn-yard attending to the horses, when their attention was called to a loud hissing noise above, and immediately a stone (No. 3) fell just at their feet, penetrating the hard-tramped earth some three or four inches, and they state that it was warm when taken from the ground. Another fell in a peach-tree near by, but the ground being newly plowed they were unable to find it.

The largest stone (No. 1) was not obtained until the following day, being dug up beside a horse-track on the streets of Buena Vista, Indiana, it having penetrated the hard gravel to the depth of four or five inches. It had a strong smell of sulphur. The last (No. 4) was dug up by Mrs. Kelly the following day in her yard.

These four aërolites, owing to their being buried deeply in the ground, are all that have been found up to this time. None have been found or were heard to fall over a greater area than four miles square.

These are all the details that I have been able to gather connected with this fall of meteoric stones. They are highly interesting, and probably as accurate as it is possible to obtain.

Nos. 1, 2, and 3 and a fragment of No. 4 were placed in my hands for examination. Nos. 1, 2, and 4 are cuboidal in shape; No. 3 was considerably elongated. They are all covered by a very black vitrified surface, equally intense on every one and on every part of each one, and when broken show the usual gray color of stony meteorites interspersed with bright metallic particles.

The mean specific gravity is 3.465; when broken up and examined under a glass four substances are distinguishable—metallic particles, dark glassy mineral, dark dull mineral, white mineral matter.

Examined as a whole, the following elements were found in it: iron, nickel, cobalt, copper, phosphorus, sulphur, silicium, calcium, aluminum, magnesium, manganese, sodium, potassium, oxygen. By the action of the magnet it was separated into

Nickeliferous iron	4.91
Earthy minerals	95.19
	100.00

The earthy minerals acted on by warm dilute hydrochloric acid, thrown on a filter and thoroughly washed, then treated with dilute caustic potash, to dissolve any silica of the decomposed portion that was not dissolved by the acid, gave

Soluble portion	62.49
Insoluble portion	37.51

The metallic portion separated from the earthy part gave

Iron	86.781
Nickel	13.241
Cobalt	.342
Copper	.036
Phosphorus	.026
Sulphur	.022

The earthy portion freed from metal gave

Silica	47.06
Oxide of iron	26.05
Magnesia	27.61
Alumina	2.35
Lime	.81
Soda	.42
Potash	.68
Protoxide of manganese	trace, not estimated.

It is clear from the analyses made out that these meteoric stones contain the constituents frequently found in similar bodies; namely, nickeliferous iron, phosphuret of iron and nickel, sulphuret of iron, olivine, pyroxene, and albite, and in about the following proportions:

Nickeliferous iron	4.989
Schreibersite	.009
Magnetic pyrites	.001
Olivine	61.000
Pyroxene and albite	34.000

I have no intention to enter into any speculations in relation to these meteoric stones, although I have accumulated some additional matter on the subject since my memoir on meteorites published in the American Journal of Science and Arts, vol. xix, pp. 152 and 322, intending to reserve their publication for a future occasion.

DESCRIPTION OF THREE NEW METEORITES.

NELSON COUNTY (KY.) METEORITE.

This came into my possession in the month of July, 1860, being obtained from a plowed field, where it may have laid for a considerable length of time.

It is a flattened mass of tough metal, a little scaly at one corner, being seventeen inches long, fifteen inches broad, and seven inches in the thickest part, shelving off like the back of a turtle, and weighs one hundred and sixty-one pounds.

It is free from any large proportion of thick rust, consequently showing no indication of chlorine. On analysis the following constituents were found in one hundred parts, No. 1 in the table below:

	1	2	3
Iron	93.10	90.12	91.12
Nickel	6.11	8.72	7.82
Cobalt	.41	.32	.43
Phosphorus	.05	.10	.08
Copper	trace.	trace.	trace.
	99.67	99.26	99.45

MARSHALL COUNTY (KY.) METEORITE.

A piece of this meteorite was sent to me from Marshall County, in this state. I have not yet seen the entire mass, which is said to weigh fifteen pounds, and to be scaly in structure. It has the usual characteristics of meteoric iron, as seen from the analysis No. 2.

MADISON COUNTY (N. C.) METEORITE.

This meteorite was presented to me in the year 1854 by Hon. T. L. Clingman, of North Carolina. It came from Jewel Hill, Madison County, of that state. There is a great deal of thick rust on the surface, with constant deliquescence from chloride of iron. Its form and surface indicate that it is entire. Its dimensions are $7 \times 6 \times 3$ inches, with a number of indentations. Its weight is eight pounds thirteen ounces. Its composition is given in the analysis No. 3.

GUERNSEY COUNTY (OHIO) METEORITES.

(FELL MAY 1, 1860.)

These meteorites were first called Concord meteorites, as the one first described was found near the village New Concord; but I have thought proper to call them the *Guernsey County meteorites*, since we are commonly in the habit of distinguishing the meteorites found in this country by the name of the county in which they fell or were found. All but one of the great number of meteoric stones that fell on this occasion were found in Guernsey County, and that exceptional specimen fell in Muskingum, on the edge of Guernsey County.

This fall of meteorites was the most remarkable ever observed in this country, and equal to, if not surpassing, the famous fall at l'Aigle, in France, with which it has many points of interest in common that will be stated in the course of this paper.

My attention was first directed to this occurrence, by a short notice of it in a newspaper, as being an earthquake that had occurred in eastern Ohio, accompanied with a shower of stones. Suspecting the true nature of the phenomenon, I immediately visited the spot where it was said to have occurred, and collected the statements of those persons who had witnessed the fall. It was ascertained that on Tuesday, May 1, 1860, remarkable phenomena transpired in the heavens, of which the following are accounts given by different observers, men of intelligence and observation.

Mr. McClenahan states that at Cambridge, in Guernsey County, Ohio (lat. 40° 4′, long. 81° 35′), about twenty minutes before one o'clock P. M., three or four distinct explosions were heard, like the firing of heavy cannon, with an interval of a second or two between each report. This was followed by sounds like the firing of musketry in quick succession, which ended with a rumbling noise like distant thunder, except that it continued with about the same degree of intensity until it ceased. It continued two or three minutes, and seemed to come from the south-west, at an elevation above the horizon of thirty to forty degrees, terminating in the south-east at about the same elevation. In the district where the meteorites fell the explosions were heard immediately overhead.

The first reports were so heavy as to produce a tremulous motion, like heavy thunder, causing the glass in windows to rattle. The sound was so singular that it caused some excitement and alarm, many supposing it an earthquake. At Barnesville, twenty miles east of Cambridge, the cry of fire was made, as the rumbling sound was thought to be the roaring of fire.

The day was cool and the sky covered at the time with light clouds. No thunder or lightning had been noticed that day, nor could any thing unusual be seen in the appearance of the clouds. Immediately on hearing the report this observer looked in the direction it came, and noticed the clouds closely, but could not see any thing unusual.

The next morning it was reported in Cambridge that aërolites had fallen on a farm in the vicinity of New Concord (eight miles west, a little south of Cambridge). Inquiries were immediately instituted, and Messrs. Noble and Hines state that they were near the house of a Mr. Amspoker at the time of the first explosion, which seemed directly over their heads. They looked up and saw two objects apparently come through the clouds, producing a twirling in the vapor of the cloud at the point where they came through, then descending with great velocity and a whizzing sound to the earth; one striking about three hundred yards to the south-west of them, and the other about one hundred yards north.

They immediately went to the spot where the first fell, and found it buried two feet in the ground. They dug it out and found it quite warm and of a sulphurous smell. The other struck a fence-corner, and breaking the ends of some of the rails penetrated into the earth sixteen or eighteen inches, passing through a heap of dry leaves. The first weighed fifty-two pounds. The other was broken up, but must have weighed about forty pounds. Another of forty-one pounds weight, not seen to fall, was discovered at the bottom of a hole two feet deep, where it had fallen on stiff turf, and was seen at the bottom of the hole, having carried the sod before it. It must have come from the south-east at an angle of sixty degrees with the horizon. Many were discovered to have fallen south-east of Cambridge, but of smaller dimensions than those already referred to. At the time of the occurrence nearly all were at dinner or in and about their houses. The stones obtained were mostly found near houses, where they were seen to fall, as the sound of their striking the ground attracted attention.

Another well-informed observer, Dr. McConnell, of New Concord (a small town eight miles east of Cambridge), furnishes the following particulars: "On Tuesday, the 1st of May, at twenty-eight minutes past twelve o'clock, the people of that vicinity were almost panic-stricken by a strange and terrible report in the heavens, which shook the houses for many miles distant. The first report was immediately overhead, and after an interval of a few seconds was followed by similar reports with such increasing rapidity that after the number of twenty-two were counted they were no longer distinct, but became continuous, and died away like the roaring of distant thunder, the course of the reports being from the meridian to the south-east. In one instance three men working in a field, their self-possession being measurably restored from the shock of the more terrible report from above, had their attention attracted by a buzzing noise overhead, and soon observed a large body descending strike the earth at a distance of about one hundred yards. Repairing thither they found a newly-made hole in the ground, from which they extracted an irregular quadrangular stone weighing fifty-one pounds. This stone had buried itself two feet beneath the surface, and when obtained was quite warm."

To this we add the following statement: "We the undersigned do hereby certify that at about half past twelve o'clock on Tuesday, May 1, 1860, a most terrible report was heard immediately overhead, filling the neighborhood with awe. After an interval of a few seconds a series of successive reports, the most wonderful and unearthly ever before heard by us, took place, taking a direction from meridian to south-east, where the sounds died away like the roaring of distant thunder, jarring the houses for many miles distant." Signed by A. G. Gault, Jas. McDonald, Nancy Mills, Ichabod Grumman, Samuel Harper, Rev. Jas. C. Murch, Mrs. M. Speer, Ang'e McKinney.

The above is from those who heard the noises, but did not see the fall; the following are a few statements of the many I collected from those who witnessed the fall of the stones. I extract from their depositions made at the time:

"I heard the reports and roaring as above described, and a few seconds afterward I saw a large body or substance descend and strike the earth four or five hundred yards from where I then stood; and then I, in company with Andrew Lister, repaired to the spot, and about eighteen inches beneath the surface found a stone weighing fifty pounds." Signed by Samuel Reblu.

"Heard the reports and roaring as above described; and the said Mrs. Fillis further says that a few seconds afterward she heard a descending buzzing noise as of a body falling to the ground. And Miss Cherry also says that she was standing near Mrs. Fillis, heard the same, and saw some substance descend and strike the earth some hundred yards distant, and that Mrs. Fillis repaired to the spot and there found a stone, eighteen inches beneath the surface, weighing twenty-three pounds." Signed by Agnes Fillis and Mary J. Cherry.

"I distinctly heard the roaring and sounds as above described, and a few seconds after the above report I saw descending from the clouds a large body that struck the earth about one hundred and fifty yards from where I then stood, and I immediately repaired to the spot, and about two feet beneath the surface found a stone weighing forty-two pounds. A second or two after seeing the first stone I saw another descend and strike the earth about the same distance from where I stood. I also took the last-mentioned stone from the earth about two feet beneath the surface. Both the above stones when taken from the earth were quite warm. I also saw a third stone descend." Signed by Samuel M. Noble.

One observer saw a stone fall within three feet of his horse's head. One of the most southerly stones struck a barn, while some people retired within doors for fear of being struck.

These, with many others of a similar nature, were the data obtained near the region of the fall of stones. It is important to remember that to these near observers no luminosity or fire-ball was visible.

In addition to the above facts we have the following from

observers at more distant points, as already published by Professors Andrews and Evans. From the data they have collected they consider the area over which the explosion was heard as probably not less than one hundred and fifty miles in diameter. "At Marietta, Ohio, the sound came from a point north or a little east of north. The direction of the sound varied with the locality. An examination of all the different directions leads to the conclusion that the central point from which the sound emanated was near the southern part of Noble County, Ohio;" its course being "over the eastern end of Washington County, then across the interior of Noble County, then over the south-western corner of Guernsey and the north-eastern corner of Muskingum, with a direction of about forty-two degrees west of north."

Mr. D. Mackley, of Jackson County, states that he was at Berlin, six miles east of Jackson, Ohio, when he saw in a north-east direction a ball of fire about thirty degrees above the horizon. It was flying in a northerly direction with great velocity. It appeared as white as melted iron, and left a bright streak of fire behind it, which soon faded into a white vapor. This remained more than a minute, when it became crooked and disappeared.

Mr. Wm. C. Welles, of Parkersburg, Virginia (lat. 39° 10′, long. 81° 24′), about sixty miles south of Cambridge, saw the meteorite as a ball of fire of great brilliancy emerging from behind one cloud and disappearing behind another. Other observers at some distance to the south of the point where the fall occurred saw this meteorite as a luminous body.

Prof. Evans, of Marietta, in his observations states:

"The successive reports heard at great altitudes in the district where the stones fell, and apparently connected with the descent of the separate pieces through the clouds, were entirely distinct from the one great detonation which was heard at great distances from that district. The former were distinctly heard only over an area of a few miles. The latter shook the buildings from Wheeling, Virginia, to Athens County, Ohio. It is ascertained by careful inquiries to have been heard from Columbiana County on the north-east to within eight miles of Chillicothe on the south-west, and from Knox County on the north-west to the borders of the third tier of counties in Virginia on the south-east—an area of about one hundred and fifty miles in diameter. At all places within this area, except those near Cambridge and New Concord, it was described as a single sound, a sudden concussion resembling thunder or the discharge of a heavy piece of ordnance, followed by a roar of about two seconds in continuance. A merchant of Marietta, happening to be at dinner, suspected it was the explosion of a powder-magazine in his store about a quarter of a mile distant. The Parkersburg News says 'the houses shook as with an earthquake.' In the counties of Washington, Morgan, Noble, Monroe, and Belmont, and in places along the Virginia side of the Ohio River from Parkersburg to Wheeling, those who were within doors very generally attributed it to an earthquake. The windows rattled, and local papers state that the door of an engine-house was jarred open at Bellair near Wheeling. The lines of direction of the sound from all sides, as distinguished by those who happened to be out of doors, cross each other in the southern (not far from the central) part of Noble County, while the inhabitants of that region thought it was overhead. Prof. Andrews, giving the results of personal inquiries, says: 'The people of the northern part of Noble County heard it in a southern or south-eastern direction, and not in a north-western direction toward New Concord.' At Zanesville, about twelve miles from New Concord, the Courier described the noise, not as a succession of sounds, but as an 'explosion.' These facts clearly indicate that the great detonation heard at these various places was one and the same sound, and that it proceeded from a point over the interior of Noble County. The most probable location is five or six miles south of Sarahsville. It was undoubtedly the *first produced*, but the *last heard* of the successive sounds described as receding to the south-east by witnesses in the neighborhood where the meteoric stones fell, and it was compared by them to the roar of thunder."

The time of the day and the number and intelligence of the observers unite to give considerable interest and value to these observations. While some of them show points of difference, natural to the observation of sudden and startling phenomena, we can yet deduce from them many conclusions with more or less accuracy, thus:

THE DIRECTION OF THE METEORITE.

My own observations of two of the stones, which fell half a mile apart, enable me to give the direction of the meteor with some degree of exactness. The first of these stones struck the end of the rails of a Virginia (zigzag) fence, half-way down, just touching the middle rail, breaking off more and more of each rail as it passed to the ground. Connecting the points of fracture by a line, this line represents a descending curve from south-east to north-west.

Again, the stone that fell at Law's (the most northerly) struck a large dead tree lying on the side of a hill, sloping north-west, passing through it as any projectile would; it then struck a small clump of elders, breaking them off at the root, falling finally at the foot of the hill. A line connecting these points shows the curve already stated. Coupling with this the observations of Mr. Callahan on the direction that one of these stones penetrated the ground, with the observed path of their distribution, no doubt can remain that the general direction of their fall was from south-east to north-west, striking the ground at an angle of about sixty degrees.

ALTITUDE OF THE METEORITE.

This is a point that can be determined but very imperfectly, if at all. It may have been when first seen forty miles above the earth, but when the explosion was heard it must have been nearer, and was even still nearer when it subdivided and was scattered ("exploded," as usually termed) over Guernsey and the edge of Muskingum counties. It is, however, but proper that I should give Prof. Evans's computation from the data he collected; they were published in the July number of the Amer. Jour. of Science and Arts, but their reproduction will not be out of place here:

"Mr. William C. Welles, of Parkersburg, Virginia (lat. 39° 10′, long. 81° 24′), a gentleman of liberal education, testifies that, being about three miles east of that place at the time of the occurrence, he happened to look up to the north-east of him and saw a meteor of great size and brilliancy emerging from behind one cloud and disappearing behind another. When about 35° east of north he thinks its altitude was 65°. Now the distance, in a direction 35° east of north, from his station to the line directly under the meteor's path, is twenty miles. Calculating from these data, I find for the vertical height, taken to the nearest unit, forty-three miles. This was at a point in Washington County near the border of Noble.

"Mr. C. Hackley testifies that he saw the meteor from Berlin in Jackson County. It crossed a cloudless space in the north-east, and he thinks its altitude at the highest point was 30°. Now the distance from Berlin to the nearest point under the meteor's path is seventy miles. These data give nearly forty-one miles for its vertical height over Noble County, a few miles to the south of Sarahsville (lat. 39° 53′, long. 81° 40′).

"Many other reliable witnesses have been found who saw the meteor through openings in the clouds from various points west of its path, and whose testimony so far agrees with the foregoing as to give results ranging between thirty-seven and forty-four miles. Care has been taken as far as possible to verify the data in each case by personal examination of the witnesses. The angles have in most instances been taken as pointed out by them from their respective posts of observation. It is unfortunate that no case has come to our knowledge in which the meteor was seen from the region east of its path. But it was a circumstance in some respects favorable to the definiteness of the observations made from the west side that the observers in nearly all cases saw the meteor only at one point, or within a very small space, on the heavens. It is impossible to reconcile the various accounts without granting that its path was very nearly as above described, and that its height did not vary far from forty miles as it crossed Noble County.

"In regard to the time which intervened at different places between seeing the fire-ball and hearing the report, the statements are so vague that not much reliance has been placed upon them. It may be remarked, however, that they will essentially agree with the foregoing conclusions, if we suppose that the loudest explosion took place in the southern part of Noble County.

"I will add under this head the statement of Mr. Joel Richardson, of Warren, Washington County, who from a place six miles west of Marietta saw the meteor as much as 15° or 20° west of north at an altitude of about 45°. The direction in this case was so oblique to the meteor's path that the data are of little value for simply determining the height; but they are important on account of their connection with the place of the meteor's last appearance. Mr. Richardson was visited by the writer, and his testimony was subjected to close scrutiny. If we take the azimuth at 15° west of north, we shall have a distance of forty-one miles to the line under the meteor's path; and these data will give about forty-one miles for its vertical height over a point not more than a mile from New Concord, at the extreme western limit of the district along which the meteorites were scattered. If we take the azimuth at 20° west of north, both the distance and the height will be greatly augmented. I have found two persons living near Bear Creek, nine miles north of Marietta, who make statements closely corroborating that of Mr. Richardson.

"D. Mackley, Esq., a lawyer of Jackson, Ohio, who at the time of the occurrence happened to be at Berlin, about six miles north-east from the former place, and seventy miles from the nearest point under the meteor's path. He took pains to note all the facts as accurately as he could at that time; and he afterward returned to the spot in order to determine more definitely the points of the compass. His testimony, in answer to my interrogatories, is substantially as follows: 'The meteor

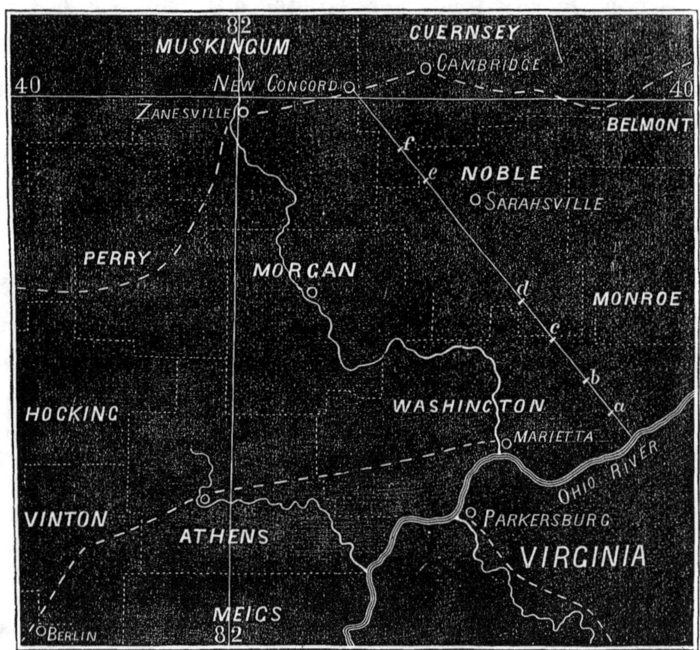

Fig. 1.

first appeared to me at a point about 55° east of north. It moved northward in a line very nearly parallel with the horizon. When it had disappeared it had described an arc of about 15°. It was in sight about six seconds. Its altitude was about 30°. In regard to its size, I have since looked at the sun through a thin cloud, and I think the apparent diameter of the meteor was one half that of the sun.'

"These data give the meteor a height of forty-one miles over the northern boundary of Noble County; a diameter of three eighths of a mile; and a relative velocity of nearly four miles a second. The results agree sufficiently well with those before given."

The accompanying map (fig. 1), made by Prof. Evans, shows the region over which the meteorite was observed to pass, and the conclusion to which he arrived is as follows: It was seen over the eastern part of Washington County (about lat. 39° 27′, long. 81° 8′), at a height of forty miles nearly. It was last seen over the north-western border of Noble County (about lat. 39° 51′, long. 81° 34′), at a height of thirty-eight miles nearly. Its velocity relative to the earth's surface, was three to four miles a second.

TEMPERATURE OF THE STONES.

Several of the largest stones were picked up ten minutes after their fall, and are described as being about as warm as a stone that had lain in the sun in the summer. One fell among dry leaves that covered it after it had penetrated the ground. The leaves, however, showed no evidence of having been heated. No appearance of ignition was discovered in places or objects with which the stones came in contact at the time of their fall; so that their temperature must have fallen far short of redness, while it may not have reached that of 200°.

SIZE AND VELOCITY.

I have no data upon which to calculate either of these. Prof. Evans, however, as just quoted, calculates from the data above given that its size was three eighths of a mile and velocity four miles a second.

While I may furnish no more reliable computations from the data obtained, I may be excused a short criticism on the above results to prevent too hasty conclusions being formed.

As regards the supposed elevation of forty miles *when the first reports were heard*, I would simply ask the question, Is it possible, with the established views of the conduction of sound by rarefied air, that any conceivable noise produced by a meteorite forty miles distant from the earth, in a medium quite as rare if not rarer than the best air-pump can produce, would reach us at all, or if so, in the manner described by observers? This question is a more important one to consider, as some observers on similar data have calculated the elevation of meteorites where they were first heard to explode at one hundred miles.

As regards the size of the meteorite I have but to refer the reader to my experiments made in 1854, and published in 1855, to show the perfect fallacy of calculating the size of luminous objects by their apparent disks, and I shall have more to say on the same subject in a future paper. It is important to note that the nearest approach of the meteor to the earth must have been in the northern part of Noble and in Guernsey counties, the point from which its most wonderful display seemed to have manifested itself; yet we hear nothing of its future career by reports from observers north of this, while its approach from the south to this point was noticed by a number of observers.

I need hardly state that my own convictions are that the meteorite terminated its career in Guernsey County, and that the group of stones which constituted it were scattered broadcast over that county. Many have been collected, and many lie buried in the soil to moulder and mingle their elements with those of this earth.

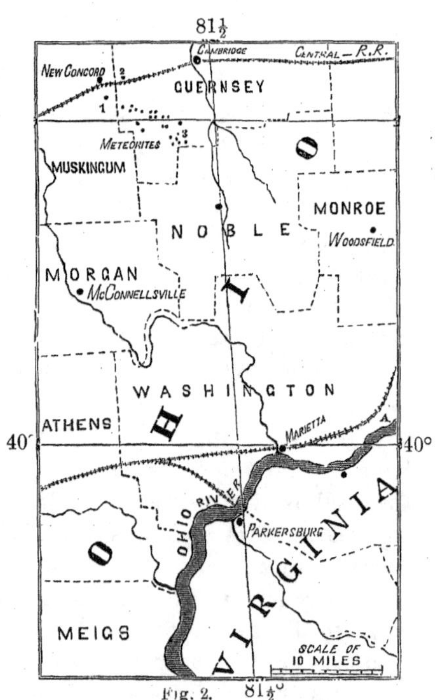

Fig. 2. 81½°

We come now to consider the stones that fell and were collected. Their number was over thirty, and their places of falling have been plotted with some care in the accompanying map (fig. 2.)

The localities of twenty-four have been fixed with precision, by the assistance of the Hon. C. J. Albright; but from the diminished scale of the map it is impossible to place a number by each dot intended to represent the locality of a meteoric stone. No. 1 on the map is the spot where the largest stone was found, weighing one hundred and three pounds; No. 2 is the next largest, weighing fifty-six pounds; and No. 3 the smallest, weighing eight ounces. The largest were at the north-west extremity and the smallest at the south-east. The space over which they were scattered was about ten miles long by three miles broad. The following is a catalogue of twenty-four:

No. 1	Weight	103 lbs.	Fell on farm of		Shenholt.
2	"	56 "		"	Law.
3	"	52 "		"	Amspoker.
4	"	50 "		"	Amspoker.
5	"	41 "		"	Torrence.
6	"	36 "		"	Reasoner.
7	"	23½ "		"	Hodges.
8	"	26 "		"	Fillis.
9	"	16 "		"	Adair.
10	"	15 "		"	Craig.
11	"	8¼ "		"	Craig.
12	"	4½ "		"	Waller.
13	"	4 "		"	Beresford.
14	"	3¾ "		"	Craig.
15	"	3¾ "		"	Stevens.
16	"	3¾ "		"	Wall.
17	"	3 "		"	Walker.
18	"	2¾ "		"	Claysville.
19	"	2 "		"	Stevens.
20	"	2 "		"	Wall.
21	"	2 "		"	Savely.
22	"	1 "		"	Carter.
23	"	1 "		"	Heskett.
24	"	½ "		"	Heskett.

Others have been found, but I have no correct record of their exact position.

Some fifteen of these stones have come under my observation. They are all irregular in shape, cuboidal, wedge-shaped, globular, and every conceivable form that irregular fragments of stone may be supposed to possess. They all have the well-known black coating, with a sharp outline between the coating and gray mass of the stone, and there is quite a uniformity in the character of the coating in both small and large stones.

When broken this meteor exhibits a gray mass, with metallic particles of nickeliferous iron,* resembling the stones I

* I have picked out pieces of iron weighing two grains, closely cemented to pyrites.

examined that fell in Harrison County, Indiana, on the 28th of March, 1859. The latter, however, is the coarser-grained of the two. Prof. Shepard, who is familiar with the meteoric stones preserved in the cabinets of this country and in Europe, says: "In its internal aspect it approaches the stone of Iekaterinoslaw, Russia (1825), though it is somewhat finer and more compact. In crust the two are identical. It is also similar to the stone Slobodka, Russia (August 10, 1818), and compares closely with those of Politz (October 13, 1819), of Nanjemoy, Maryland (February 10, 1828), and of Kuleschowka, Russia (March 12, 1811); but the crust is less smooth on the Ohio stone than in that of the latter. In fact, its character is that of a large portion of the known meteoric stones."

The general thickness of the crust is about from one thirtieth to one fourtieth of an inch.

The cut (fig. 3) is a representation of the largest stone that has been found, now in the cabinet of Marietta College, and described by Prof. E. B. Andrews. We reproduce the figure from Prof. Andrews's article here cited.

Fig. 3.

Several specimens have been examined. They all show the presence of the same minerals, with a slight variation in their proportions, as might be expected in a mass not homogeneous. Its composition is fairly represented as follows: Specific gravity,

3.550, varying slightly in different specimens. In one hundred parts there are

Nickeliferous iron	10.7
Earthy minerals	89.3

The nickeliferous particles separated by a magnet from the crushed stone, and well washed, presented the following constituents in one hundred parts:

Iron	87.011
Nickel	12.360
Cobalt	.421
Copper	minute quantity, not estimated
Phosphorus	.012
Sulphur	1.080

The sulphur comes from magnetic pyrites that the meteorite contains, and that it is not easy to separate mechanically from the small particles of nickeliferous iron.

The earthy part, when freed as thoroughly as possible from nickeliferous iron (which can be done pretty effectually by the magnet), was treated with warm dilute muriatic acid thrown on a filter first washed thoroughly with water, then with a solution of potash to dissolve the last portion of the silica of the decomposed portion of the mineral. The result was, in one hundred parts,

Soluble portion	63.7
Insoluble	36.3

The earthy material analyzed as a whole was found to contain

Silica	47.30
Oxide of iron	28.03
Alumina	.31
Magnesia	24.53
Lime	.02
Soda } Potash }	1.04
Manganese	trace.

From these results it is very clear that the mineralogical constitution of these meteoric stones is about as follows, in one hundred parts:

Nickeliferous iron	10.690
Schreibersite	.005
Magnetic pyrites	.005
Olivine	56.884
Pyroxene	32.416

This sums up the history of this meteoric shower, with as full an account as possible of the stones that fell at that time.

In the first part of this paper it was stated that this fall was quite as remarkable as that near l'Aigle, in France, in 1803. Although it does not equal this latter in the number of stones that were collected, it exceeds it in the size of the stones that fell. The largest of the l'Aigle stones weighed seventeen and a half pounds, while the largest in the present case was one hundred and three pounds.

There are many points of coincidence in the phenomena and circumstances attending the two falls. Were I to copy Biot's description of the phenomena of the fall at l'Aigle, as detailed to the Academy of Sciences nearly sixty years ago, it would be but a repetition of what has been written in the first part of this paper.

The date of fall at l'Aigle was the 26th of April; the date of the Guernsey fall May 1st. Time of the day of the former, one o'clock; of the latter, twenty minutes of one; the direction of both falls, from south-east to north-west. The extent of surface covered by the first, seven and a half miles wide by two and a half broad; by the latter, ten miles long by three wide; and both were seen by a large number of persons.

THREE NEW METEORITES.

---

1. Lincoln County Meteorite. (Fell August 5, 1865.)

This meteorite was examined several years ago, having been sent to me for that purpose by Prof. J. M. Safford, State Geologist of Tennessee. The result of my examination was embodied in Professor Safford's report of the geology of Tennessee for 1855, but has never received a special notice in any scientific journal; and as it is not too late to make up that deficiency, the following is sent for publication, embracing Prof. Safford's account of its fall, with the chemical examination. The following particulars in regard to its fall were furnished by Rev. T. C. Blake, of Cumberland University:

"It fell two miles west of Petersburg and fifteen north-west of Fayetteville, in Lincoln County, about half past three o'clock P. M., August 5, 1855, during or just before a severe rain-storm. Its fall was preceded by a loud report resembling that of a large cannon, followed by four or five lesser reports. These were heard by many persons in the surrounding country. Immediately after this mass or fragment was seen by James B. Dooley, Esq., to fall to the ground. It approached him from the east, appeared while falling to be surrounded by a 'milky' halo two feet in diameter, and fell one hundred and fifty or two hundred yards from him, burying itself about eighteen inches in the soil. When first dug out it was too hot to be handled.

"This specimen has an edge broken off, revealing the character of the interior. Within it is of an ashen-gray color, varied by patches of white, yellowish, and dark minerals. With the exception of the broken edge it is covered, and when first obtained was entirely covered, as most meteorites of this kind are, with a very 'black, shining crust, as if it had been coated with pitch.'

"One end or face, which may be regarded as the base, has an irregular rhomboidal outline, averaging two and three fourths by two and a half inches. Placing the stone upon this end, the body of it presents the form of an irregular, slightly oblique, rhomboidal prism. The upper end, however, is not well defined, but runs up to one side in a flattened protuberance, giving the entire specimen a form approaching roughly an oblique pyramid. The length from the base to the apex is four and a half inches. Three adjacent sides are rough, being covered with cavities and pits. The other sides are smoother and rounded.

"The specimen acts upon the needle; fragments of it readily yield particles of nickeliferous iron by trituration in a mortar. The specific gravity of the entire specimen is 3.20. Its weight in its present condition is three pounds fourteen and a half ounces.

"The minerals found in the meteorite are: pyroxene, principal portion of the mass; olivine and orthoclase, disseminated through the mass; nickeliferous iron, forming about one half per cent. of the mass. In addition to these there are specks of a black, shining mineral not yet examined."

The general analysis is as follows:

Silica	49.21
Alumina	11.05
Protoxide of iron	20.41
Lime	9.01
Magnesia	8.13
Manganese	.04
Iron	.50
Nickel, minute quantity.	
Phosphorus, minute quantity.	
Sulphur	.06
Soda	.82
	99.23

The minute quantity of nickel that was separated did not permit of my examining for cobalt, but there is no doubt that this metal was present.

2. Oldham County Meteorite.

The announcement of the discovery of this iron meteorite with the one that follows was made in a note in the American Journal of Science and Arts. It was discovered in the month of October, 1860, by Mr. William Daring, near Lagrange, in Oldham County, Ky. There is nothing known with reference to the time of its fall. It came into my possession shortly after its discovery.

It was entire and weighed *one hundred and twelve pounds.* Its extreme dimensions were: length, twenty inches; breadth, ten and three fourths inches; and thickness, six and a half inches. Its shape was elongated and flattened. Its specific weight is 7.89, and an analysis furnished

Iron	91.21
Nickel	7.81
Cobalt	.25
Copper, minute quantity, not estimated.	
Phosphorus	.05
	99.32

3. Robertson County Meteorite.

This mass of meteoric iron came into my possession during the month of December, 1860, being sent by Prof. Lindsley, of Nashville, Tenn. It was discovered by Mr. D. Crockett, near Coopertown, in Robertson County, Tenn. The time of its fall is not known.

Its weight was *thirty-seven pounds.* Its form was wedge-shaped; and its extreme dimensions were: length, ten inches; breadth, nine and a half inches; and thickness, five and a half inches. Its specific gravity is 7.85. On cutting through the mass a module of sulphuret of iron was discovered about one fourth of an inch in diameter, and there are doubtless others in its interior. The iron on analysis furnished

Iron	89.59
Nickel	9.12
Cobalt	.35
Copper, minute quantity, not estimated.	
Phosphorus	.04
	99.10

A NEW METEORITE FROM WAYNE COUNTY, O.

REMARKS ON THE METEORITE FROM ATACAMA, CHILI.

METEORIC IRON OF WAYNE COUNTY, OHIO.

The existence of a mass of meteoric iron from Wayne County, Ohio, has been known to me for some years; but I have delayed noticing its existence, hoping to obtain the mass, and thus give a more complete description of it than I am able to do.

My attention was first called to it by Prof. James C. Booth, of the United States Mint at Philadelphia, it having been brought to him by Peter Williams, of Wooster, Wayne County, Ohio, who supposed it to be a mass of silver or some other precious metal. Prof. Booth saw at once that it was meteoric iron, and tried to procure it from Mr. Williams; but from some notion of its possessing considerable intrinsic value he retained it, and since that time both the iron and Mr. Williams have been lost sight of.

Prof. Booth detached a small portion of it, part of which specimen he placed at my disposal, with the following memorandum: "Meteoric iron, given me in 1858 by Peter Williams, of Wooster, Wayne County, Ohio. It was a rounded mass, weighing about fifty pounds, and found by him in a woods near the above place while gathering bowlders to pave a town. It exhibits the usual figures on application of acid to a smooth surface."

As it is a well-authenticated meteorite, it is proper to make a record of it. Its specific gravity is 7.901, and it is composed of

Iron	93.61
Nickel	6.01
Cobalt	.73
Copper, very minute, not estimated.	
Phosphorus	.13
	100.48

There was a very small quantity of manganese, that has been estimated along with the nickel.

THE NEW ATACAMA METEORITE.

A fragment of the meteorite lately described by Prof. Joy (Amer. Jour. of Science and Arts, March, 1864, p. 243) has been sent to me by Prof. C. F. Chandler, and I have thus been afforded an opportunity of carefully examining it. I had at first supposed that it might be in some way related to the well-known Atacama iron; but it is very clear, by the most casual inspection, that it has no connection with that iron; at the same time it resembles so closely another meteoric mass from that region—in fact, is so identical with it in all particulars—that if it had not hailed from another locality it would be pronounced a portion of the meteorite from Sierra de Chaco, Atacama, described in 1863 by Prof. Rose (see Buchner, *Geschichte der Meteoriten*, p. 131).

Prof. Joy omitted to mention in his paper that the meteorite was said to have been found in the Janacera Pass.

The meteorite from Sierra de Chaco was, at the time it was described, unique in its physical characteristics; the close resemblance to it therefore of the one under notice, and its coming from Atacama, has induced me to investigate as far as possible the relative position of Sierra de Chaco and Janacera Pass.

The best authority on the geography of Chili in this country is doubtless Capt. Gilliss, of the United States Observatory at Washington; in answer to my inquiries on the subject he gives the following information: "I do not know any pass in Chili named Janacera; there is a river Jarquera, which has its origin near one of the passes in Atacama, and very probably there is a pass of the same name. The river Jarquera is to the northward and eastward of Chaco, the former being within the chain of the Andes, and Chaco most probably is in the western or coast range. They are from one hundred and twenty to one hundred and fifty miles apart."

As it is important to locate this meteorite correctly, I have written to Prof. Domeyko on the subject. The village of Chaco is situated near latitude 25° 20′ S. and longitude 69° 20′ W. from Greenwich, and its height above the sea is 8,778 feet.

The meteorite in question is so intimate a mixture of metallic and stony matter that it is difficult to say whether to rank it among the stony or metallic meteorites. Treated with a mixture of nitric and chlorhydric acids, and slightly warmed, the metallic portion is rapidly dissolved without the form of the mass being altered. Its mineral constituents are readily separated by the combined aid of chemical and mechanical means, and besides the iron I have been enabled to separate small but distinct particles of chromic iron, small spherical masses of olivine as beautiful in color and as transparent as that from the Pallas meteoric iron, and also a pyroxenic mineral; and perhaps with a larger amount of material to work upon other minerals might have been recognized.

I have nothing to add to the careful chemical examination by Prof. Joy, having detached mechanically most of the minerals that he deduced from analysis.

NEWTON COUNTY (ARK.) METEORITE:

CONTAINING ON ITS SURFACE CARBONATE OF LIME.

The first notice of the meteorite of Newton County was made in 1860 by Prof. Cox, who was engaged in the geological survey of Arkansas. The original has not been obtained. The only fragment of it, being in the hands of Judge Green, was given to Prof. Cox, who has kindly presented it to me. The weight of the fragment is twenty-two and a half ounces, and was evidently broken off from one corner of the mass, as it presents three of the original surfaces.

This meteorite is of the mixed variety, and can not be classed with either the metallic or the stony meteorites. It is one of the most interesting that has been discovered in the United States, differing from any other yet found in these regions.

The stony matter is very distinctly crystallized, and some of the minerals can be easily detached and examined separately. The metallic portion constitutes somewhat over one half of the mass, and owing to the diffusion of the stony matter has a coarsely reticulated structure.

When broken under the hammer, and the iron separated by the magnet, it is obtained in coarse grains, varying from three to four grains down to very small fragments. The exterior is of a rusty color, roughened by projection of nickeliferous iron, and over several parts of the surface there is a white incrustation.

Specific gravity taken on different pieces varies from 4.5 to 6.1. By mechanical means and the aid of the magnet the following minerals were separated: Nickeliferous iron, chrome iron, sulphuret of iron, hornblende, olivine, and carbonate of lime.

Nickeliferous Iron.—I may as well mention the manner in which I separate the iron from the stony matter of meteorites.

In most instances it is necessary to sacrifice a fair portion of the specimen. The mass is crushed in a steel mortar. The magnet is then able to take out the iron from the mass of stony matter, especially if the crushing operation is repeated two or three times. The iron is then introduced into an iron or, better still, a silver capsule or crucible, and a strong solution of potash added. Heat is applied until all the water is driven off, and the residue is heated to redness. On cooling water is applied, and the excess of potash washed out, as well as some silicate of potash that is formed. After thoroughly washing the particles of iron they are moistened with a little alcohol, and dried on blotting-paper with a gentle heat; and by holding a magnet a little distance from them the particles of iron will adhere to the magnet, almost perfectly free from earthy matter.

The iron, if of a coarse reticulated structure, as the one in question, may require to be crushed in the steel mortar after treatment by potash, to detach particles of silicate remaining in small crevices; and in this variety I sometimes repeat the treatment by potash. In this way the foreign matter associated with the iron can be reduced to one half per cent. Of course this process sacrifices more or less of the iron, especially if the iron be in very small particles. But this sacrifice is of secondary importance compared with the necessity of having the metallic matter in a pure state. Thus purified the iron was found to be composed of

Iron	91.23
Nickel	7.21
Cobalt	.71
Copper, Phosphorus, } too small to be estimated.	
	99.15

In the analysis, after separating the iron by the acetate of soda, the nickel and cobalt were separated by nitrite of potash; which method I have used frequently, and with the best results. Liebig's method for accomplishing the same end has been much improved by the modification lately devised by Prof. Gibbs, of dissolving the oxide of mercury in the cyanide of mercury. But having every arrangement necessary for executing successfully the method by the nitrite of potash, I have not yet tried Prof. Gibbs's modification, but shall do so shortly.

Chrome Iron.—This is found in small quantity in minute particles, some of them showing distinct faces of crystals; but I failed to find any complete octahedron. The quantity was too small for analysis, but was readily recognized by the blowpipe.

Sulphuret of Iron.—This also is discernible only in minute quantity, and could not be collected for analysis. I would remark, with reference to the sulphuret of iron found in meteorites, that it can not be classed with the terrestrial magnetic pyrites, whose formula is considered $Fe_7 S_3$, having always found the sulphur too small for this formula; in which conclusion I believe that I am sustained by Rammelsberg and others. My results point to the formula FeS; and if the compositions of these two kinds of pyrites are correctly made out, then the meteoric variety has no terrestrial representative.

Hornblende.—This mineral is easily separated, and is of a greenish-gray color, more or less soiled by iron. With some care it can be detached unmixed with other constituents. It has a very distinct cleavage in one direction and an imperfect one in another. On analysis it gave

Silica	52.10
Alumina	1.02
Protoxide of iron	16.49
Protoxide of manganese	1.25
Magnesia	29.81
Alkalies (potash, soda, lithia)	.24
	100.91

The oxygen relations of the silica and protoxides furnish the formula $\dot{R}_4 \ddot{Si}_3$ — the formula of hornblende. In structure and composition it is not unlike some varieties of anthophyllite.

Olivine.—This mineral is diffused through the mass. Some of the smaller pieces are almost colorless; others again are more or less yellow, being stained with oxide of iron. Some of the fragments are iridescent, like varieties of oligoclase, which I at first took it to be. Sufficient of it was detached in a pure state for analysis, and was found to be composed as follows:

Silica	42.02
Alumina	.46
Protoxide of iron	12.08
Magnesia	47.25
	101.81

There was a minute quantity of the manganese estimated with the oxide of iron and magnesia. This analysis overruns the 100. This is accounted for in part by the quantity used for analysis not being more than 0.160 grammes. The oxygen ratio of the silica and protoxides show the composition $\dot{R}_3 \ddot{Si}$, which is that of olivine.

Carbonate of Lime.—The observation of this constituent in a meteorite is something entirely new, yet it is found on the exterior surface of the meteorite in question in various places. There is no doubt in my mind, however, that this ingredient was not a part of the mass when it fell; but that it has been exposed to certain conditions since its fall by which carbonate of lime has been incrusted on its surface.

It is much to be regretted that the entire original mass is not accessible to furnish facilities for determining whether it is an incrustation or not, and if the former, whether the incrustation was formed prior or subsequent to its fall.

In relation to the presence of carbonates in meteorites we have the first and only announcement, up to the present time, in connection with the meteorites which fell at Orgueil in 1863. Messrs. Des Cloizeaux, Pisani, Daubrée, and Cloez discovered minute rhombohedral crystals of double carbonates of magnesia and iron.

The above statements exhaust about all that I have to say at the present time on the meteorite under investigation. There may be one or two other minerals in its composition, but I could not separate them in a manner to pronounce as to whether they were different from those already described or not.

COLORADO METEORIC IRONS.

---

1. Russel Gulch Iron.

I have known of the existence of a new meteoric iron from Russel Gulch in Colorado for about two years, but it was only recently that it passed into my hands. I first heard of it in the possession of Mr. Fisher, of New York, who subsequently turned it over to Prof. C. F. Chandler, of Columbia College, New York, who kindly submitted it to me, as I am furnished with the necessary means for cutting up and scrutinizing thoroughly this class of bodies.

The mass of iron is accompanied with the following label: "Meteoric iron found in Russel Gulch, February 18, 1863, by Mr. Otho Curtice. Weight twenty-nine pounds. Brought to New York, February, 1864."

The mass measures in its extreme length, breadth, and thickness $8\frac{1}{2} \times 7\frac{1}{4} \times 5\frac{1}{2}$ inches. It is perfect in all parts except at one extremity, and, as stated above, weighs twenty-nine pounds.

The iron is one of medium hardness, with the density 7.72, and when cut through was found to contain a few small nodules of iron pyrites. It is attacked readily by nitric acid, and gives bold Widmannstättian figures without very sharp angles. It resists the action of the air and moisture very well, and is consequently but little altered on the surface. No siliceous minerals could be traced in any of the crevices. On analysis its composition was found to be

Iron	90.61
Nickel	7.84
Cobalt	.78
Copper, minute quantity.	
Phosphorus	.02
	99.26

I have not made any further observations in relation to the presence of copper in meteoric iron since 1852, when I called attention to it. Since then I have become more confirmed in the opinion, then first expressed, that copper would be found in all meteoric irons; this has been the result of examinations of many well-known meteoric irons and all new ones that have come under my examination.

One or two grains of the iron is all that is necessary for the examination, if it be done carefully; but four or five grains had better be used. Dissolve the iron in chlorhydric acid, and if necessary add a little nitric acid; it is as well at all times to add a drop or two at the end of the operation. Evaporate away the excess of acid, add water, precipitate with sulphureted hydrogen until there be an excess of gas in the solution; throw on a filter and wash with water containing a little HS, dry the filter, burn in a porcelain crucible, treat the residue in a little *nitro-muriatic acid*, and evaporate to dryness, with the addition of a drop or two of sulphuric acid; treat the residue with water, when the introduction of a clean plate of iron will cause a deposition of the copper with all its characteristic properties.

2. Bear Creek Iron.

Of this there are two short notices in the November number of the Amer. Jour. of Science and Arts, pages 260 and 286. The specimen of it in my possession has enabled me to make a thorough examination of the constituents. The piece I have has a portion of the exterior attached.

As has already been stated by Prof. Shepard, it is coarsely crystalline, and laminated from the effects of decomposition between the crystals; the surface contains considerable pyrites, although Prof. Shepard did not discover any in his specimen. I was enabled to separate and analyze magnetic pyrites, schreibersite, and nickeliferous iron. Of the magnetic pyrites sufficient was separated to make a quantitative determination, which was as follows:

Sulphur	35.08
Iron	61.82
Nickel	.41
Insoluble residue	1.81
	99.12

The schreibersite was not obtained in sufficient quantity for a complete analysis; about fifty milligrammes of the pure mineral gave all the constituents usually found in this interesting mineral.

The nickeliferous iron, constituting of course the great bulk of the mass, was composed as follows:

Iron	83.89
Nickel	14.06
Cobalt	.83
Copper, minute quantity.	
Phosphorus	.21
	98.99

The laminæ of iron are often very brilliant, having the luster of silver, and caused me to suspect more nickel than was found. It was supposed that in the decomposition of the crystals the iron would disappear more rapidly than the nickel, and that by a process of cementation the nickel would accumulate in the laminæ; but from careful examination of the process of decomposition there is no doubt that the interior of the mass will not differ materially in its composition from the analysis already given of the nickeliferous iron. Besides the minerals already mentioned, and which properly belong to the original mass, there is much oxide of iron, containing some nickel arising from the decomposition of the surface.

COHAHUILA (MEXICO) METEORITES OF 1868.

The region of Mexico bordering on Texas seems to have been most profusely furnished with these celestial visitors. In 1854 I first drew the attention of the scientific public to the meteoric irons of this region, at which time I described one brought from there by Lieut. Gouch, referring at the same time to one mentioned by Mr. Weidner near the south-western edge of the Balsin de Mapini, on the route to the mines of Panal, weighing not less than one ton; also to another mentioned by Dr. Berlandier, in his journal of the commission of limits, that at the Hacienda of Venegas there was (1827) a piece of iron that would make a cylinder one yard in length with a diameter of ten inches. It was said to have been from the mountains near the hacienda (see my article on the subject, Amer Jour. of Science and Arts, 1854). In the description there given it was stated that the specimen examined came from sixty miles north of Santa Rosa, and therefore in one or two collections in which it is to be found it is called incorrectly *Santa Rosa meteorite.* I was allowed to cut off but a small piece of it from the original specimen, which is in the Smithsonian Institution, and consequently I was able to supply but two or three specimens. For the discovery and collection of the specimens now under consideration we are indebted to Dr. H. B. Butcher, and I will give a full detail of the discovery as communicated in letters to his father by Dr. Butcher, to whom the scientific world are certainly indebted for the labor, expense, and danger incurred in procuring them. I must not, however, fail to state that I am indebted to Dr. Feuchtwanger for first informing me of the fact of their arrival in this country, and for the exhibition of a small fragment to the members of the American Scientific Association at Chicago in 1868.

In a letter dated September 8, 1868, Dr. Butcher writes, from information received from the son of Dr. Long, who had resided many years at Santa Rosa, that in the fall of the year

1837 there appeared over the town a most brilliant meteor, having a north-west direction. He describes it as most beautiful, lighting up the whole horizon, with a trail of brilliant light following in its progress. Shortly after its disappearance among the distant mountains they heard a rumbling sound, immediately followed by a tremendous explosion.

From the report he thought it fell and exploded as it reached the earth, somewhere between Santa Rosa and the mountains, a distance of some thirty-five miles, and the next day he started with friends to examine the route, hoping to find it. After two days' severe and rough riding they abandoned the search and returned to town. Shortly afterward an Indian brought a piece weighing ten or twelve pounds into Santa Rosa, supposing it to be silver, having found it some ninety miles north-west of the town, being in the same direction in which Dr. Long and his friends had been exploring, the doctor having been deceived as to distance, he only going to the base of the mountain instead of crossing it and then following the valley for some forty miles farther, where I think his search would have been a success.

Dr. Butcher now undertook the search, after which he writes: "I have returned fully successful, and am making preparations to send on the iron. In making my arrangements I hired eight Mexicans and two Indians as guides, and started into the mountains in a north-west direction, the same as taken by Dr. Long, and found the iron about ninety miles from Santa Rosa. As no vehicle could go into the mountains by the route we entered, I spent two days in exploring a new road whereby the ox-teams could bring them out and get them to Santa Rosa. They consist of eight pieces, varying from two hundred and ninety pounds, which is the smallest, to six hundred and fifty-four pounds, which is the largest, making a total of nearly four thousand pounds. Before the explosion the weight must have been much greater, as it is not probable that I have secured the whole, and we know some was taken away by the Indians, who thought they found large masses of silver, and carried their specimens to Santa Rosa. It appears there is on record a statement of the meteor having passed over the city in 1837, and one of my guides relates as a fact that at that time (1837) a Lepan Indian was riding one of their small ponies through the valley, when his stirrup struck against one of the masses, causing a ringing sound like silver. He dismounted, and was confirmed in his opinion of silver, and took away a piece ten or twelve pounds in weight, which he carried to Santa Rosa to sell. I have received from various sources information relative to this meteor, and all confirm me in the opinion that the autumn of 1837 is about the time of its fall. My party were in considerable danger while in the mountains, as we were encamped two miles from the regular trail when some three hundred Indians went through with a large number of their stolen horses."

Whether or not the time above specified is that of the fall of one or more of these irons is a matter of little moment; the probabilities are, however, strongly in favor of it; nevertheless, it forms one of the most interesting groupings of meteoric irons known in any part of the world, especially as the masses are solid and compact, and not fragile and half stony, as the Atacama iron, that may have been broken artificially after its fall, and the fragments scattered by Indians and explorers in search of silver. Each one of these masses merits a separate examination, which I hope to be able to give, sooner or later, to satisfy my mind on one or two points connected with their common physical structure and chemical composition. But I will not delay this paper until then.

Six of these masses have been brought to this country, weighing respectively 290, 430, 438, 550, 580, and 654 pounds. They are irregular compact masses, without any evidence of stony minerals. They belong to the softer irons, not very difficult to cut with the saw; as yet there has been but about one ounce detached from one of the masses, which has enabled me to make out the following description Specific gravity 7.692. It contains

Iron	92.95
Nickel	6.62
Cobalt	.48
Phosphorus	.02
Copper, very minute quantity.	

This composition differs somewhat from the meteoric iron called Santa Rosa; but recently I have reason to believe that the quantity of nickel given in the Santa Rosa is too small, some portion of it having remained with the iron. Future examinations may prove that the Santa Rosa belongs to the group of irons under notice.

THE WISCONSIN METEORITES:

WITH SOME REMARKS ON WIDMANNSTÄTTIAN FIGURES.

These meteorites were first brought to my notice by Mr. I. A. Lapham, of Wisconsin; and his attention was called to them by Mr. C. Daflinger, Secretary of the German Natural History Society of Wisconsin. They were discovered in the town of Trenton, Washington County, Wisconsin, and I have called them the "Wisconsin Meteorites." Up to the present time fragments have been found, indicating that these meteorites were of the same fall, and separated at no great elevation. They were found within a space of ten or twelve square yards, very near the north line of the forty-acre lot of Louis Korb, in latitude 43° 22′ north, and longitude 88° 8′ west from Greenwich, and about thirty miles north-west of Milwaukee.

They were so near the surface as to be turned up with the plow; they weigh sixty, sixteen, ten, and eight pounds respectively, and present the usual pitted and irregular surfaces.

The largest of the meteorites in its extreme dimensions is fourteen inches long, eight inches wide, and four inches thick, weighing sixty-two pounds. Its specific gravity is 7.82, and composition:

Iron	91.03
Nickel	7.20
Cobalt	.53
Phosphorus	.14
Copper	minute quantity.
Insoluble residue	.45

A polished surface when etched gives well-marked Widmannstättian figures. There is something, however, peculiar about the markings on this iron, which is doubtless common to other irons, but which has heretofore escaped my observation; and I can not discover, in a hasty investigation, that it has been noticed by others. My attention was called to this peculiarity by Mr. Lapham, on a slice of the meteorite I sent him etched. Should these markings be entitled to a separate notice, I propose calling them *Laphamite markings.* The little drawing accompanying this, which is on a somewhat exaggerated scale, will show what they are.

The Widmannstättian figures are *a*, bright metallic, with convex ends and sides; *b c*, of a darker color, are the other markings, usually smaller and with the sides and ends concave. The material of which these dark figures are composed seems to have enveloped the lighter colored portion, which serves to make the dark lines so beautifully conspicuous. A good pocket-glass will show that the dark figures are striated, with lines at right-angles to the bounding surfaces. When the figure is nearly square the lines extend from each of the four sides, but when much elongated, as at *c*, they are parallel with the longer sides. Often these lines do not reach the middle of the figure, where only a confused crystallization can be detected. In the interior of the elongated figures the lines are quite irregular, often running together and showing a striking resemblance to woody fiber. The nature of these markings may be easily understood. They indicate the axes of minute columnar crystals, which tend to assume a position at right-angles to the surface of cooling.

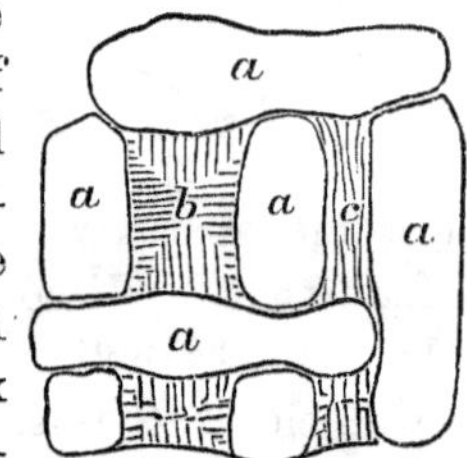

These markings may have been observed by others; and as soon as the subject can be examined on other irons, a better conclusion can be formed.

FRANKLIN COUNTY (KY.) METEORIC IRON:

WITH REMARKS ON THE PRESENCE OF COPPER AND NICKEL IN METEORIC IRONS; THE METHOD OF ANALYZING THE SAME; AND THE PROBABILITY OF THE LEAD IN THE TARAPACA IRON HAVING BEEN ORIGINALLY FOREIGN TO THAT MASS.

---

1. The Franklin County Meteoric Iron.

The Franklin County meteoric iron was first brought to my attention in a blacksmith-shop in Frankfort, Ky. It was carried there to be tested in regard to its quality as iron; being supposed by its discoverer to indicate an iron-mine. Mr. Nelson Alley became possessed of it, and kindly presented it to me.

It came from a hill eight miles south-west of Frankfort, latitude 38° 14′ north, longitude 80° 40′ west from Greenwich, and was discovered in 1866. It passed into my possession in 1867, and was then described by me, but the manuscript was lost after its leaving my hands, and the original notes were misplaced; the notes have been recently discovered, and the iron again analyzed.

Its form is somewhat globular, with a highly crystalline structure. Its weight was twenty-four pounds; and this appears to have been its original weight, only a few flakes having become detached by the rusting through of some of the fissures. Specific gravity 7.692. Its composition when perfectly freed from rust and earth is

Iron	90.58
Nickel	8.53
Cobalt	.36
Copper	minute quantity.
Phosphorus	.05
	99.52

Having, as it will be seen, the usual composition of meteoric irons. While on the subject of this iron I will add some remarks.

2. On the Presence of Cobalt in Meteoric Irons.

My attention has been directed again and again to meteoric irons whose analyses are given without mention of the presence of cobalt; and in some instances with the distinct statement that it is absent, as in the recent examination of a meteoric iron from Auburn, Macon County, Alabama, by Prof. Shepard, who states that "neither cobalt, tin, nor copper was detected in this iron." I can not but suggest the importance of making a most critical examination of these irons before pronouncing this fact; for in every analysis that I have made of meteoric irons (over one hundred different specimens) with this in view, cobalt has been invariably found, along with a minute quantity of copper. A great many of the analyses made were of irons that had been previously examined without a recognition of the cobalt.

The presence of these ingredients, even in small quantities, is a matter of considerable mineralogical interest, as is the case of the presence of small quantities of other elements in many minerals; a fact that I will have occasion to refer to at some future time in connection with leucite and other silicates.

As a guide to those who may wish to know the manner of my examinaton of meteoric iron, I will give a little in detail the method adopted in separating the metals.

Method of analysis.—A small piece of the iron is selected perfectly free from crust and earthy matter. I sometimes plunge the fragment into nitric acid somewhat diluted, warm the acid and continue the action for a few seconds, withdraw the iron, wash well and dry it. The piece selected for analysis should be about one gramme, or a little over (except where the copper is sought for quantitively, and then at least ten grammes should be used for the copper estimate alone). Treat the iron in a porcelain capsule, or glass flask, with a mixture of hydrochloric and nitric acids, consisting of four parts of the former to one of the latter, and about as much water as acid; dissolve over a water-bath. If a capsule be used, invert a funnel over the mixture, the edges of the funnel entering the capsule but not touching the mixture. Continue the action on the water-bath until the solution is complete; evaporate to dryness (having washed what may adhere to the inner surface of the funnel into the capsule), then add a little more hydrochloric acid and evaporate again nearly to dryness. This is done to insure driving off the last portion of nitric acid and rendering the iron easily soluble. Add to the contents of the capsule an ounce or two of water, and if there be a residue it must be collected on a weighed filter, dried, weighed, and reserved for future examination; if the quantity be too small for examination, a larger portion of the iron must be examined with special reference to this residue—which most commonly is a silicate, but may contain carbon or chromic iron (chromite).

If, however, there is no residue, proceed to the next step at once without filtering; if the solution has been filtered the following steps are the same. Examine first for sulphur. This is done by adding a few drops of chloride of barium; if there is a precipitate it is collected on a filter, and the sulphate of baryta obtained furnishes the amount of sulphur present. Next pass a stream of sulphureted hydrogen through the filtrate to complete saturation, previously adding a drop or two of sulphuric acid; much sulphur will be deposited and a very minute quantity of copper (not traceable by the color of the precipitate, but only recognized by the most delicate tests after the sulphur of the collected precipitate is burnt away); the solution thrown on a filter leaves the precipitated sulphur, the little trace of copper, and all the excess of the baryta that had been added, if the excess had been slight. The filter is then ignited in a porcelain crucible; the residue treated with a few drops of nitric and sulphuric acids, which suffices to dissolve the copper, and when evaporated to dryness leaves the copper in the form of sulphate, slightly acid, as there is no necessity of heating so high as to drive off the very last trace of sulphuric acid. The presence of the trace of copper is easily shown by adding a drop or two of water to dissolve the sulphate, and with the end of a glass rod placing a little of the solution on a clean and bright surface of iron, as the blade of a knife, for instance. In no examination of a meteoric iron have I failed to detect copper by this means.

When more iron is used, and there is consequently more copper, it can be separated and weighed. Tin and lead will also be found in the precipitate if they be present; but I have never detected either, except lead in the case of the Tarapaca iron, which I have every reason to believe was originally foreign to the iron.

The sulphureted hydrogen precipitate, which I have always obtained, is so minute in a gramme of iron that it may be dispensed with, and the iron, nickel, and cobalt be separated, which is accomplished as described a little further on. If, however, sulphureted hydrogen has been used, the iron in solution is in the form of protoxide, and must be converted into the peroxide, which is accomplished by adding a little chlorate of potash and hydrochloric acid, that have been made to react on each other by heating, before adding it to the boiling solution of iron, etc.

The solution of iron should have a bulk of ten or twelve ounces; to it is added a solution of carbonate of soda in sufficient quantity to nearly neutralize the free acid; the iron is now precipitated by acetate of soda with all the well-known precautions. I wash this precipitate only partially, and detach it from the filter by washing it into a beaker and re-dissolving it by hydrochloric acid, and precipitate it a second time by acetate of soda; I then subject it to complete washing, and estimate the iron in the way usually employed. This second precipitation is necessary to separate an appreciable quantity of nickel remaining in the acetate of iron after the first precipitation; and after considerable experience I must say that it is the only method of separating with any degree of accuracy iron from nickel.

The solution separated from the iron, and containing nickel and cobalt, is concentrated down to four or five ounces, then treated with caustic potash or soda, thrown on a filter and washed with hot water, with all the usual precautions when nickel is precipitated by an alkali. The precipitate is dried, ignited and weighed; after which it is dissolved in nitric acid over a water-bath and the acid solution evaporated nearly to dryness; about two or three drachms of water are added, then a concentrated solution of nitrite of potash, then an excess of acetic acid; this is now set aside for forty-eight hours, during which time the nitrite of cobalt is completely separated; it is next thrown on a filter, washed and estimated in the manner proposed by the author of this method. I usually employ a concentrated solution of the sulphate of potash to wash with. This method, to say the least of it, has in my experience proved fully equal to the cyanide method, and is much more simple;

and by it I have never failed to detect and estimate cobalt in every meteoric iron that has come under my examination.

In examining for *phosphorus* the following method is adopted: To three or more grammes of the iron, in a porcelain capsule, add nitric acid diluted with water; then invert a funnel over the mixture to protect from loss during the action, evaporate to dryness over a water-bath, and then on a sand-bath to a temperature of 500° to 600°; the iron is thus converted into an oxide with little or no nitric acid remaining, and the phosphorus is transformed into phosphoric acid that is now combined with oxide of iron. The residue is detached as thoroughly as possible from the capsule and mixed with twice its weight of carbonate of soda, or, better still, with a mixture of carbonates of soda and potash; a little carbonate of soda added to the capsule and rubbed with a pestle detaches the last portion of oxide of iron, or rather leaves so small an amount as to make no error in the future steps of an analysis where the original quantity of phosphorus is so small.

The mixture of oxide of iron and phosphate is now to be heated in a platinum crucible to the point of fusion of the carbonates for about twenty minutes; then heat the mass with water, when the excess of carbonates will be dissolved, and what phosphate may have been formed. The phosphates will represent all the phosphorus in the iron. Now neutralize the carbonate with hydrochloric acid, and estimate the phosphorus in the ordinary way by a magnesia salt.

With regard to the detection of chromium and other special constituents of some meteoric irons, especially those containing some siliceous minerals intimately mixed in the iron, it is not the province of this paper to discuss.

3. Lead in Meteoric Irons.

The only instance of the finding of lead in meteoric irons is that of the Tarapaca iron, found in 1840, in Chili, which was examined by Mr. Greg; the metallic lead was detected by him in small masses of varied dimensions.

I have examined several specimens cut from the original mass of iron, two of which are in my possession, and my conviction is that the metallic lead was altogether foreign to the iron when it originally fell, and has been doubtless derived from lead with which the mass was probably treated by the original discoverers for the purpose of extracting some precious metal, they being ignorant of its true nature. My reasons for coming to this conclusion are, that the lead is found in cavities near the surface of the iron, these cavities having channels of more or less size leading to the exterior of the mass; the iron is honey-combed in its character in many places, which is evident to the eye, and is also indicated by its specific gravity, 6.5. In pieces of the iron detached from the interior of the mass, and examined with the utmost care by a magnifying glass to see that there is no possible fissure in it, no lead has been found. These pieces are exceedingly difficult to obtain, and can only be had in very small pieces.

The crust of the iron having the most cavities furnishes most lead, and is in some parts covered by a fused yellow crust of oxide of lead; this last fact has no significance, however, in the present consideration of the matter. Without venturing to insist too sharply on the view here taken, after the careful examination of so distinguished an observer as Mr. Greg, I recommend this view of the subject to those having larger specimens of the iron than myself.

STEWART COUNTY (GA.) METEORITE.

(Fell on October 6, 1869.)

In October, 1869, I learned through the public press that certain meteoric phenomena had occurred in Stewart County, Georgia, and that one or more stones had fallen. Inquiries were immediately instituted by me, and through Prof. Willet I obtained for examination the only stone found, one that was seen to strike the ground.

The stone, as it reached me, was nearly intact, and weighed twelve and a quarter ounces; it must originally have weighed twelve and a half ounces. It is of an irregular conical shape, having a flattened base, and is covered with a dull, heavy black coating. The specific gravity is 3.65. The fractured surface has a grayish aspect, and when examined closely, especially by the aid of a glass, exhibits numerous greenish globules with a whitish granular material between; through the mass are dark particles consisting principally of nickeliferous iron, with some pyrites and a few specks of chrome iron. The nodules are sometimes three or more millimetres in diameter, and of an obscure fibrous crystalline structure, the crystals radiating usually from one side of the nodule; they have a dirty bottle-green color, a greasy aspect when broken, and are more or less opaque.

Some of these little nodules were separated in a tolerable state of purity, amounting to one hundred and twenty-one milligrammes. On analysis they afforded:

		Oxygen.	Ratio.
Silica	48.62	25.90	2
Alumina	8.05	3.79	
Protoxide of iron	11.21	2.51	1
Magnesia	30.18	11.80	
	98.06		

The hardness of the mineral is about 6, and it is quite tough. The formula would be $\dot{R}\ddot{Si}$, with a part of the silica replaced by alumina, a not unfrequent case in minerals such as hornblende, hypersthene, etc. As it is impossible to derive any light from its crystalline structure, the above analysis warrants me in concluding that it is either *bronzite* or *hornblende*, but I am more inclined to the former supposition, as it appears to take the place of the enstatite in many meteorites.

Nickeliferous iron constitutes about seven per cent. of the mass, and a portion separated in as pure a state as possible afforded on analysis:

Iron	86.92
Nickel	12.01
Cobalt	.75
	99.68

These are the proportions after allowing iron for a small amount of sulphur present in a minute quantity in the nickeliferous iron, which could not be separated mechanically. I did not test for copper or phosphorus. The quantity of iron separated from the stone did not warrant my making special analyses for substances the quantity of which present could only be exceedingly minute.

The stony matter freed from the iron was treated with nitromuriatic acid and water, and heated for some time over a water-bath, renewing the water and acid once or twice; the solution was filtered and the residue washed; the residue was then treated with a warm solution of caustic potash, filtered, and again washed. The filtrate was neutralized by hydrochloric acid and added to the first filtrate, and the whole evaporated to dryness over a water-bath, warmed gently over the lamp, and treated with water and a little hydrochloric acid, thrown on a filter, the silica collected and estimated; the last filtrate was treated with a solution of chloride of barium to ascertain the quantity of sulphuric acid present (due to the pyrites in the original mass); it was found to indicate 6.10 per cent. of magnetic iron pyrites. The solution freed from the excess of baryta was now analyzed in the ordinary way.

The insoluble portion of the meteorite was fused with carbonate of soda and a small fragment of caustic potash, and its ingredients ascertained.

A separate portion of the stony part of the meteorite was examined for alkalies.

The various analyses referred to above gave, omitting the nickeliferous iron:

The part soluble in acid	58.05
The part insoluble in acid	41.95

	Soluble part.	Insoluble part.
Silica	41.08	56.03
Alumina	.32	5.89
Protoxide of iron	18.45	15.21
Magnesia	41.06	21.00
Lime	......	.10
Soda, with a little K̇ and L̇i	......	2.97
	100.83	101 20

The soluble part consists principally of olivine. The insoluble is doubtless the bronzite already referred to, with a little albite or oligoclase.

Chrome iron was detected by fusing some of the stony part of the meteorite with carbonate of soda and a little niter and separating in the usual way. The quantity was quite minute. The composition of the stone as made out would be

Nickeliferous iron	7.00
Magnetic pyrites	6.10
Bronzite, or hornblende Olivine Albite, or oligoclase Chrome iron	86.90
	100.00

PHENOMENA ATTENDING THE FALL OF THE STONE.

Mr. J. B. Latimer, of Bladen's Creek, Stewart County, has kindly furnished the following particulars of the *flight* of the body through the air, and of the several *explosions*, which occurred nearly vertically above him:

"The morning of the 6th of October last (1869) was quite clear, scarcely any cloud being visible, quite calm; about ten A. M. the atmosphere grew somewhat hazy, no clouds; at about fifteen or twenty minutes before twelve M. a roaring, rushing sound was heard in a north-westerly direction, about 80° above the horizon. In a moment or two it was almost directly overhead, at which point a loud explosion occurred, followed in rapid succession by six other reports, but less in volume than the first—making seven in all. The explosions appeared about as loud as a twelve-pound cannon at a distance of ten or twelve miles. These explosions did not occur all at the same point in the heavens, but seemed to emanate from some body moving rapidly to the south-east. After the explosions a peculiar whirring sound was heard, apparently produced by some large irregular body moving very rapidly. This also went in a south-easterly direction. This sound was heard several seconds; many have compared it, and aptly too, to an imperfect steam-whistle. I have no precise idea of the time consumed in all this demonstration. Some persons say several minutes, but I think ten or fifteen seconds would about cover the time.

"As the larger body was going out of our hearing, some moments after the explosions, a smaller one passed to the south-west with just such a noise as is always produced by a flying fragment of a shell after its explosion, or of any angular body cast violently through the air. This piece descended to the earth, distinctly traced in its passage by many persons, and struck in the yard of Capt. E. Barlow—which point of contact is, on an air-line, about two and a half miles from a perpendicular beneath where the explosions occurred. This is the only one known to have fallen in this section.

"The explosions, together with the rushing sound afterward, were heard over a region about thirty miles north-east and south-west, and fifty or sixty miles north-west and south-east. No shock was felt—at least no tremor of the earth.

"Two men say that they were looking in the exact direction of the explosions at the time they occurred and saw a quantity of vapor, much like the volume of steam escaping from the pipe of an engine at each successive stroke, which vapor or mist was violently agitated and increased in bulk with each successive report, but disappeared soon after the cessation of the reports. This corroborates the testimony of some of my own laborers, who say that immediately after the explosions something like a thin cloud cast its shadow over the field they were in."

Hon. John T. Clarke, of Cuthbert, Ga., who has interested himself in collecting the history of the meteorite, and through whose influence it has come into the possession of Mercer University, writes me the following particulars of its *fall:*

"It fell about 11½ A. M., on the 6th of October last (1869), in Stewart County, Georgia, on the premises of Elbridge Barlow, Esq., about twelve miles south of west from Lumpkin. Capt. Barlow picked it up a few moments after it fell. His account of it is this: While standing in the open yard, the sky being bright and clear, he heard first a succession of about three explosions, resembling bursts of thunder or discharges of artillery, followed by a deep roaring for several seconds, and then by a rushing or whizzing sound of something rushing with great speed through the air near by. The sound ceased suddenly. The noise from first to last was some half a minute. Two negroes were washing near the well in the same yard, about sixty yards from where Barlow stood. They heard the noise, and supposed it to be the falling in of the plank well-curbing, banging from side to side in its descent, and so spoke of it to one another before it fell. While they were speaking thus it struck the ground about twenty steps from them, in full sight, knocking up the dirt. They called Capt. Barlow, and showed him the spot. It was upon very hard-trodden ground in the clean open yard. The earth was freshly loosened up very fine in a circle of about one and a half feet in diameter; and, upon scraping the loose dirt away with the hands the stone was found about ten inches below the surface. From the direction in which the ground was crushed in it must have come from the north-west, and at an angle of about thirty degrees with the horizon. The stone when it was picked up was covered all over with the black shell which it bears now, except a triangular spot on one corner, about one inch each way, where the corner appeared freshly knocked off, and about four other spots near a quarter of an inch in diameter, where the shell was slightly knocked off. The other bruises which you will find upon it have been made since by persons who have handled it. To enable you to distinguish the original breaks upon it I have marked each of them with a red cross. The stone still has a strong odor, which I will not undertake to describe. Capt. Barlow says it smelled stronger when he first picked it up. He does not remember that it had any noticeable heat. It was not cold, as a stone found so deep in the ground should be.

"The stone weighs now twelve and a quarter ounces; about half an ounce has been pecked off from it. Its color within is strikingly like very light granite; and with the exceptions above noted it is entirely covered with a smooth, almost black shell, a trifle thicker than common letter-paper, so that externally it looks very much like a lump of iron-ore. It is an irregular, seven-sided figure, its longest side being about two and three quarter inches long. If put into a spherical form it would make a ball about one and three quarter inches in diameter. So far as I have been able to ascertain no other parts have been found.

"The noise attending this phenomena is variously described by different persons, and from different places. Two intelligent ladies, residing four miles south of Lumpkin, nearly east of where the stone fell, and about ten or twelve miles off, describe it thus: While sitting in the house they heard, as it were, the sound of a great fire suddenly bursting forth from some confinement into the open air. They rushed out of doors and heard the roaring sound continue for several seconds. They located the source of the noise in the direction of Barlow's.

"In Cuthbert, about eighteen miles from Barlow's, nearly south-east, a gentleman engaged in a workshop heard a lumbering noise, which he took to be several heavy pieces of machinery in an adjoining room falling down one after another. On going in he found no one, and that he had mistaken the cause of the noise. Many persons here heard sounds like repeated thunder followed by roaring. Some say that they first heard several rapid, cracking explosions, like that of volleys of small arms, followed immediately by the louder burst of artillery. Most persons here thought the noise came from the south-east, passed over the place in a north-westerly direction, and died away in the distant north-west.

"The foregoing statements have been selected from many in circulation, showing how differently the senses were affected at different points. The facts are purposely presented in their nakedness. If you can find them available in aid of a scientific investigation of the origin of this phenomenon, I shall have accomplished more than I expect."

The above accounts agree as to the main facts: the point of greatest discrepancy—the direction of flight. It is probable that the meteorite came from some point in the *north* quarter; the statement of Mr. Latimer, over whom it exploded, and that of Mr. Barlow as to the direction in which the earth was penetrated, concur in this regard. Persons in Cuthbert, who represent it as coming from the south, may have been misled by an echo, mistaking this for the original sound.

DANVILLE (ALA.) METEORITE.

(Fell November 27, 1868.)

Although the meteorite of Danville, Ala., fell in November, 1868, and an analysis has been made of it during the past summer, it is only recently that I obtained a complete account of the phenomena attending its fall.

On Friday evening, November 27, 1868, about five o'clock, Mr. T. F. Freeman, of Danville (about lat. 34° 30′ and long. 87° W. Greenwich), on stepping from his house, was startled by a loud report, so much like artillery that for the moment its origin was attributed to the firing of a small piece of artillery kept in the village; but on inquiry it was ascertained that no firing had taken place there, but that the sound was heard at the village, and attributed to very heavy artillery at Decatur, Trinity, Hillsboro, or some other point to the northward of Danville. During the war artillery had been often heard in the valley of the Tennessee, and various speculations were indulged in as to what was meant by this cannonade at such a time of day and in such a direction.

The following day Mr. Wm. Brown, living three miles west of Danville, brought to the village a piece of rock, which he said fell near him and some laborers who were picking cotton. He dug it up at a depth of about one and a half to two feet. It weighed about four and a half pounds, and had the characteristic aspects of a meteoric stone. But it was broken by the party obtaining it, and all but half a pound, now in my possession, has been scattered, and probably lost or thrown away.

Several other stones fell in the same vicinity. Some negroes working in a cotton-field on the plantation of Capt. McDaniel, half a mile from Danville, heard a body fall with a whizzing, humming sound, and strike the ground near them with tremendous force; but they were alarmed, and did not approach the spot that night. A rain fell during the night, and no trace of it could be found the next day. Various other stones were heard to fall in different parts of the adjacent country. Two brothers by the name of Wallace were plowing in their field, about one and three fourths miles north-west of Danville. They distinctly heard two or three fainter reports after the first loud one, and heard the sound of two falling bodies whizzing down, one to the right and the other to the left of them.

With the above data, and the known geography of the country, its direction must have been north-east and south-west; but it is impossible to say from which of these quarters it came.

The portion of the meteorite that I possess has a large portion of it covered with the usual black crust. Its general aspect is rough and dull; a portion of the outer surface, not covered with the black coating, is nevertheless a surface that it had when it reached the ground, for on this surface are streaks and little patches of a bright, pitchy matter, which was once fused, and was derived either from another part of the coating that was thrown off in a melted state from the coated portion, and whipped around (as it were) on to the unfused surface as the stone fell through the air, or from an incipient fusion that was begun on the denuded surface, and arrested by the termination of the fall. Where the black crust reaches the denuded places it appears to be rounded off, as if it had been melted matter passing from another portion of the stone, and rolled over the surface of the borders.

The broken surface has a dark-gray color, and is somewhat oölitic in structure, but not as much so as many other meteoric stones. There are veins and patches of a slate-colored mineral running through it. Pyrites and iron are also to be seen diffused through the stone; thin flakes of the iron giving that slickenside-like appearance to a fracture not unfrequently seen in this class of bodies. There seems to be more of iron in the slate-colored mineral than in the other parts. There are a few patches of white mineral, which I take to be enstatite. The specific gravity of the stone is 3.398.

For further examination a portion of the meteorite was separated mechanically into three parts, the pyrites, the metallic iron, and the earthy minerals. As in the case of most meteorites, the earthy minerals were so intermixed that it was impossible to separate the different varieties, three of which were easily traceable by the eye.

The iron, separated with great care from the pulverized meteorite, constitutes 3.092 per cent. of the entire mass, and an analysis furnished

Iron	89.513
Nickel	9.050
Cobalt	.521
Copper	minute quantity
Phosphorus	.019
Sulphur	.105
	99.208

The sulphide of iron detached very carefully from the mass of the meteorite gave

Iron	61.11
Sulphur	39.56
	100.67

Which corresponds with the protosulphide of iron, FeS. Whether it contains any of the sulphide known as troilite I am not prepared to say.

The stony minerals were freed as much as possible from iron and pyrites, and one gramme treated with ten grammes of hydrochloric acid on a water-bath, and evaporated nearly to dryness, then filtered, and the filtrate well washed. After which the residue in the filter was warmed with a solution of caustic soda, to dissolve any silica belonging to the portion dissolved by the acid. It was then filtered again and washed. The result was

Soluble portion	60.88
Insoluble portion	39.12

The treatment by a solution of caustic soda or potash is of importance for a correct result, as otherwise a portion of the silica of the decomposed minerals will be estimated with the portion that is undecomposed.

The insoluble portion of it was analyzed; for, although the analysis made in this way can not furnish any positive indication in regard to the true mineral constitution of the meteorite, it is nevertheless an important guide. It was found to consist of

Silica	50.08
Alumina	4.11
Protoxide of iron	19.85
Magnesia	20.14
Lime	3.90
	98.08

From all the circumstances connected with this mineral, its physical characters, etc., it is doubtless a pyroxene of the augite variety.

The soluble portion, owing to the unavoidable presence of a little iron and pyrites, simply furnished results on analysis that showed it to be mostly olivine. The stony matter, as a whole, freed as much as possible from pyrites and nickeliferous iron, gave

Silica	45.90
Protoxide of iron	23.64
Magnesia	26.52
Alumina	1.73
Lime	2.31
Soda	.51
Potash	.64
Oxide of manganese, a minute quantity, not estimated.	
Oxide of chrome, " " " "	
Phosphorus, " " " "	
Lithia—marked reaction with the spectroscope.	
Sulphur	1.01

The excess in the footing up of the analysis above one hundred per cent. is due to the fact that a part of the iron, estimated as protoxide, is combined with sulphur, forming sulphide of iron.

This meteoric stone is similar in every respect to that which fell, March 28, 1859, in Harrison County, Ind. (which locality I see referred to in catalogues of meteorites as Harrison County, Ky.) This meteorite is therefore composed of nickeliferous iron, olivine, pyroxene, protosulphide of iron, with minute quantities of schreibersite, chrome iron, and probably albite.

In concluding these observations on the Danville meteorite I can not but feel more and more convinced of the importance of a thorough re-examination of the mineral nature of the meteoric stones; and in the present case I am not at all satisfied that the mineral characteristics are perfectly made out.

SEARSMONT (MAINE) METEORITE:

(FELL MAY 21, 1871.)

MINERALOGICAL AND CHEMICAL COMPOSITION.

Immediately after the fall of this meteoric stone a portion of it was placed in my hands for examination. The circumstances accompanying its fall, as well as its physical characters, have been described by Prof. Shepard.

It resembles very closely the Mauerkirchen stone that fell in 1768, the crust of the specimens corresponding quite closely to that in thickness and appearance; the Mauerkirchen stone, however, has not well-marked globules like that of Searsmont; in this respect it corresponds more nearly with the Aussun, as already stated by Prof. Shepard.

The specific gravity of the specimen examined was 3.701. The nickeliferous iron and stony matter were separated mechanically for analysis. One hundred parts of the meteorite gave

Stony matter (including a little sulphuret of iron)..	85.38
Nickeliferous iron	14.62

The iron afforded

Iron	90.02
Nickel	9.05
Cobalt	.43

Phosphorus and copper were not estimated. The stony part, treated with a mixture of hydrochloric and nitric acids, gave

Soluble in the acid	52.3
Insoluble ··	47.7

The soluble portion afforded

Silica	40.61
Protoxide of iron	19.21
Magnesia	36.34
Sulphuret of iron	3.06

Leaving out the sulphuret, which is obviously only a mechanical mixture, this soluble part is evidently an olivine, which is almost invariably the case with soluble portions of meteoric stones. The insoluble part was composed as follows:

Silica	56.25
Protoxide of iron	13.02
Alumina	2.01
Magnesia	24.14
Alkalies, Na O, K O with trace of Li O	2.10
Chrome iron, small black specks, not estimated.	

The above analyses give for the composition of the stone:

Nickeliferous iron	14.63
Magnetic pyrites	3.06
Olivine	43.04
Bronzite or hornblende, with a little albite or orthoclase, and chrome iron	39.27

With the bronzite there may also be enstatite, which would be confounded with the former if existing in the stone.

METEORIC IRONS OF NORTH MEXICO:

PRECISE GEOGRAPHICAL POSITION, WITH DESCRIPTION OF A NEW MASS—THE SAN-GREGORIO METEORITE.

Some of the remarkable masses of meteoric iron in Northern Mexico have been known to travelers for a number of years, but no very precise information concerning them had been given until the year 1854, when the first mass brought from that locality was placed at my disposal by Lieut. Gouch, of the United States army, and was described in a memoir on meteorites published in the Amer. Jour. of Science and Arts, April, 1854; it is now in the Smithsonian Museum, and weighs two hundred and fifty-two pounds.

On the return of Mr. Bartlett, of the Boundary Commission, I learned of two other masses in that region, and Lieut. John G. Parke, of the United States army, placed a fragment of one of them in my possession; the fragment of the other mass was lost. I figured and described both of those meteorites in the memoir just alluded to; the first, which I called the *Tucson meteorite*, is now in the Smithsonian Institute, and weighs, I believe, several thousand pounds; the second one I called the *Chihuahua iron*, and is still at the Hacienda de Conception, where it was first found. Still later, in the year 1868, Dr. H. B. Butcher placed under my examination eight masses of meteoric iron that had been brought to the United States from the same region of Mexico; these I examined, and published a full account of in the Amer. Jour. of Science and Arts. These masses are now in Philadelphia, still owned by Dr. Butcher, and vary in weight from about three hundred to eight hundred pounds. Dr. Butcher having returned to Mexico, I requested him to get all possible information in regard to the geographical position of these bodies; this he has succeeded in accomplishing. At the same time he has sent a fragment of another mass, still larger than any yet known, which will be called the *San-Gregorio meteoric iron.* Its description is as follows.

THE SAN-GREGORIO METEORITE.

This immense mass of meteoric iron is situated on the western border of the Mexican Desert, a map of which is given on the next page. Some idea of its form may be had in the accompanying sketch. It measures six feet six inches in its greatest length, is five feet six inches high, and four ft. thick at its base; on one part of its surface 1821 is cut with a chisel, and above this date is the following inscription:

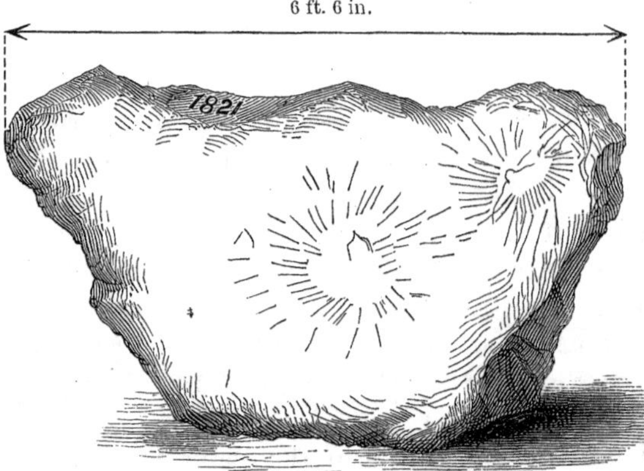

"*Solo dios con un poder este fierro destruerá, por que en el mondo no habra quien lo puedo deschacer.*"

It lies within the inclosure of a hacienda, having been hauled to the ranch many years ago by the Spaniards, who thought that it could be made use of as iron for farming utensils. It is said to have fallen quite near its present site, and from its huge bulk and weight, which is calculated to be about five tons, it could not have been transported very far. Nothing more is known of its history. Small specimens were detached by Dr. Butcher, one of which I have examined. I find it to be of the softer meteoric irons, with a specific gravity of 7.84. The fragment I possess is too small for the study of the true character of its Widmannstättian figures. On analysis it furnished the following composition:

Iron	95.01
Nickel	4.22
Cobalt	.51
Copper, minute trace.	
Phosphorus	.08

This San-Gregorio iron makes the fifth that has come under my observation and examination from this famous Mexican locality, the geography of which I will now describe, referring to the accompanying diagram for details.

The Bolson de Mapini, or Mexican Desert, occupies the western portion of the province of Cohahuila and the eastern portion of the province of Chihuahua. It is four hundred miles from east to west, and five hundred miles from north to south, bounded on the north by the river Rio Grande. Some of the villages and haciendas are specified in the diagram, and the numbers 1, 2, 3, etc., are the localities of the different meteoric masses discovered.

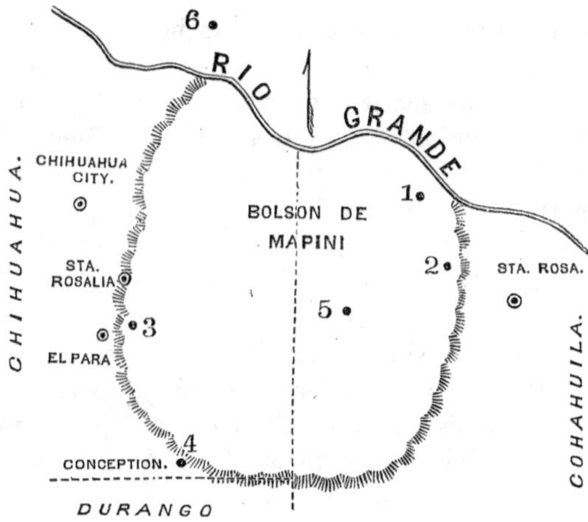

No. 1.—The locality of the Cohahuila meteorite, described by me in the Amer. Jour. of Science and Arts, April, 1854; it is now in the Smithsonian Institution.

No. 2.—The locality of the Cohahuila meteorite of 1868, described by me in the Amer. Jour. of Science and Arts, April, 1870; it is now in the possession of Dr. Butcher.

No. 3.—The locality of the San-Gregorio meteorite just described; it is still in the place where it was first observed.

No. 4.—The locality of the mass described and figured in my memoir on meteorites (Amer. Jour. of Science and Arts, April, 1854), and called the *Chihuahua meteorite;* it is still in place at the Hacienda de Conception, ten miles from Zapata, its greatest height being forty-six inches, breadth thirty-seven, and in the thickest part eight feet three inches in circumference.

Signor Urquida calculated its weight to be about four thousand pounds.

No. 5.—The locality of a huge meteorite lately discovered, of which no specimen has yet been detached, and is said to be larger than any one yet found in that locality.

No. 6.—The locality of the large mass described and figured by me in 1854 as the *Tucson iron*, and now in the Smithsonian Institution at Washington, having a large hole in the center, and sometimes called the *Signet meteorite*, also the *Ainsa meteorite*. I do not know its exact weight, but suppose that it must weigh two or three thousand pounds.

The question naturally arises, what can be the cause of the number of meteoric masses in the circumscribed region, and whether each one represents a separate fall? My study of them leads me to the belief that they are the products of two falls. First of all, No. 6, the Signet or Ainsa meteorite, has peculiar physical and chemical characters that separate it entirely from the others. Nos. 1, 2, and 3 I have examined chemically, and find them very closely allied in composition, also in physical properties, as the softness of the iron, and freedom from rusty crusts over the exterior; in fact, the pieces I have examined were more or less bright on the exterior surface. The Widmannstättian figures I have not had an opportunity to compare, since, with the exception of No. 1, I have had only small pieces that were detached from the surface by a cold-chisel, which are unfit for the study of these figures. Thus far in my investigation there appear strong reasons for supposing that at some epoch, probably far remote, the meteoric masses 1, 2, 3, 4, and 5 were the products of the fall of one meteoric mass, moving from the north-east to the south-west, the smaller masses falling first at 1 and 2, and the larger masses farther on. The distances of these bodies from each other are: from Nos. 1 to 2, about eighty-five miles; from 2 to 5, about one hundred and thirty-five miles; from 5 to 3, about one hundred and sixty-five miles; from 3 to 4, about ninety miles. Of course there is no great stress laid upon these deductions, but it would not be surprising if farther investigation should sustain this view.

Since my first publication on these meteorites, Burckhardt, of Bonn, has made some observations on them, but his publications are not within my reach at the present time.

VICTORIA METEORIC IRON:

(Seen to fall in South Africa in 1862)

WITH SOME NOTES ON CHLADNITE OR ENSTATITE.

The Victoria Meteoric Iron, although found about ten years since, has never been described; and yet it is one of the most interesting of this class of meteorites. I have succeeded in collecting the following facts in connection with it. It was seen to fall in the year 1862 by a Dutch farmer in Victoria West, Cape Colony, South Africa, and was given by him to Mr. Auret, the civil commissioner of that district, who presented it to the South African Museum at Capetown.

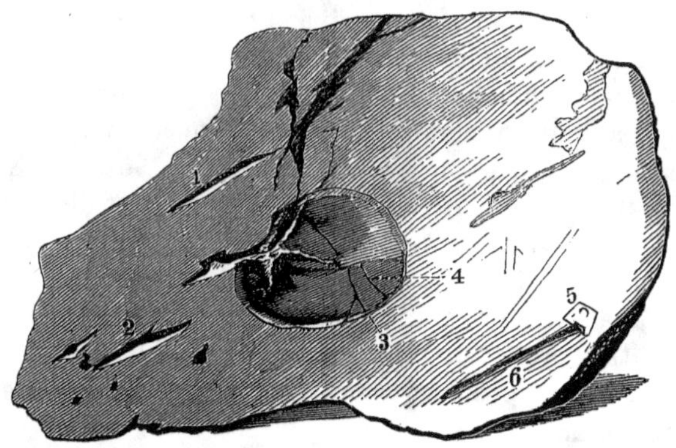

Although it is an iron that has a tendency to decompose rapidly, it has not undergone the decomposition that would have inevitably taken place had it long remained exposed to atmospheric influences. This fact, coupled with another that the farmer could have no object to deceive any one with reference to a body which certainly bears evidence that it fell at some time from the heavens, and as those who know the farmer have every confidence in his statement, we are led to conclude that it is to be placed along side of the Agram, Brannau, and

Dickson County irons. The mass was pear-shaped, and weighed abour six pounds eight ounces. One end was smooth and rounded; the smaller end was jagged, as if torn or parted from a larger meteorite.

A small fragment of it was first brought to Europe in 1868, by which its true meteoric character was established. In 1870 the trustees of the South African Museum had it cut, retaining one half of it, and sending specimens to the British, Calcutta, Vienna, and Berlin museums; also to Mr. Nevill of Godalming; and a mass weighing about twelve ounces to myself.

From the specimen in my possession I here describe its characteristics. The iron is compact, with a tendency to fissure near some portions of its surface. The amount of oxide on the surface is small, the cut surfaces showing bright metal quite up to the exterior surface. The Widmannstättian figures developed are of that class where the lines are delicate and straight, inclined at a considerable angle to each other, a form I have seen common to irons rich in schreibersite. This last mineral is diffused through the iron in masses with absolutely straight boundaries, some of them five eighths to three fourths of an inch long, by one eighth of an inch broad (1 and 2 on figure), and others much longer (6) and narrower; others again triangular (5) and arrow-shaped; and on my specimen a layer of it (3) coating an oval cavity that must have been an inch and a half in its longest and one inch in its shortest diameter, the schreibersite having a thickness of about one twentieth of an inch; the rest of this cavity being filled by pyrite (this is distinctly seen in the photograph). My specimen shows one fourth of this cavity, and there are doubtless others in the original mass. The specific gravity is 7.692. On analysis it was found to contain:

Iron	88.83
Nickel	10.14
Cobalt	.53
Copper	minute quantity
Phosphorus	.28
	99.78

Enstatite of Chladnite.

This mineral now occupies so important a relation to the mineral constitution of meteoric stones that it is well to give an account of its discovery, and the subsequent investigations of different observers. Its discovery is beyond doubt due to Prof. C. U. Shepard, who first described it in the American Journal of Science, Sept., 1846, p. 381, calling it chladnite.

It constituted nearly the entire mass of the Bishopville meteorite that fell in 1843. Prof. Shepard did not make out its composition correctly, his analysis being imperfect. The composition given by him was

Silica	70.41	3 of oxygen.
Magnesia	28.25	1 "
Soda	1.39	

Making it out to be a tersilicate of magnesia. Although the constitution was incorrectly determined, Prof. Shepard clearly showed that it differed in character from any then known mineral.

Eight years after the mineral was first made known, a small fragment of the meteorite coming into my possession, a re-examination was made of its chemical constitution, and the errors of the first analysis discovered. But not having enough of the meteorite for analysis, the simple statement was presented to the American Association for the Advancement of Science, in April, 1854, "that from some investigations just made chladnite is likely to prove to be a pyroxene." This was noticed in the proceedings of the association for that year, and referred to in the American Journal of Science, March, 1855, p. 162. Ten years later a specimen of the Bishopville meteorite of good size being placed at my disposal, the mineral was separated in a very pure state, and found to be composed as follows:

Silica	59.97
Magnesia	39.33
Peroxide of iron	.40
Soda, with feeble potash and H	.74
	100.44

This minute quantity of peroxide of iron came from a little metallic iron that was present. The analysis afforded the oxygen ratio 2:1, corresponding to the formula $\dot{M}g^3, \ddot{S}i^2$ (or $\dot{M}g\ \ddot{S}i$), corresponding to the general formula of pyroxene. The details of the examination then made are to be found in the American Journal of Science, September, 1864, where it is further stated that chladnite approaches those forms of pyroxene known as white augite, diopside, white coccolite, etc., these last-named minerals having part of the magnesia replaced by lime. It is identical with the enstatite of Kenngott, a pyroxenic mineral from Aloysthal in Moravia.

From these observations it will be seen that the Bishopville meteoric stone, however different in external characteristics from other similar bodies, is after all identical with the great family of pyroxenic meteoric stones.

Enstatite.

This form of pyroxene was first noticed by Kenngott as a new species, in a communication made by him to the Vienna Academy in 1855 (see Vien. Acad. Ber., xvi, p. 162; Jahresbericht for 1855, page 928). Its composition there given is

Silica	57.09
Alumina and oxide iron	5.13
Magnesia	35.85
Water	1.92
	99.99

As the crystallographic character of this mineral entitles it to separation from pyroxene, or, in other words, as it is entitled to be ranked as a new species, the prior right of discovery belongs to Prof. Shepard, and the name first given by him, chladnite, has the priority; but as it has for so long time borne the name of enstatite among mineralogists, any attempt to change it would only bring confusion. This is the more to be regretted since the name of chladni would be a most appropriate affix to a mineral, the true and pure type of which is so pre-eminently that in meteorites.

In this connection I would refer to the simple chemical relation of three of the most characteristic minerals of meteoric stones; these minerals forming at least ninety per cent. of the earthy minerals in the aggregate mass of all meteoric stones. The three minerals are:

Enstatite,	$\dot{R}\,\ddot{Si}$	$\dot{Mg}\ \ddot{Si}$
Bronzite,	$\dot{R}\,\ddot{Si}$	$(\dot{Mg}\,\dot{Fe})\,\ddot{Si}$
Chrysolite,	$\dot{R}_2\,\ddot{Si}$	$(\dot{Mg}\,\dot{Fe})\,\ddot{Si}$

In these minerals the protoxide of iron replaces but a small portion of the magnesia in the last two; so they are virtually silicates of magnesia, containing one or two atoms of silica with one atom of magnesia.

INVERTED MICROSCOPE:

A NEW FORM OF MICROSCOPE; WITH DESCRIPTION OF A NEW MICROMETER AND GONIOMETER.

The instrument forming the subject of this article was invented by me in the summer of 1850, and first brought to the notice of the *Société de Biologie* of Paris in the month of September of the same year, and with additional improvements in the micrometer movement was laid before the American Scientific Association in 1851. Besides the mention made of this instrument in the minutes of the proceedings of those scientific bodies, no account of it has been published giving a full detail of the objects sought after and effected by this new form of microscope; and this I now hasten to do, in justice to myself, since seeing in the last edition of Quekett's work on the microscope a short description of this instrument under the title of Nachet's Chemical Microscope. How it is that my name has been entirely omitted in connection with it is a mystery to me; it must have arisen through Mr. Nachet's neglect to mention who the inventor of it was while exhibiting it at the World's Fair in London in 1862, or through the forgetfulness of Mr. Quekett after being informed on the subject. This omission is still more glaring, from the fact that the instrument as then exhibited, with one or two very unimportant modifications, is the same in all its mechanical details as was constructed for me, from my plans, by Mr. Nachet, of Paris, and used in the laboratory of Messrs. Wurtz and Verdiel.

I am sorry to be obliged to preface the description of the microscope with this reclamation; but after considerable experience I feel that the instrument is an important one for general as well as chemical purposes, and that it will in time be considered a decided advancement in the construction of microscopes. With these views in the matter I am unwilling to yield what little credit might be due the inventor of it.

The great development made in microscopic research during the last twenty or thirty years is due in great part to improvements in the construction of achromatic object-glasses; still the mechanical arrangements of the instrument have contributed their share to facilitate observation and diminish the fatigue dependent upon this character of research. In fact, observers have not hesitated to make use of different descriptions of mounting in their varied field of research, and now we have instruments for general purposes, but the construction of which is imperfectly adapted to certain special researches; as, for instance, the dissection of animal tissues. This last circumstance has given rise to the invention of various forms of dissecting microscopes, such as the Pancreatic Microscope of Oberhauser, and more recently the simple and better instrument for arriving at the same end constructed by Nachet, of Paris.

These remarks are made to show how the use of the microscope might be extended by paying proper attention to its mechanical arrangements, and it is from this cause I have been led to seek out a form of instrumant by means of which microchemical research might be facilitated and enlarged. The instrument about to be described is calculated to produce these results.

The great obstacles to chemical research beneath the microscope are twofold: first, the necessity of manipulating in the limited space between the object-glass and the stage; and secondly, the exposure of the most essential parts of the instrument to the vapors emanating from the re-agents employed, and the condensation of vapor on the under-surface of the object-glass, thereby obscuring the view. A less important obstacle is the impossibility of heating a liquid or other substance while beneath the microscope.

The only way by which these difficulties can be surmounted is to place the object-glass beneath the stage and the object above it, with an optical arrangement of such a nature as to permit observation. It was with this view that M. Chevalier made a chemical support to go with his general instrument; but those familiar with it know how awkward it is for manipulation, although exceedingly ingenious, and doubtless as perfect as could be for attaching to his instrument. Feeling then the want of something more effective, I was led to the construction of the inverted microscope entirely with reference to its chemical uses, other purposes to which it might be applied being of secondary consideration; but I would here remark that since its completion its value even in this latter respect yields to no other form of instrument, and has induced me to change

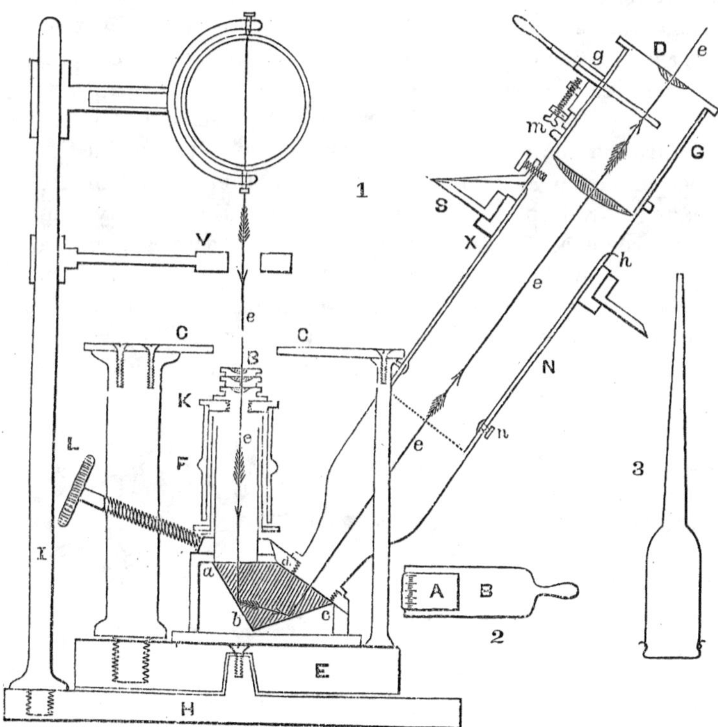

its original designation of Chemical Microscope to that of Inverted Microscope, as the former name might mislead as to the extent of its uses.

It was important for the arrangement in question so to have the relative position of the stage and eye-piece that the eye, while on a level with the latter, could readily see the former and guide the required manipulations.

Without entering into any detail of the steps taken in the construction of the instrument, I will at once proceed to a description of it that will be readily understood by referring to the figure. The most important part is a four-sided prism, with the angles *a*, *b*, *c*, *d*, respectively 55°, 107½°, 52½°, 145°, the angles being of such dimensions that a ray of light passing into the prism in the directions shown by the arrows, and perpendicular to the surface *a d*, after undergoing total reflection from the inner surfaces *a b* and *b c* (on both of which the light strikes at an angle much less than forty-five degrees), will pass out perpendicular to the surface *c d*. If the line *c* be followed, it will be readily seen how a ray of light passing through the object-glass B descends into the prism and passes out of it upward through the eye-glass D, the tube of which is inclined to the perpendicular 35°. The other parts of the instrument are understood by simply looking at the figure. E is a heavy support that revolves on another support H, which carries a column I, on which are placed the mirror, diaphragm, etc. The prism used has each side nearly an inch in length, and little less in width, which is about the most convenient size. The arrangement for adjusting the focal distances is somewhat peculiar, and is readily understood by reference to fig. 4. There are three tubes (the outer one of which is F) that slide on each other; the inner is fastened to the plate O; the second tube has a projecting collar, on the under surface of which rest the extremities of two springs *y*, and on the upper surface two points of the lever X, which is moved by means of the screw T. The plate O is fastened on to the top of the prism by the binding-screw L (fig. 1), that readily allows of the plate being detached at pleasure, which it is necessary to do at times in order to wipe the upper surface of the prism. The way in which the observer operates is to screw one or other of the object-glasses to a small cap, K (fig. 4), that simply rests on the upper end of the outer tube F, which is readily moved up and down by the finger for the coarser adjustment, while the minute adjustment is obtained by moving the screw T.

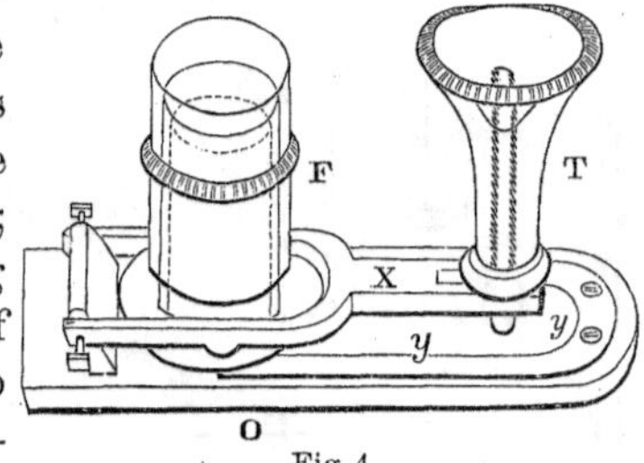

Fig. 4.

This description suffices to make it clearly understood how

the instrument is used, and the conveniences arising therefrom. In examining an object with this microscope the object is arranged in the ordinary way; when liquid it is placed in a watch-glass, or such glass cells as are convenient to use. In employing re-agents they can be added and their effects watched immediately, for it is readily seen how the eye guides the manipulations on the stage, and looks into the instrument almost at one and the same time—a circumstance that facilitates and renders highly satisfactory all such operations, as nearly two years' experience has convinced me.

With this arrangement we need not hesitate to employ hydrofluoric acid among our re-agents, as Prof. Bailey has already done, for the purpose of settling, in a most ingenious manner, that the markings on certain microscopic siliceous animalculæ are elevations, and not depressions, as they disappear last under the action of this acid.

Under the supporting ring V are placed the diaphragms, palarascope, achromatic condenser, etc. I have also arranged a small ring of ivory, through the edge of which two wires pass, that can be made the poles of a galvanic battery, and thereby subject any thing to an electric action while on the microscope. The extremities of the wires may be united with a spiral of small platinum wire, which would become heated by the passage of the electricity, and in this state can be brought immediately over the object under examination.

There is another and very simple method which I have adopted for heating or evaporating liquids while on the stage of the microscope. It consists of a thin plate of brass, about five inches long and an inch wide, with a hole in the center. About an inch and a half each side of the hole there are two screws projecting about the tenth of an inch. When required it is placed on the stage with the projecting screws downward, that prevent the plate from touching the stage, and the part of the plate projecting beyond the stage is heated by a small lamp. The heat is readily propagated along the plate, and imparted to the glass that may be placed along it.

In using this instrument for chemical purposes it is very necessary to be able to apply the re-agents conveniently, and for this purpose I place such of them as are used in two ounce vials, in the neck of which there is a small drop tube, as represented in fig. 3, over the top of which is stretched a piece of sheet india-rubber, and by pressing and relaxing it the liquid is drawn in, and by pressing the same the smallest possible quantity can be discharged on the object subject to examination. The tube also serves as a stopper to the vial, for the length of the capillary extremity is such that it reaches nearly to the bottom.

The acids and ammonia used are always diluted to about one half their ordinary strength, to prevent any unnecessary disengagement of vapors.

A movable stage, under many circumstances, is very convenient, and I have adopted one of a very simple character, and quite equal to any of those where the motion is produced by screws or pinions. It is a circular plate of metal or glass, about three fourths of an inch less in diameter than the fixed stage of the instrument, and an eighth of an inch thick, with a hole in the center of nearly an inch diameter. This is laid in the stage of the instrument, the glass sustaining the object placed on it, and when required the former is moved by the fingers, which can readily impart to it the most delicate motion, as they are in part supported by the edge of the fixed stage. For this suggestion I am indebted to Prof. Riddell, and both he and myself, after much experience, feel convinced of its usefulness.

In observing with high powers, as the object-glass is beneath the glass supporting the object, and as this glass is usually of a certain thickness, we have to change our method of observation—for all powers resorted to in chemical examination this difficulty never occurs, and in using high powers it is easily obviated. Where the object is already mounted and dry, the thin glass can be readily turned downward; but where it is moist—as, for instance, in examining fresh *Desmidiæ* and *Diatomaciæ*—the following plan is resorted to, namely, to use a cell made of a thin piece of brass or glass, perforated with a hole about half an inch in diameter; it is best to give the hole a considerable bevel in one direction, as it facilitates the cleaning of it; over the small end of the hole a piece of thin glass is stuck with balsam or other cement. When used the object to be examined is placed within, and a cover of thin glass placed above. When brass is used to make the cell it may be as thin as the twentieth of an inch; and I have two such in my possession, made for me by Prof. Riddell, and they are certainly the most convenient things of the kind I have ever used. And here I may remark that for all observation with high powers the Inverted Microscope is decidedly superior to the ordinary forms of mounting; for in the latter case, when an object-glass of a one-twelfth or one-sixteenth inch focus is used, the focus is too short to admit of the use of cells; whereas in the inverted form, as the object is looked at from beneath, the cell may be as thick as one pleases. Another thing that I have discovered connected with this class of observations is that the *Desmidiœ* and *Diatomaciœ* can be observed to much greater advantage from beneath than from above, for reasons that will be obvious to persons accustomed to observe these classes of objects.

Another advantage possessed by this instrument, calculated to extend its use for general purposes, is its great capacity for every variety of illumination, without sacrificing the ease and freedom from fatigue belonging to the use of this form of microscope; for when placed on a table, rather higher than the one commonly used, and a foot or two from the edge, the observer can recline on his arms, and observe for hours without the slightest sensation of fatigue.*

---

* As Prof. Riddell, of the Medical Department of the University of Louisiana, has been using my microscope for general purposes for more than a year, I requested of him his opinion as to its advantuges, which is expressed in the following letter:

Prof. J. LAWRENCE SMITH: *Dear Sir.*—In reply to your note respecting your Inverted Microscope, I have to say that having formerly been in the habit of using the mountings of Pritchard, Dollard, Raspail, Chevalier, and Nachet, and having the past year constantly used my best lenses (Spencer's make) in the inverted microscope, I am fully satisfied of the practical superiority of the latter for general purposes. With it observation can be made with more ease and comfort, the light admits of more convenient and efficient management, chemical re-agents can be applied to the object with the greatest facility, without endangering the instrument, and the slides can be moved or changed with the utmost facility, and with perfect safety to the object-glass and the slides themselves. The instrument is so firm as to manifest no vibration with the highest powers, and admits of the attachment of every collateral appliance. I shall never willingly return to the habitual use of any other known form of microscope, especially with high powers. The excellence of your form of microscope depends on having a good reflecting prism below the object-glass. The one used by me, made by Oberhauser of Paris, seems to be perfection itself, and seems neither to absorb or distort the luminous rays in the slightest degree.

Respectfully yours, J. L. RIDDELL.

Many little additional conveniences will suggest themselves to almost all microscopists who may use this instrument; but the great principle belonging to it is what I desire to make public, and any other adjuncts that may be described are such as belong to all forms of microscopes.

NEW FORM OF EYE-PIECE MICROMETER FOR MEASURING OBJECTS UNDER THE MICROSCOPE.

Facility of measuring objects under the microscope is a great desideratum; and for this reason I now make known what I have been using for this purpose for upward of two years, as it furnishes all that can be desired. The eye-piece micrometer ordinarily in use consists of a glass with divisions drawn in it, contained in a special eye-piece adapted for its use; and whenever the measurement of an object is required we replace the eye-piece used by the micrometer eye-piece, and move the object on the stage, so that its image falls on the marking of the micrometer. With all its advantages this form has many inconveniences, among which I will mention the necessity of using an eye-piece, which is not always the best for examination, the constant interposition of the micrometer in the field of observation, and the necessity of moving the object so as to superpose its image on the micrometer.

The eye-piece micrometer of Mr. George Jackson, described in Quekett's work on the Microscope, is an improvement on the one just mentioned, but does not do away with all the objections.

By the present arrangement the micrometer can be used with any eye-piece; it can be withdrawn at pleasure, and placed over the image of the object without regard to its position in the field; for this purpose the tube of the microscope is in two parts, G and N (fig. 2); the former has a collar *h*, and projects a couple of inches into N, turns freely in it, and is retained by a small screw *n* passing through N, and playing on a groove in G. On the upper part of G there is a small rectangular opening in a little mechanical arrangement, as seen in *m*. The various eye-pieces are so mounted that when placed in the tube G the planes of their foci correspond with the opening at *g*, and at the same time there is an opening in their mounting, which is made to come opposite to that of *g*.

The micrometer is seen in fig. 2, and consists of a brass mounting B, with a small plate of glass A, having near the outer edge a fine graduate scale (the one used is ten millimetres divided in one hundred parts) made in the direction of the breadth and not of the length of the micrometer, which little circumstance is of vast importance; for, made as it is, it can sweep the field of the microscope; whereas, were it graduated longitudinally, it would simply move in the radii of the field, and therefore could not be brought on the object in many of its positions.

The manner of using the micrometer can be understood in a few words. In examining with any eye-piece, if it be required to measure an object, the micrometer B is introduced into the opening *g*, and if not seen distinctly, by turning the screw *p* it is readily adjusted, and by pushing it backward or forward, or turning the tube D, the graduated scale can be readily brought over the image of the object, either longitudinally or otherwise; and, knowing the value of each division, the dimensions of the object is readily made out. The manner of ascertaining the value of these divisions is learnt in almost every work on the microscope. This method of mine is now adopted by M. Nachet, of Paris, in the construction of his large microscope.

A NEW FORM OF GONIOMETER FOR MEASURING ANGLES OF CRYSTALS UNDER THE MICROSCOPE.

The measurement of the angles of crystals beneath the microscope is at best a very imperfect operation, for we can only measure plane angles, the angles between the faces not being measurable; yet, imperfect as it is, it is an important adjunct in certain researches, as may be seen by referring to the following, which is an abstract from Lehmann's work on Physiological Chemistry, where he speaks of detecting a minute quantity of urea in albuminous fluids: "If the residue of the fluids from which the coagulated matters have been filtered be extracted with cold alcohol, and the solution rapidly evaporated, so as to cause the chloride of sodium (taken up by the cold alcohol) to separate as much as possible in crystals, or then bringing a drop of the matter-liquid in contact with nitric acid under the microscope, we shall observe the commencement of the formation of rhombic octahedra, and the hexagonal tablets, in which, if the investigation is to be unquestionable, the acute angles (=82°) must be always measured. After the determination of the nitrate we may also obtain the oxalate, and submit it to microscopic examination. *A good crystalline determination yields the same certainty as an elementary analysis, which in these cases would never, or extremely seldom, be possible.*"

Two of the best goniometers used for this purpose are those of Ross and Leesom, the latter being doubtless the best, and based on the use of the double refracting spar. After trial, however, I find it neither as accurate nor as economical in its construction as the following: Around the tube N (fig. 1) there is a collar X fastened to it. On the collar there is a graduated circle S, about three inches in diameter, turning freely on X. On the tube G a small index *t* is fastened. These are all the additional parts necessary, as the micrometer just described is used to aid in the measurement, which is accomplished as follows: The angle of the crystal to be measured is brought as near the center of the field as the eye can readily judge of (a little deviation will not sensibly affect the measurement). The micrometer is then introduced in the opening *g*, and turned about until the lines are parallel to one side of the angle, or until one of the long marks corresponds with that side. This done, without disturbing the tube G, the graduated circle S is turned until the index *t* points to zero. Now look again into the instrument, and turn G until the markings on the micrometer are parallel with the other side of the angle; read the number of degrees on the circle, and this will be the angle or its complement. It frequently happens that the micrometer has to be moved in or out to make the lines on it accord with the second side; but as this motion is altogether a parallel one, the accuracy of the measurement is not at all affected. The simplicity of the mechanical arrangement can now be very readily seen. The same advantage in using the micrometer with every eye-piece belongs to the goniometer.

FLAME-HEAT IN THE CHEMICAL LABORATORY,

ESPECIALLY THAT FROM GAS, WITHOUT THE AID OF A BLAST.

There is probably no more important era in the operations of the chemical laboratory than that of the introduction of the lamp as a source of heat for a large number of chemical operations, and that without the aid of a blast. Berzelius was doubtless the first to accomplish much in this direction, which he did by the agency of the lamp that so commonly bears his name, and which, more or less modified, is still in use where the ordinary illuminating gas is not to be had.

Although illuminating gas has been in use for about seventy years, it is only within a comparatively recent date that it has been pressed into service and used as a heating agent in the laboratory. The reason of this arose from the fact that when burnt in the ordinary manner it deposited soot on the vessels heated by it. This difficulty has been overcome by burning the gas from small orifices made in a tube bent in the form of a circle, the holes being from one to two centimetres apart, and sometimes combining two or more rings in concentric circles. This method, however, has not been generally adopted.

We must date the successful introduction of gas for heating purposes to the use of a mixture of gas and air passed through wire-gauze and ignited above the gauze, giving a flame without light and with great heat. The invention of this method is claimed by several, and doubtless was discovered by different individuals at about the same time, without a previous knowledge of each other's results; this method is still more or less employed for certain purposes.

The next step in this direction, and doubtless the most important up to the present time, is to burn the mixture of gas and air without the agency of wire-gauze; it was first made known to the public in the burner commonly called the Bunsen burner, doubtless from its being either invented or brought into extensive practical use by the distinguished chemist of Heidelberg. Its form is too well known to require more than a mere mention here, and it is now made of all sizes, from those capable of burning four cubic feet of gas and under to those which can burn fifteen or twenty cubic feet from a single burner or from a combination of several smaller ones. To this burner some material additions have been made by different individuals. J. J. Griffin (the chemical instrument dealer in London) was, I believe, the first to introduce the use of the rosette and the register for the supply of air. The most remarkable results accomplished by this method of burning gas and air are those obtained by G. Gore, of Birmingham (all of whose results I have verified), where gold, copper, cast-iron, etc., were fused in crucibles without the agency of any artificial blast. Mr. Gore evidently realized fully the true principle of burning this mixture so as to obtain a maximum effect; the burner, however, with its furnace arrangements, is unavoidably of a form and on a scale limiting its application.

The usual form of the Bunsen burner, with the rosette and register (when required), bids fair to hold its own against any other form for general purposes, and whatever modifications may be made on it should be of such a character as not to intrench on its simplicity. One or two of these modifications are now in daily use in my laboratory, for which there is no claim to any special originality, nor are they intended to supplant the ordinary form.

As simple an instrument as the Bunsen burner appears to be, its principles and effects are well worthy of being carefully studied.

As the gas passes from the small orifices* in the lower part of the burner, and mixes with the air drawn in at the lower opening and passes out at the open end of the tube, it usually contains not quite enough oxygen for its complete combustion, and requires free access of air to the outer portion of the flame

* The outlet for gas may be in the form of crossed slits or two small holes (one thirty-second inch in diameter each) for the small-size burner, the length of the tube being about four to four and a half inches; the next larger has four openings (about one twenty-fifth inch diameter each), and the tube about five inches long.

to complete the combustion; yet even with this the flame is hollow in its lower portion, having a cool center, its most intense heat being at about three or four inches above the end of the tube in the smaller Bunsen burners, and eight or ten inches in the largest size. If a proper access of air is not allowed to the flame, as sometimes happens in some of the furnace connections occasionally used with Bunsen's burner, acetylene is formed from the imperfect combustion, which is recognized by its disagreeable odor, or by collecting some of the gas formed during the combustion; the presence of acetylene may be rendered evident by a small amount of a solution of ammoniacal cuprous chloride.

The best heating effects of the gas used in the ordinary round Bunsen burner, when employed in the heating of crucibles and other vessels, are not obtained; yet in the great majority of cases the small loss of gas is not worth considering, especially as to obtain better results in most cases would only complicate this beautifully simple instrument.

To get the best effects of heat, we must imitate the principle applied in the Argand burner; namely, to flatten down the exit of the mixed gases. It was by following out this principle that Mr. Gore was enabled to make a burner having a number of radial flat orifices as represented in the figure (1); the air from without having free access to the flame along the entire length of the slit openings, the number of slits used are more numerous than those represented in the figure. With the flame from this burner introduced into a certain form of refractory cylinders, cast-iron can be melted in a crucible without the aid of a blast, as has already been stated, the little chimney to the furnace being two inches in diameter and four feet long. This burner and its furnace is of but limited application, and the amount of gas consumed considerable.

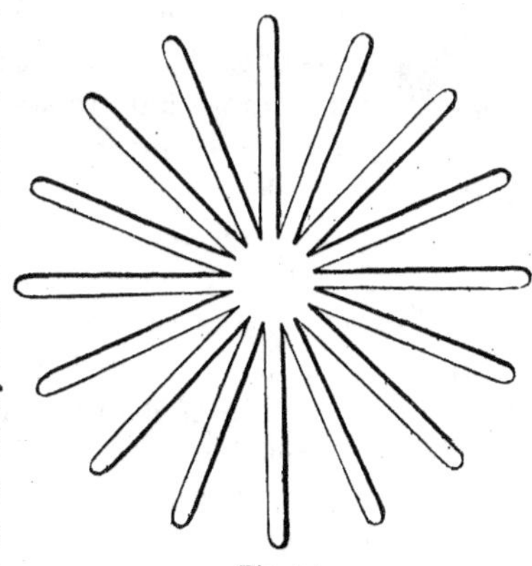

Fig. 1.
The openings at the exit of Gore's burner.

The principle, however, of the above burner is introduced in constructing a more simple form, and the flattened orifice is now used in the construction of what I conceive to be the best form of furnace for heating glass tubes for organic analyses and other purposes. Such furnaces are made by Weisnig, of Paris, and Desaga, of Heidelberg.

The use of the flattened burner is not fully appreciated; its advantages are, that there is no cold point in the flame, and the burner can be brought much nearer to the object to be heated—within twenty to twenty-five millimetres for the small-sized burners. In this burner, as usually made, the opening is too broad, experience having convinced me that a slit two millimetres across and about forty millimetres in length is the most effective one for a small-size burner, consuming about five and a half cubic feet per hour. This burner is represented in fig. 2, which can be used with the ordinary tube by detaching the tube with the flattened orifice.

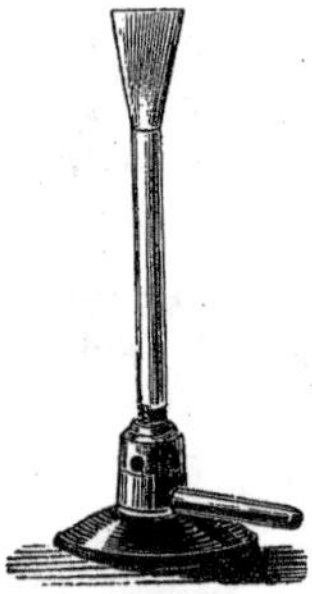

Fig. 2.

By taking a burner of this description and putting two pieces on each side of the center, as represented in fig. 3, a very efficient burner is made for heating platinum crucibles in silica fusions, etc., and with such a burner, consuming five and a half to six cubic feet of gas per hour, I conduct most effectually all silica fusions in one hour or less, taking care to protect the crucible from the current of the air by a properly-constructed short conical chimney.

Fig. 3.

As was stated in the commencement of this article, it was not intended to describe the methods of burning gas in furnaces and by means of a blast, but to confine the remarks to the simpler forms in every-day use, which can be made to accomplish all the ordinary requirements of the laboratory, and when a higher heat is required the furnace must be our recourse, whether burning gas, charcoal, or coke. The burner represented in figure 2 is the one I now employ in heating the crucible in my method of alkali determination with carbonate of lime and sal ammoniac.

FREEZING WATER BY THE AIR-PUMP

WITHOUT SULPHURIC ACID OR OTHER DESICCATING AGENT.

In attempting to freeze water under the air-pump, without the aid of a desiccating agent, the cooling of the water to the point of congelation is prevented by the heat received from the containing vessel. I have lately found that by obviating this difficulty water may be readily frozen by its own evaporation.

It was first shown by Count Rumford that water does not wet a sooted surface, but forms in globules like quicksilver. Three drops of water were placed in a sooted watch-glass; the spheroidal globule lay on the soot, exposing a large surface for evaporation, at the same time that the water was insulated from any source of heat. Arranged in this manner and placed under an air-pump, two or three minutes were sufficient to freeze the water. The glass was sooted over an oil-lamp with great care; the experiment fails if the globule of water touches the glass even by a small point.

In place of the sooted watch-glass make a shallow cavity in the end of a large cork, and over a lamp burn it, sooting it at the same time. By putting three drops of water into the cavity thus prepared, and subjecting it to the action of the air-pump under a pint receiver, the water froze solid in a minute and a half, and in two and three fourths minutes twenty grains of water congealed, though at 73° Fah. when introduced. Under a receiver of three quarts' capacity twenty grains of water froze in four minutes. I could not succeed in freezing the same amount in the sooted watch-glass.

By placing corks, prepared as above, over a saucer of sulphuric acid, the same results are obtained more rapidly. I put half a drachm of water, at 65° Fah., in each cavity, and exhausted the receiver till the mercurial gauge reached four tenths of an inch, which was effected in one minute. In a minute and a half the water on one cork began to freeze, and in five minutes they were all frozen. An ounce of water in a large flat cavity froze in three and a half minutes.

A flat-bottom porcelain capsule was prepared for an experiment on a large scale by sooting it in the following manner: After coating it with soot over a lamp, and allowing it to cool a little, a small quantity of oil of turpentine was carefully poured upon the edge and passed over the entire surface; the vessel was then warmed to drive off the redundant turpentine. The surface was again coated with soot and again with turpentine, and this process was repeated a third time; finally another coating of soot was added, when it was ready for use. Two ounces of water were placed in this capsule under a receiver, and the air-pump worked for one minute. After standing six minutes the surface was frozen.

This experiment, as well as similar ones, was attended with violent ebullition on the part of the liquid, throwing the water against the sides of the receiver, which was owing to the rapid formation of vapor on the under surface of the liquid.

DETERMINATION OF ALKALIES IN SILICATES

BY IGNITION WITH CARBONATE OF LIME AND SAL AMMONIAC.

In the following description of a method of separating and determining the alkalies *I aim to give the minutest details*, and it is to be regarded as an appendix to the one on page 200. Numerous analyses have given me the experience here presented; and I am convinced that analytical chemists, if they follow out the directions, will not resort to any other known method. If there be a better method, it is yet to be discovered. The presence of boracic, hydrofluoric, and phosphoric acids in the minerals in no way interferes with the process. Even in silicates *soluble in acids* I prefer this method, in common with other analysts, for its ease and accuracy. I made the researches during the latter part of 1852, and the details were published early in 1853.* Since then I have employed the process many hundreds of times with the most accurate results. Some minor points were not completed satisfactorily until several years after the first notice of the method; these have since been perfected, and I now know of nothing further that is needed.

The purpose of this article is to give all the improvements, with a minute detail of the manipulations and of the precautions necessary, all of which are simple and easily executed. In the two articles on the subject of alkali determination in minerals, published in 1853, the whole subject was reviewed, and it is needless to return to it now. I then considered the processes by caustic baryta and its salts, and by hydrofluoric acid, and also detailed some experiments on the separation

---

* Shortly after my first publication in this country M. St. Claire Deville made known his method of analyzing the silicates by fusion with carbonate of lime, but the nature of his process and the objects to be arrived at were quite different from those attained in this process.

of the different alkalies from each other, and on the microscopic examination of the same, etc. It was proved that, after the caustic alkalies, the most powerful agent to attack silicates at a high temperature is caustic lime, a fact not new to chemists. But for the purpose of accomplishing conveniently by this method *a quantitative determination of the alkalies in silicates*, certain methods of manipulation and facts with regard to quantity of material, admixture, etc., had to be discovered; and in them resides the success of my process—converting the most difficult parts of the analysis of a silicate into the easiest.

The methods of analysis by caustic baryta and by means of its carbonate are now no longer used, for various reasons fully detailed by Rose in his Analytical Chemistry. The method still extensively employed is that with hydrofluoric acid, proposed by Berzelius; and when used with the necessary precautions it has seemed to decompose all silicates; still, according to Rose, there are siliceous compounds that can not be completely decomposed by hydrofluoric acid.*

Dismissing all criticism, I at once proceed to the method which is the subject of this article—viz., the *decomposition of silicates by ignition with carbonate of lime and sal ammoniac.* A mixture of carbonate of lime and sal ammoniac is used in the decomposition simply for the purpose of bringing the caustic lime to act in a most thorough manner upon the silicates at red heat.†

Pure carbonate of lime.—The first requisite is pure carbonate of lime. This is made in my laboratory as follows: Take as good marble (calcite) as can be conveniently found, and dissolve it in hydrochloric acid (it is not necessary that the acid be perfectly pure); add an excess of the marble and warm the solution; to it add lime-water, or some milk of lime made from pure lime, until the solution is alkaline to test-paper; the lime is added to precipitate any magnesia, phosphate of lime, etc., that may have existed in the marble. Filter this solution and precipitate with carbonate of ammonia after heating to at least

* The process used by Deville in fusing with carbonate of lime is in most cases better than that by hydrofluoric acid, and one that I should use in preference to all others except the one now under notice.

† Chloride of calcium at a red heat will dissolve more or less caustic lime.

160° Fah.* The carbonate of lime thus precipitated is to be thrown on a filter and well washed with distilled water. Thus prepared, the carbonate of lime is a dense powder and perfectly pure, or if it contain any impurity it will be a trace of carbonate of baryta or strontia, which in no way interferes with its use.

Sal ammoniac.—To obtain this re-agent in the most convenient form take some fragments of clean sublimed sal ammoniac, dissolve them in water with a gentle heat, filter, evaporate the filtrate over a steam-bath or a sand-bath, or by means of any other convenient gentle heat, and as the crystals deposit themselves stir the solution to keep them small; when half or two thirds of the sal ammoniac is deposited pour off the liquid without waiting for it to cool, throw on a cotton filter, and dry the crystals at the temperature of the atmosphere. In this way sal ammoniac is obtained that can be easily pulverized.

Vessel for the decomposition.—The ordinary platinum crucible can be used for this purpose, and for many years was employed by me. It was found, however, that while it is the best hitherto contrived both for precision and ease, there was yet a very minute quantity of alkalies lost by volatilization; and I made further researches to overcome this small loss. This I have successfully accomplished, and for some time I have used an improved form of crucible. The one for half to one gramme of silicate is of the following form and dimensions; viz., an elongated, slightly conical crucible with rounded bottom and cover (either with or without the central wire by which to hold it); length, ninety-five millimetres; diameter of opening, twenty-two millimetres; diameter of smaller end just at the turn of the bottom sixteen millimetres; weight about thirty-five to forty grammes. These are now made by Messrs. Johnson, Matthey

* This precaution must not be overlooked, as it is desirable to obtain the precipitated carbonate of lime as dense as possible. If the carbonate of ammonia be added to the cold solution, the precipitate, at first gelatinous, will ultimately become much more dense and settle readily; the same is true if the mixture be heated after the addition of the carbonate; but in neither case will it be as dense as when the carbonate is added to the hot solution of chloride of calcium. The reaction in the analysis is in no way affected by the form of the carbonate of lime; but by using the denser form the mixture occupies less space in the crucible.

& Co., Hatton Garden, London, to whom I have furnished all the necessary directions. This shape is given it in order that the portion of the crucible containing the mixture may be heated strongly, while the upper portion is below a red heat.

Manner of heating the crucible.—The ordinary crucible, if used, may be heated in the manner commonly employed for the fusion of silicates. If the new form of crucible, however, is employed, then the upper part is grasped by a convenient metallic clamp in a slightly-inclined position, and a moderate blast from the table blow-pipe made to play upon it for about twenty-five or thirty minutes; but as gas is to be found in every well-mounted laboratory, Bunsen burners of all dimensions are used, and when properly applied can be made to give all gradations of heat. A simple, cheap, and convenient furnace, with a properly-arranged draught, can be made to accomplish all fusions of silicates without the aid of any manual labor; and I therefore employ such an apparatus. (A description of it is given at the close of this article.)

Method of analysis.—We have now the pure carbonate of lime, granular sal ammoniac, and the proper crucible. The silicate should be well pulverized in an agate mortar,* and half a gramme or one gramme of it is taken. The former amount is most commonly used, it being sufficient and best manipulated in the crucible; a gramme, however, may be conveniently employed. The weighed mineral is placed in a large agate mortar, or better in a glazed porcelain mortar, of half to one pint capacity. An equal quantity of the granular sal ammoniac is weighed out (a centigramme more or less is of no consequence) and put into the mortar with the mineral, and the two are rubbed together intimately. After this add eight parts of the carbonate of lime in three or four portions, and mix intimately after each addition. Empty the contents of the mortar completely upon a piece of glazed paper that ought

* While in all mineral analyses thorough pulverization of the mineral is usually essential, still it is a singular fact very good analyses can be made with this method even when the powder is tolerably coarse: and in some experiments with lepidolite I used powder of which much of it was in particles of from one fortieth to one thirtieth of an inch, and obtained excellent results. Notwithstanding this, thorough trituration of the mineral is recommended.

always to be under the mortar, and introduce into the crucible. The crucible is now tapped gently upon the table and the contents settled down.

The crucible is then clasped by a metallic clamp in an inclined position, or it is placed in the upper part of the support referred to in the latter part of this article, leaving outside about three fourths of an inch or one inch. By means of a small Bunsen burner the heat is brought to bear just above the top of the mixture, and gradually carried toward the lower part, until the sal ammoniac is completely decomposed, which takes about five minutes. Heat is then applied in the manner suggested, either with the blast or with the burner referred to, acting by its own draught, and the whole kept up to a bright-red heat for from forty to sixty minutes. It is well to avoid too intense a heat.

The crucible is now allowed to cool, when the contents will be found to be more or less agglomerated in the form of a semi-fused mass. A glass rod or blunt steel point will most commonly detach the mass, which is then dropped into a platinum or porcelain capsule of about one hundred and fifty centimetres capacity, and sixty or eighty centimetres, and distilled water added. After some time the mass will slack and crumble in the manner of lime. Still better, this may be hastened by bringing the contents of the capsule to the boiling-point either over a lamp or water-bath; at the same time water is put into the crucible to slack out any small particles that may adhere to it, and subsequently this is added to that in the capsule, *washing off the cover of the crucible also.*

After the mass is completely slacked the analysis may be proceeded with. As a general thing, I prefer to allow the digestion to continue six or eight hours, though this is not necessary.

If the contents of the crucible are not easily detached, do not use very much force, as the crucible may be injured by it; but fill the crucible to about two thirds its capacity with water, bring it almost to the boiling-point, and lay it in the capsule with the upper portion resting on the edge. The lime will slack in the crucible, and then may be washed thoroughly into the dish; also, as before, the cover is to be washed off.

We have now by this treatment with water the excess of

lime slacked into a hydrate, and some of the lime combined with the silica and other ingredients of the silicate in an impalpable form. In solution there is an excess of chloride of calcium formed in the operation, and all the alkalies originally contained in the mineral, as chlorides. All that now remains to be done is to filter and separate the lime as carbonate, and nothing is left but the chlorides of the alkalies. To do this proceed as follows: throw the contents of the capsule on a filter, the best size of which for the quantity above specified is one three to three and a half inches in diameter; wash well, to do which requires about two hundred centimetres of water; the washing is executed rapidly. The contents of the filter (except in those cases where the amount of the mineral is very small, and there is no more for the estimation of the other constituents) are of no use, unless it be desired to heat again to see if any alkali still remains in it.

The filtrate contains in solution all the alkalies of the mineral, together with some chloride of calcium and caustic lime; to this solution, after it has been thrown into a platinum or porcelain capsule, is added a solution of pure carbonate of ammonia (an amount equal to about one and a half grammes is required). This precipitates all the lime as carbonate; it is not, however, filtered immediately, but is evaporated over a water-bath to about forty centimetres, and to this is added again a little carbonate of ammonia and a few drops of caustic ammonia to precipitate the little lime that is redissolved by the action of the sal ammoniac on the carbonate of lime; filter on a small filter (two-inch), which is readily and thoroughly washed with but little water, and the filtrate allowed to run into a small beaker. In this filtrate are all the alkalies, as chlorides and a little sal ammoniac. Add a drop of carbonate of ammonia to make sure that no lime is present. Evaporate over a water-bath in a tared platinum dish, in which the alkalies are to be weighed; the capsule used is about sixty centimetres' capacity, and during the evaporation is never filled to more than two thirds of its capacity.

After the filtrate has been evaporated to dryness the bottom of the dish is dried, and on a proper support heated very gently by a Bunsen flame to drive off the little sal ammoniac. It is well to cover the capsule with a piece of thin platinum to prevent any possible loss by the spirting of the salt. After the sal ammoniac has been driven off by gradually increasing the heat the temperature of the dish is brought up to a point a little below redness, the cover being off. (The cover can be cleaned from any sal ammoniac that may have condensed upon it by heating it over the lamp.) The capsule is again covered, and when sufficiently cooled, and before becoming fully cold, is placed on the balance and weighed. This weight gives as chlorides the amount of alkalies contained in the mineral.

If the chloride of lithia be present, it is necessary to weigh rapidly, for this salt, being very deliquescent, takes moisture rapidly.

It not unfrequently happens that the chlorides at the end of the analysis are more or less colored with a small amount of carbon arising from certain constituents in carbonate of ammonia; the quantity is usually very minute, and in no way affects the accuracy of the analysis. In selecting pure carbonate of ammonia for analytical purposes it is well to take specimens that are not colored by the action of light.

It only remains now to separate the alkalies by the known methods. Under this head I have made several observations that at some future time may be published as soon as the results are sufficiently definite.

A SPECIAL ARRANGEMENT FOR HEATING THE CRUCIBLE BY GAS.

The support and burner, where gas is to be had, are simple in their character, and have been contrived after a great variety of experiments with gas furnaces. The figure here given illustrates the stand, burner, crucible, etc., and is about one sixth the natural size:* *h* is the stand with its rod *g*; *d* is a brass clamp with two holes at right-angles to each other, having two binding screws; it slides on the rod *g*; the second hole is for a round arm attached to *b*, the binding-screw *e* fixing it in any position. *b* is a plate of cast-iron five to six mm. thick, ten to eleven cm. long, and four and one half cm. broad, having a hole in its center large enough to admit the crucible to within about fifteen mm. of the cover without binding. *a* is the crucible

---

* This apparatus can be obtained from Johnson, Matthey & Co., Hatton Garden, London.

already referred to, which is made to incline a few degrees downward by turning the plate of iron that supports it. *c* is a chimney of sheet-iron, eight to nine cm. long, ten cm. high, the width at the bottom being about four cm. at one end and about three cm. at the other end. It is made with the sides straight for about four cm., then inclines toward the top so as to leave the width of the opening at the top about one cm. A piece is cut out of the front of the chimney of the width of the diameter of the hole in the iron support, and about four cm. in length, being semi-circular at the top, fitting over the platinum crucible. Just above this part of the chimney is riveted a piece of sheet-iron in the form of a flattened hook *n*, which holds the chimney in place by being slipped over the top of the crucible support; it serves as a protection to the crucible against the cooling of the currents of air. *f* is the burner, which has been described in the article on flame-heat. The upper opening of it is a slit from one and a half to two mm. in width and from three and a fourth to four and a half cm. long, and when used is brought within about two cm. of the lowest point of the crucible, the end of the flame just playing around the lower end of the crucible; the gas enters the lower part of the burner by two small holes of one sixteenth of an inch, furnishing at one inch pressure about five and a half cubic feet of gas per hour; the precaution must be observed, already referred to, in heating the crucible at first gently above the mixture. It is surprising to see the effect produced by this simple burner as here used; eight grammes of precipitated carbonate of lime can be decomposed to within two or three per cent. in one hour, and when mixed with silica or a silicate in a very much shorter space of time, although in my analysis I employ one hour, as it requires no attention after the operation is once started. This form

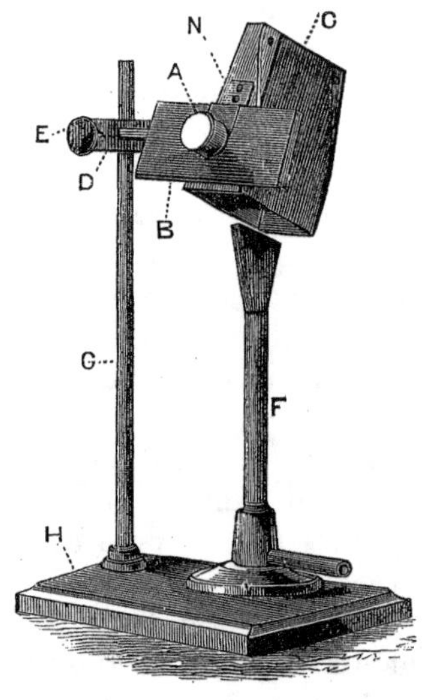

of furnace and crucible is found to be convenient for other operations.

Although the details here given are long, the time occupied in the analysis is short, and the necessary precautions are of a simple character, so much so that results obtained by students in beginning chemical analysis have been found by me far more reliable and less variable on the alkalies of the silicates than on any of the other constituents. Good alkali determinations can be made in three hours or less from the commencement of the operation, hastening the evaporation by more direct application of the heat, which of course requires more close watching.

It is a common practice of mine, when a silicate presents itself of which there are no physical means of ascertaining its nature, to make at once an alkali determination, which not unfrequently indicates immediately what it is, if it be a known silicate containing an alkali; and if an unknown compound, the analysis made is one step in the examination.

www.ingramcontent.com/pod-product-compliance
Lightning Source LLC
LaVergne TN
LVHW020109110826
845151LV00001B/106

* 9 7 8 1 4 2 5 5 4 3 7 6 1 *